AUSTRALIAN PALMS

AF703618

AUSTRALIAN PALMS

BIOGEOGRAPHY, ECOLOGY AND SYSTEMATICS

JOHN LESLIE DOWE

Australian Centre for Tropical Freshwater Research,
James Cook University, Townville, Qld, Australia,
and the Montgomery Botanical Center, Coral Gables, FL, USA

CSIRO
PUBLISHING

© John Leslie Dowe 2010

All rights reserved. Except under the conditions described in the *Australian Copyright Act* 1968 and subsequent amendments, no part of this publication may be reproduced, stored in a retrieval system or transmitted in any form or by any means, electronic, mechanical, photocopying, recording, duplicating or otherwise, without the prior permission of the copyright owner. Contact **CSIRO** PUBLISHING for all permission requests.

National Library of Australia Cataloguing-in-Publication entry

Dowe, John L.

Australian palms : biogeography, ecology and systematics /
John Leslie Dowe.

9780643096158 (pbk.)

Includes index.
Bibliography.

Palms – Australia.
Palms – Australia – Identification.
Palms – Norfolk Island – Identification.
Palms – Lord Howe Island – Identification.
Palms – Christmas Island (Indian Ocean) – Identification.

584.50994

Published by
CSIRO PUBLISHING
36 Gardiner Road, Clayton VIC 3168
Private Bag 10, Clayton South VIC 3169
Australia

Telephone: [+613] 9545 8555
Local call: 1300 788 000 (Australia only)
Fax: +61 3 9662 7555
Email: csiropublishing@csiro.au
Website: www.publishing.csiro.au

Front cover (clockwise from top left): *Licuala ramsayi, Livistona lorophylla, Calamus moti, Livistona muelleri, Carpentaria acuminata, Archontophoenix alexandrae, Nypa fruticans* and *Normanbya normanbyi.* Photos by the author.

Back cover (clockwise from top left): *Livistona victoriae*, Bungle Bungle Ranges, Western Australia; *Licuala ramsayi; Calamus moti; Linospadix apetiolatus;* and *Caryota albertii.* Photos by the author.

Set in Palatino 9.5/13
Edited by Adrienne de Kretser, Righting Writing
Cover and text design by James Kelly
Typeset by MPS Limited, A Macmillan Company
Printed by Ingram Lightning Source

CSIRO PUBLISHING publishes and distributes scientific, technical and health science books and journals from Australia to a worldwide audience and conducts these activities autonomously from the research activities of the Commonwealth Scientific and Industrial Research Organisation (CSIRO).

The views expressed in this publication are those of the author(s) and do not necessarily represent those of, and should not be attributed to, the publisher or CSIRO.

Jan26_RP_ILS

Foreword

Despite the long and venerable tradition of rivalry between Australia and the UK, it is obvious that us Pommies are on a losing wicket when it comes to plant diversity. The thrills of my own encounters with wild Australia still resonate today – the first tastes of sclerophyll vegetation on the Sydney sandstones, a formative road trip through the Atherton Tablelands, and the lush subtropical forests of the Lamington National Park, for example. I have scarcely dipped my toe in the water, but such experiences can only leave one awestruck and distinctly envious of the natural wonders that Australians can enjoy on their doorsteps.

The palm family makes only a modest contribution to the Australian flora, but is exceptional nonetheless. No other regional palm flora includes representatives of all five palm subfamilies. It encapsulates flavours of both the Malesian and Pacific palm floras while including several specialities of its own. Australia contains memorable palm-dominated landscapes, such as the forests of *Licuala ramsayi* in famous localities such as Mission Beach, *Livistona mariae* on the Finke River system in the Northern Territory, or the vast stands of Kentia palms, *Howea forsteriana*, on the off-shore island of Lord Howe. Significant among the endemics is *Oraniopsis*, an ancient-looking palm whose affinities as a representative of a lineage otherwise found only in South America, Madagascar and Juan Fernandez were determined only 25 years ago. Similarly dramatic, the foxtail palm *Wodyetia* was only described in 1983, having just been discovered among spectacular granite boulders in the Melville Range in Queensland. That such finds can be made so recently hints at the possibility of more great botanical discoveries to come in Australia.

It is perhaps surprising that a comprehensive guide to the palms of Australia has been lacking until now, but readers will not regret the wait. In typical style, John Dowe has tackled the subject in great depth and detail, producing an account so thorough that it is unlikely to be superseded for many years to come. It has arrived at a time when the need for palm information has never been greater, to service both horticultural appetites and much-needed conservation agendas. This rich source of knowledge and beautiful images is a milestone in the botanical history of Australia and a fitting tribute to the wonders of its palms.

William J. Baker
Head of Palm Research
Royal Botanic Gardens, Kew

Contents

Preface

This book provides an updated and thorough systematic and taxonomic treatment of the Australian palm flora. The only other similarly broad treatment of Australian palms was prepared over 130 years ago, by German botanists Hermann Wendland and Oscar Drude in 1875 in their *Palmae Australasicae*, published in the journal *Linnaea*. Based on the knowledge of the time, that early work described 26 species and examined the palm flora from aspects of biogeography, relationships and taxonomy. It provided the basis for much of our understanding of Australian palms well into the 20th century. Treatments such as Bentham's *Flora Australiensis* of 1878 and F.M. Bailey's *Queensland Flora* of 1902 provided timely synopses, but did not address broader aspects of biogeography, relationships and classification. Early 20th-century palm botanists such as Odoardo Beccari and Max Burret provided some broader examination of Australian palms, but most often in a context of revisions and new species descriptions. In the later decades of the 20th century, interest in Australian palms was renewed through the work of A.N. Rodd, A.K. Irvine, J.L. Dowe, D.L. Jones and F.B. Essig; revisions of *Archontophoenix*, *Caryota*, *Hydriastele*, *Licuala*, *Linospadix*, *Livistona* and *Ptychosperma* have been completed in recent decades.

Sixty species of Australian palms are recognised and discussed in this book. The book takes note of recent advances in biogeographic and phylogenetic research, which allow the Australian palm flora to be incisively placed within a regional and global context. These aspects are examined in detail. *Australian Palms* includes expanded and thorough descriptions of genera and species, with additional information on distribution, ecology and typification. The descriptions of both genera and species were obtained following extensive field-work and examination of over 1600 specimens in herbaria in Australia, south-east Asia, Europe and the US. The author has studied and collected all mainland palm species in their natural habitats, and has studied those from off-shore territories as cultivated specimens in Australian botanic gardens such as The Palmetum in Townsville, Flecker Botanic Gardens in Cairns and Mt Coot-tha Botanic Gardens in Brisbane, and in Florida at the Fairchild Tropical Botanical Gardens and the Montgomery Botanical Center. This work is the culmination of over 20 years of research into Australian palms.

STRUCTURE OF THE BOOK

This book is divided into two sections. The first section provides a broad introduction, with detailed summaries of botanical history, historical biogeography, distribution and ecology. The second section deals with systematics, classification and taxonomy, providing assessment and description of taxa from the level of subfamily to subspecies, arranged according to the most recent classification of palms in *Genera Palmarum: The Evolution and Classification of Palms* (Dransfield *et al.* 2008). Each species is clearly illustrated with images of habit, leaves, flowers, fruit and relevant diagnostic features. The final chapter offers a key for the field identification of Australian palm species. The book concludes with a glossary.

METHODS

The taxonomic history of each species was investigated and nomenclature and typification were reviewed. Protologues for all taxa were sourced and scrutinised for validity of publication, typification and adherence to the appropriate articles in the *International Code of Botanical Nomenclature* (McNeill *et al.* 2006). Type specimens were located at numerous herbaria, and images of at least one sheet (of those with multiple sheets) are included. Where types had been lost, misapplied or inappropriately designated, new typification is presented. In some cases, typification is designated for the first time. Species descriptions are based on the examination of about 1600 herbarium specimens and living palms in their natural habitats and cultivation. Descriptions are based on mature plants. For many measurements, such as stem height and leaf length, only the upper range is provided. Otherwise, the range of sizes recorded for some organs, where size is more critical, is provided. Phylogenetic assumptions and reconstructions are based on the most recent analyses as referenced in the text, mostly accepting those presented in *Genera Palmarum* (Dransfield *et al.* 2008). Author abbreviations follow Brummitt and Powell (1992), herbarium acronyms are according to Holmgren and Holmgren (1998), journal abbreviations follow Bridson and Smith (1991) and publication abbreviations follow Stafleu and Cowan (1976–98).

Specimens examined in compiling species descriptions and associated with typification may be viewed as an Excel spreadsheet at http://www.montgomerybotanical.org/Pages/Research.htm.

Acknowledgments

I sincerely wish to thank all those who have contributed in many ways to the preparation and completion of this book. Chapters and sections were reviewed by Anders Barfod (Aarhus University, Denmark), William J. Baker (Royal Botanic Gardens Kew, UK), Bob Congdon (James Cook University, Townsville, Australia), Don Hodel (University of Southern California, USA), Nanette Hooker (James Cook University, Townsville, Australia), Neils Lunoe (Cairns, Australia), Larry Noblick (Montgomery Botanical Center, Florida, USA), Mike Pole (Queensland Herbarium, Australia), Alison Shapcott (University of the Sunshine Coast, Australia), Gary Wilson (Australian Tropical Herbarium, Australia), Stuart Worboys (Cairns, Australia) and Scott Zona (Florida International University, USA).

Images of palms were generously provided by William J. Baker (Royal Botanic Gardens Kew), Jeff Clausen (Christmas Island), Ian Hutton (Lord Howe Island), John Lok (New Zealand), Jeanne Price (Cooran, Qld), John G. Pritchard (Iron Range, Qld), David Tanswell (Brisbane, Qld), Jo Wilkens (Melbourne, Vic.) and the late Robert Tucker. They are individually acknowledged on the pages where their images are reproduced.

The following generously provided or assisted with images of herbarium specimens, portraits, literature and fossils: Helen Cohn (National Herbarium of Victoria), Ronald Clark (Herrenhausen Gardens), Piero Cuccuini (FI), Ewan Fordyce (University of Otago), Paul Forster (Queensland Herbarium), Wayne Gebert (National Herbarium of Victoria), Greg Jordan (University of Tasmania), Norbert Kilian (Berlin Herbarium), Pina Milne (National Herbarium of Victoria), Tony Orchard (ABLO 2009), Neville Pledge (South Australian Museum), Mike Pole (BRI), Ian Raine (GNS Science, New Zealand), A'mhara Russell (James Cook University), Ota Sida (PR), Karin Schwabe (Dresden Botanic Garden), John Simes (GNS Science, New Zealand), Mikael Siversson (Western Australian Museum, Perth), Marianna Terezow (GNS Science, New Zealand), Roy Vickery (BM), Lynley Wallis (Flinders University), Karen Wilson (NSW), Ged Wiren (Auckland Museum, New Zealand) and Scott Zona (Florida International University).

Access to living collections was provided by Chris Cole (Townsville Botanic Gardens, Qld), Mary Collins and Marilyn Griffiths (Fairchild Tropical Gardens, Florida, USA), Patrick Griffith and Larry Noblick (Montgomery Botanical Center, Florida, USA), Edi Heide and Barbara Lamont (Cairns, Qld), Will Kraa (Brisbane, Qld), Julie Roach (Townsville City Council, Qld), Stan Walkley (Burpengary, Qld) and David Warmington (Cairns Botanic Gardens, Qld).

Assistance with discussion and information was provided by John Conran (AD), John Dransfield (RBGKew), Andrew Ford (CNS), Betsy Jackes (JCT), Kath Prickett (Auckland Museum), Dian Latifah (JCT), Mike Macphail (ANU), Mike Pole (BRI), John and Jeanne Price (Cooran, Qld) and Tim Willing (Broome, WA).

Field work was funded by the Palm and Cycad Societies of Australia and the Brisbane Palm and Cycad Society. Maps were prepared by Mirjam Maughan, Australian Centre for Tropical Freshwater Research, James Cook University, Townsville.

Access to herbarium data and collections was provided by Gordon Guymer (BRI), Pina Milne (MEL), Frank Zich (CNS), Betsy Jackes (JCT), Scott Zona (FTG), William J. Baker and John Dransfield (K).

Jenny Martin is lovingly thanked for her support and comfort, especially during field work.

1. Introduction

This account deals with the 60 species in 21 genera of palms, family Arecaceae, that are indigenous to the Australian geopolitical region, which includes the Australian continent and the off-shore territories of Christmas Island, Lord Howe Island and Norfolk Island. For continental Australia, 54 species in 17 genera are currently recognised, while there are four species in three genera on Lord Howe Island and one species on both Christmas Island and Norfolk Island (Du Puy and Telford 1993; Green 1994) (Fig. 1.1).

Considering the great diversity and number of palm species in nearby areas such as New Guinea, with c. 280 species in 31 genera (Barfod *et al.* 2001; Baker and Dransfield 2006) and the comparatively small island of New Caledonia with 38 species in 10 genera (Pintaud and Baker 2008), the Australian palm flora is not a prominent element in the overall context of the Australian flora. The limited distribution and/or absence of palms from suitable habitats in tropical northern Australia are notable, as other areas of the world with similar climates and environments often have a relatively greater abundance. Two somewhat different scenarios may be the cause of this:

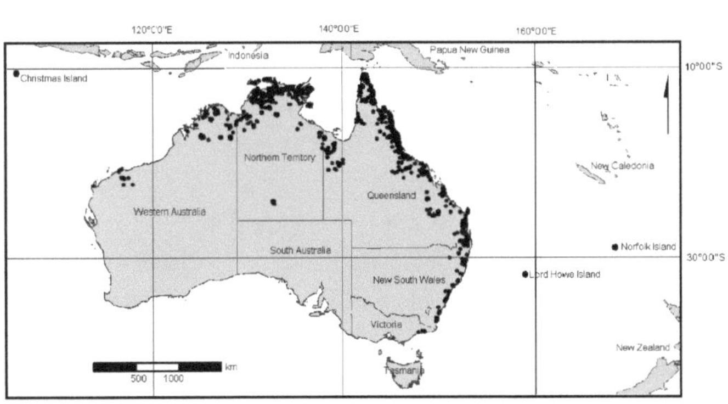

Figure 1.1 Distribution of the Arecaceae in Australia and off-shore territories.

- a formerly diverse palm flora has largely been extirpated across much of the continent through long- and short-term climate change;
- the low vagility of palms into Australia from nearby palm-rich areas.

These aspects will be discussed further in Chapter 3, 'Historical biogeography', and Chapter 4, 'Distribution and ecology'.

Despite the relatively small number of species in Australia, diversity at higher classification levels is considerable, with all the five subfamilies in the Arecaceae represented. Two of the subfamilies are represented by a single species in Australia: the Nypoideae (a monospecific subfamily) with *Nypa fruticans* (mangrove palm) and the Ceroxyloideae (globally eight genera, 42 species) with *Oraniopsis appendiculata*. The Calamoideae also has relatively moderate representation, with only one genus with eight species in Australia (globally 21 genera, c. 620 species). The most diversified subfamilies in Australia include the Arecoideae with 13 genera and 26 species (globally 107 genera, c. 1250 species) and the Coryphoideae with five genera and 24 species (globally 46 genera, c. 450 species). The largest genus in Australia is *Livistona*, in the Coryphoideae, with 18 species of 36 species world-wide. Other significant genera include the endemic *Archontophoenix* with six species, and *Linospadix* with five endemic species (plus two species in New Guinea), in the Arecoideae. Aspects of relationships between taxa and other phylogenetic themes will be presented in Chapter 5, 'Systematic arrangement' and subsequent chapters.

A number of genera with species in Australia have their centre of distribution and diversity in New Guinea and central or western Malesia. These are represented in Australia by a single or a few outlier species: these genera include *Arenga*, *Caryota*, *Corypha*, *Hydriastele*, *Licuala* and *Ptychosperma*. An unusual feature of the Australian palm flora is the relatively high number of endemic monotypic genera, including *Laccospadix*, *Oraniopsis*, *Carpentaria*, *Normanbya*, *Wodyetia*, *Hedyscepe* and *Lepidorrhachis*. This may be an artefact of taxonomy rather than an expression of biological diversity, as the relationships of some are presently unresolved. Other genera are *Howea*, with two species endemic to Lord Howe Island, and *Rhopalostylis* with two species, one which occurs on Norfolk Island and the Kermadec Islands (the latter is politically part of New Zealand) and one which is endemic to New Zealand.

In Australia, most palm species occur in tropical regions that have high rainfalls (Fig. 1.2). In many habitats, palms do not form a significant element in the vegetation and only few grow gregariously. In Australia, 60% of the palm species occur in rainforest habitats. This is lower than the approximate 75% of palm species that occur in rainforest globally (Dransfield 1978). The lower percentage of rainforest palms is explained by the presence in Australia of *Livistona* which, with 18 species, is the continent's most diverse palm genus. Only one species of *Livistona* occurs in rainforest (*L. australis*) (Fig. 1.2); the others occur in monsoon forest, woodland, savanna, littoral forest and riparian forest, in monsoonal or semi-arid regions where rainforest is absent (Fig. 1.2). Where palms do occur in low-rainfall habitats such as open woodlands and savannas, they most often occupy niches where subsurface water is available for most of the year, or in environments that are seasonally inundated and have a high watertable for some months (Fig. 1.2). There are no palm species in Australia that are obligate xerophytes, although some may experience periods of seasonal drought and possess xerophytic characteristics such as wax on the leaf surface (Fig. 1.3). These species are often associated with habitats that have strongly seasonal rainfall patterns, occurring in northern monsoonal areas or inland Australia. Environmental requirements and ecological amplitude of Australian palms will be examined in Chapter 4, 'Distribution and ecology'.

Forty-four palm species occur in Queensland, of which 26 are endemic to that state. The Northern Territory has 14 species of which four are endemic, and Western Australia has five species of which three are endemic. The palm diversity in Queensland is high for two main reasons: there are vast areas with monsoonal or ever-wet tropical climate that sustain species-rich forest types in general, and there are forests in the extreme north of Cape York Peninsula that harbour a New Guinean element. Ten thousand years ago, when the Torres Strait land-bridge still existed, these forests were coherent with rainforest in New Guinea (Hope 1994). The other Australian states that have palms are New South Wales with four species, all shared with Queensland, and Victoria with one species which also has distribution in New South Wales and Queensland. Eleven species are shared with New Guinea (see Chapter 3); five of these have further distribution in south-east Asia or the south-west Pacific. The latter species are outlier populations of genera that are more diverse in areas to the north or north-west of Australia.

Some low- to moderate-altitude rainforest habitats of north Queensland, such as the Wet Tropics Bioregion and monsoonal Cape York Peninsula, have the highest diversity of palm species in Australia, with up to 12 species occurring together in some localities and occasionally forming palm-rich forests (Fig. 1.3). Other areas of significant diversity include the low- and moderate-altitude rainforest habitats in the MacPherson/Macleay Overlap region of southern Queensland and northern New South Wales, where four species may occur together (Fig. 1.3), and the Top End of the Northern Territory where three species may occur together in any one location. Conversely, some palm species exhibit extreme isolation from other species, such as *Livistona alfredii* in the Pilbara of Western Australia and *Livistona mariae* in the Finke River system of central Australia (Fig. 1.4). These isolated species are sometimes referred to as relicts. They have been interpreted as representing the remnants of a flora that occurred in those areas when the climate was wetter (Lothian 1959; Latz 1975) and that have persisted and adapted despite increasing environmental dryness. Alternative views suggest that these populations may be the result of recent long-distance dispersal and are not relict species but are recently evolved entities with some adaptive traits that have allowed them to survive in 'difficult' environments.

Palms also occur in some restricted habitats, such as mangroves (e.g. *Nypa fruticans*) (Fig. 1.5), riparian forests (e.g. *Archontophoenix myolensis*) (Fig. 1.5) and high-altitude habitats (e.g. *Laccospadix australasicus* and *Linospadix palmerianus*) (Fig. 1.5). Many of these species have specialised morphological adaptations or fruit dispersal characteristics associated with those environments.

Figure 1.2 Palm habitats. **Top left:** *Licuala ramsayi* var. *ramsayi* and *Calamus* spp., lowland rainforest, Licuala State Forest, Mission Beach, north-east Queensland. **Top right:** *Livistona australis*, moderate-altitude rainforest, Broken River, Eungella Range, central Queensland. **Bottom left:** *Livistona drudei*, near-coastal savanna woodlands, Clemant State Forest, north-east Queensland. **Bottom right:** *Livistona decora*, seasonally inundated riparian habitat, Healeys Lagoon, Haughton River catchment, north-east Queensland.

Figure 1.3 Palm habitats. **Top left:** *Livistona lanuginosa*, tall open woodland with c. 550 mm rainfall per annum, Deep Creek, Dandenong Park, Burdekin River catchment, north-east Queensland. **Top right:** Leaves of *Livistona lanuginosa* have a coating of wax on the surface, a xerophytic adaptation. **Bottom left:** *Archontophoenix alexandrae*, *Calamus* spp., *Licuala ramsayi* var. *ramsayi* and *Ptychosperma elegans*, Lamb Range, north-east Queensland, c. 600 m altitude. **Bottom right:** The moderate-altitude rainforest habitats in the MacPherson/Macleay Overlap region of southern Queensland and northern New South Wales with *Archontophoenix cunninghamiana*, *Calamus muelleri* and *Linospadix monostachyos*, Tamborine Mountain, south-east Queensland, c. 700 m altitude.

Figure 1.4 Remote and isolated palm species. **Top:** *Livistona alfredii*, Millstream-Chichester National Park, Fortescue River, Pilbara, Western Australia. **Bottom:** *Livistona mariae*, Palm Valley, Finke River, Northern Territory.

■ ECONOMIC BOTANY

Australian palms are significant in ornamental horticulture and many species are used in urban and domestic landscapes within Australia and in other countries (Jones 1996). Among the best-utilised and popular Australian palms are the King palms (*Archontophoenix* spp.), the Darwin palm (*Carpentaria acuminata*), the Kentia palms (*Howea* spp.), the Mission Beach fan palm (*Licuala ramsayi* var. *ramsayi*), the cabbage palms (*Livistona* spp.), the solitaire palm (*Ptychosperma elegans*), the Macarthur palm (*Ptychosperma macarthurii*) and the foxtail palm (*Wodyetia bifurcata*).

■ ENDANGERMENT AND CONSERVATION OF AUSTRALIAN PALMS

The conservation status of each Australian palm species has been provided as part of general status ratings for the indigenous flora, and assessments are discussed in a number of publications (Rodd 1998; Bostock and Holland 2007). Conservation status of species deemed endangered, vulnerable or rare has been discussed and conservation action plans have been implemented (Leach 1992; Liddle *et al.* 1996, 2006; Dowe 1998, 2007c). Conservation status ratings, based on Australian and IUCN criteria, are provided in the Distribution and Ecology section for each species.

Figure 1.5 Palm habitats. **Top left:** *Nypa fruticans* in the upper tidal zone of mangrove habitat, in a substrate of anaerobic mud, Herbert River delta, north-east Queensland. **Top right:** *Archontophoenix myolensis* is a riparian species that has specialised leaf and fruit adaptations, Warril Creek, Myola, Atherton Tableland, north-east Queensland. **Bottom left:** *Laccospadix australasicus* in high-altitude rainforest, Paluma Range, north-east Queensland, c. 900 m altitude. **Bottom right**. *Linospadix palmerianus* in high-altitude rainforest, Bartle Frere, Bellenden Ker Range, north-east Queensland, c. 900 m altitude.

2. Early documentation of Australian palms

This chapter examines the early documentation and first records of Australian palms, with an emphasis on pre-1900 literature and pertinent summaries of the known extent of the palm flora. The first Australian palms were scientifically described in 1810, and the most recent in 2005. Species of *Livistona* were the first palms documented by early explorers and among the first palms described in Australia. Several distinct periods of taxonomic activity can be identified for Australian palms, each corresponding to the production of flora treatments or the activities of specialist palm botanists.

■ EUROPEAN PLANT COLLECTORS AND BOTANISTS IN AUSTRALIA BEFORE SETTLEMENT

Although specimens of Australian plants had been returned to Europe by explorers such as William Dampier by the late 1600s (George 1999), it was not until Cook and Banks' *Endeavour* voyage in 1770 that the first Australian palms were documented. The entry dated 27 April 1770 in Joseph Banks' *Endeavour Journal* reads:

> *The trees were not very large and stood separate from each other without the least underwood; among them we could discern many cabbage trees but nothing else which we could call by any name (Banks 1962).*

This impression of the coastal vegetation of the area now known to be near Wonoona Beach, southern New South Wales, was gained during an unsuccessful attempt by Banks to venture on to the Australian mainland for the first time (Banks 1962). The first documented observation of an Australian palm, *Livistona australis*, described it as 'cabbage trees'.

Farther north, Banks (1962) provided a summary of the palms he had seen on the *Endeavour's* three-month voyage along the east coast of Australia:

> *Palms here were of three different sorts. The first which grew plentifully to the Southward had leaves pleated like a fan; the Cabbage of these was small but exquisitely sweet and the nuts which it bore in great abundance a very good food for hogs. The second was very much like the real cabbage tree of the West Indies, bearing large pinnated leaves like those of a Cocoa nut; these too yielded cabbage if not so sweet as the other sort yet the quantity made ample amends. The third which as well as the second was found only in the Northern parts was low, seldom ten feet in height, with small pinnated leaves resembling those of some kinds of fern.*

The last was the cycad *Cycas media* R.Br., that with 'leaves pleated like a fan' referred to *Livistona australis* R.Br. and that with 'large pinnated leaves' was *Archontophoenix* spp. and/or *Ptychosperma elegans* (R.Br.) Blume, which Banks failed to recognise as separate species.

The botanist on the *Endeavour*, Daniel Solander, left a handwritten manuscript describing plants seen during the voyage, with intended names and descriptions of three Australian palms; it was never formally published (Solander 1768–71). Solander's manuscript names were *'Corypha nuda'*, *'Corypha juncea'* and *'Palma oleracea/Caryota oleracea'*. A specimen consisting of a single leaf with the original inscription *'Corypha nuda'* (identified as *Livistona australis* at a later date) and inscribed as *'Nova Cambria*: Botany Bay J.B & D.S.' is conserved in the Natural History Museum, London (Fig. 2.1). This specimen of *Livistona australis* and another of *Ptychosperma elegans*, collected by Banks and Solander from the Endeavour River, are the

Figure 2.1 **Left:** The oldest known botanical specimen of an Australian palm: *Livistona australis* (R.Br.) Mart., conserved in the herbarium of the Natural History Museum, London. The original label data include 'Nova Cambria, Botany Bay, J.B. & D.S. Corypha nuda'. With permission of Natural History Museum, London. **Right:** Portrait of Robert Brown (1773–1858), DCL, LLD, FRS, PLS. Oil on canvas by Stephen Pearce (1819–1904), 1856 [AO 028]. © Board of Trustees of the Royal Botanic Gardens, Kew.

earliest surviving botanical specimens of Australian palms. Solander's unpublished manuscript name *'Corypha nuda'* was informally used by Robert Brown for a palm species he observed at Hervey Bay and Sandy Cape, 31 July 1802 (Vallance *et al.* 2001), but no specimens are extant and no description using that name was left by Brown. It must be noted that Brown's informal use of Solander's epithet does not confer any nomenclatural legitimacy on Solander's manuscript name, and it remains a name of no botanical standing.

■ EXPLORERS, NATURALISTS AND BOTANISTS AFTER SETTLEMENT

The first published systematic work on the Australian flora was by French explorer and botanist Jacques-Julien Houton de Labillardière (b.1755, d.1834) (Labillardière 1804–06). The work included 270 species described from c. 4000 specimens collected in western and south-western Australia and Tasmania during the d'Entrecasteaux Expedition of 1791–94, but Labillardière did not describe any Australian palms. It was not until the work of Robert Brown (b.1773, d.1858) (Fig. 2.1) a decade later that the first Australian palms were described (Brown 1810). Brown was employed by Banks as botanist on Matthew Flinders' *Investigator* voyage of 1801–03 (Vallance *et al.* 2001). Also on the voyage were horticulturist Peter Good, illustrator Ferdinand Bauer (Norst 1989; Mabberley 1999) and landscape artist William Westall (Findlay 1998). Prior to Brown's departure for Australia, he had familiarised himself with the *Endeavour* voyage specimens and Solander's unpublished manuscript, as well as the collections, made since settlement in 1788, that had been sent to London (Vallance *et al.* 2001). During the *Investigator* voyage of 1801–03, Brown made collections along the eastern and northern coasts of the continent. In *Prodromous Florae Novae Hollandiae*, Brown (1810) described four palms: *Livistona humilis*, *L. inermis*, *Corypha australis* (= *Livistona australis*) and *Seaforthia elegans* (= *Ptychosperma elegans*), which were the first palm species

Figure 2.2 Top: The first published illustration of an Australian palm. Engraving of *Livistona inermis*, based on the original painting by William Westall, 'View in Sir E. Pellew's Group Gulph of Carpentaria Discovered by Captain Flinders 1802', in Matthew Flinders (1814), *A Voyage to Terra Australis*, Vol. 2. With permission of James Cook University Library, Townsville. **Bottom:** The first published scientific illustrations of Australian palms appeared in C.F.P. von Martius (1838), *Historia Naturalis Palmarum*, Vol. 3, including *Livistona humilis*, Tab. 110 (left) and Tab. 111 (right).

botanically classified for Australia. The genera *Livistona* and *Seaforthia* (= *Archontophoenix* and *Ptychosperma*) were established at that time by Brown.

Flinders' (1814) account of the *Investigator* voyage included engravings based on Westall's original works, one of which depicted *Livistona inermis* on the shore of North Island, type locality for the species (Fig. 2.2). Bauer completed detailed botanical illustrations of the species but these remained unpublished until recently (Watts *et al.* 1997). In an appendix to Flinders' *Voyage to Terra*

Australis, Brown (1814) provided brief summaries of certain plant families in Australia, including the palms:

> **Palmae.** *Only six species of this order have been observed in New Holland, and two of these the fructification is at present unknown. The New Holland Palms exist chiefly within the tropics, but one species is found in 34°S. lat.; it seems, however, that this is nearly the southern limit of the order in that country, no species having been seen on any part of the South coast. In New Zealand a species of Areca was observed by Sir Joseph Banks, in about 38°S. lat., which is probably nearly the limit of Palms, in the southern hemisphere. In the northern hemisphere their extent is not materially different from this: in North America, indeed, they do not appear to grow beyond 36° lat.; but in Europe Chamaerops humilis extends as far as the neighbourhood of Nice. It is remarkable that no species of Palm has been found in South Africa, nor was any observed by Mr Lechenault, on the West coast of New Holland, even within the tropic.*

Brown communicated with Carl von Martius, the foremost palm botanist of the era, and provided herbarium specimens that were used by Martius (1838) in compiling the most ambitious account of the world's palms, the epic *Historia Naturalis Palmarum*. In addition to citing Brown's specimens, Martius acknowledged Bauer's illustrations as the basis for figures in *Historia* (Fig. 2.2). Although not acknowledged directly, elements from Westall's paintings were incorporated in some figures in *Historia*. These were the first published scientific illustrations of Australian palms.

■ ALLAN CUNNINGHAM

In 1816, Allan Cunningham (b.1790, d.1839) (Fig. 2.3) arrived in the colony of New South Wales as a botanical collector for Kew Gardens (McMinn 1970). A protégé of Banks, he contributed perceptive accounts of the Australian vegetation, particularly the rainforest flora. Cunningham accompanied Phillip Parker King on a number of voyages during which many new plant species were collected (King 1827; Curry *et al.* 2002). These were productive botanical ventures, with Cunningham observing and collecting a number of palms for the first time.

In an appendix to King's *Narrative of a Survey*, Cunningham (1827) provided a summary of the Australian palms:

> **Palmae**. *On considering the vast expanse of the continent of Terra Australis, and that great extent of coast which passes through climates favourable for the production of certain genera of this remarkable natural family, it is singular that so few of the order should have been discovered: a fact in the history of the Australian vegetation, which (upon contemplating the natural economy of many genera of plants) can only be considered for, by the great tendency to drought of at least three-fifths of its shores. To Corypha, Seaforthia, and Livistona, the only three genera that have been enumerated in the productions of the Australian Flora, may now be added Calamus; of which a species (discovered without fructification by Sir Joseph Banks, during the celebrated voyage of Captain Cook), has at length been detected bearing fruit in the vicinity of Endeavour River. The existence of this palm, or rattan, on the East Coast, to which it is confined, seems almost to be limited to an area within parallels of 15° and 17° South; should, however, its range be more extensive, it is southerly one or two degrees, in which direction a remarkable primary granitic formation of the coast continues, throughout the whole neighbourhood of which is a peculiar density of dark moist forest, seemingly dependent on, and evidently indispensable to the life of this species of Calamus; but at the termination of this geological structure, it most probably ceases to exist. A dioecious palm of low stature, and in habit similar to Seaforthia, was detected in the shaded forests investing the River Hastings, in latitude 31° South, bearing male flowers; but as it may prove to be a dwarf state of a species of that genus, which has lately been observed, with all its tropical habits, in a higher latitude, it cannot now be recognised as a sixth individual of the family whose fructification has been seen. Although this order has been observed to be sparingly scattered along the line of East Coast almost to the thirty-fifth degree of latitude, its range on the opposite shores of the continent is very limited. Upon the Northwest Coast, the genus Livistona alone has been remarked, in about latitude 15° South; beyond which, throughout a very extensive line of depressed shore, towards the Northwest Cape, no palms were seen. If the structure of a coast, and its natural disposition to produce either humidity or*

drought be consulted, (appoint with respect to this order, as well as certain other tropical tribes, appearing very important) those portions of the western shores recently seen, indicates no one character that would justify the supposition of the existence of the Palmae in this corresponding extremes of the respective parallels that produce them on the opposite or East Coast. Another remark relative to the economy of this family is, that in New Holland it seems confined to the coasts, Corypha australis, so frequent in particular shaded situations in the neighbourhood of Port Jackson, having never been detected in the vicinity of, or upon the mountains, much less in the distant country to the westward of that extensive boundary.

The reference to 'a dioecious palm of low stature' is the first documentation of *Linospadix monostachyos*. It is not a dioecious species, as suggested by Cunningham; the maturation of male and female flowers, both of which occur on a single inflorescence, are separated by many months and only one type of flower is functionally obvious at a time.

■ CARL FRIEDRICH PHILLIP VON MARTIUS: *HISTORIA NATURALIS PALMARUM*

The herbarium specimens collected by Brown, Cunningham and other contemporary collectors were studied by botanists in Britain and Europe. The first palm specialist to comprehensively study Australian palms in this manner was Carl F.P. von Martius (b.1794, d.1868) (Egge 1979), who described a further three Australian taxa, *Linospadix monostachyos*, *Calamus australis* and *C. caryotoides*, in *Historia Naturalis Palmarum* (Martius 1838). These were described from collections made by Cunningham. Martius acknowledged that the illustrations of some Australian palms in *Historia Naturalis Palmarum* were based on original artwork produced by Ferdinand Bauer from the voyage of the *Investigator*.

■ FERDINAND VON MUELLER

The next phase of Australian palm taxonomy was included in the work of Ferdinand von Mueller (b.1825, d.1896) (Fig. 2.3). Mueller was Government Botanist for Victoria for over 40 years, from 1853 to 1896, and established the National Herbarium of Victoria. Mueller was a productive palm taxonomist and among the foremost authorities on the Australian flora (Willis 1949).

Mueller published most of his palm taxonomy in the journal *Fragmenta Phytographiae Australiae*, where 19 species were described over a 30-year period. A small number of Mueller's other palm species were described in other publications. In *Second Systematic Census of Australian Plants*, Mueller (1889) listed 25 species of palms for Australia and territories.

■ HERMANN WENDLAND AND OSCAR DRUDE: *PALMAE AUSTRALASICAE*

By the last decades of the 19th century, there were sufficient herbarium collections and information for a comprehensive account of Australian palms to be attempted. Hermann Wendland (b.1825, d.1903) (Fig. 2.3), based at the Royal Gardens of Herrenhausen in Germany, had developed an interest in palms and was providing regional taxonomic accounts. Mueller and others sent specimens of Australian palms to Wendland. In conjunction with Oscar Drude (b.1852, d.1933) (Fig. 2.3), Wendland compiled the first comprehensive taxonomic account of Australian palms, *Palmae Australasicae* (Wendland and Drude 1875). It included a biogeographic assessment, descriptions of 17 new species and the designation of many new combinations.

■ GEORGE BENTHAM: *FLORA AUSTRALIENSIS*

The period 1860–80 saw a general consolidation of the classification and taxonomy of the Australian flora. The result was the first comprehensive account, *Flora Australiensis*, by English botanist George Bentham (b.1800, d.1884) (Fig. 2.4), with assistance from Mueller, serially completed over 17 years from 1863 to 1879. Bentham used the herbarium resources at Kew, the British Museum and Paris as well as loans from Australia (George 1981). Bentham's (1878) treatment of the Palmae included 10 genera and 21 spp. for Australia and territories. The taxonomy of Martius, Mueller and Wendland and Drude formed the basis of Bentham's work, which was synoptical rather than revisionary. He provided no new taxa and only one new combination.

■ FREDERICK MANSON BAILEY

When statehood was achieved by the Australian colonies, each government engaged a botanist to undertake taxonomy of a more regional nature. Frederick Manson

Figure 2.3 Top left: Portrait of Allan Cunningham (1790–1839), by an unknown artist. Oil painting on wood panel, c. 1835. With permission of the National Library of Australia. **Top right:** Ferdinand von Mueller (1825–1896). With permission from the archives of the Royal Botanic Gardens Melbourne. **Bottom left:** Hermann Wendland (1825–1903). With permission of the Director, Herrenhausen Gardens, Hannover, Germany. **Bottom right:** Oscar Drude (1852–1933). With permission of Dresden Botanic Garden, from B. Ditsch (1999), *Führer durch den Botanischen Garten der Technischen Universität Dresden*.

Figure 2.4 Top left: Portrait of George Bentham (1800–84), CMG, LLD, FRS, PLS. Oil on canvas by E.M. Merrick, undated. [MO 021]. © Board of Trustees of the Royal Botanic Gardens, Kew. **Top right:** Frederick Manson Bailey (1827–1903). With permission of the Queensland Herbarium. **Bottom left:** Odoardo Beccari (1843–1920). With permission of the Director, Museo di Botanica, Museo di Storia Naturale di Firenze, Italy. **Bottom right:** Karl Ewald Maximilian Burret (1883–1964). With permission from the Archives of the Botanic Garden and Botanical Museum Berlin-Dahlem, Germany.

Bailey (b.1827, d.1915) (Fig. 2.4), an English migrant, was appointed Queensland Government Botanist in 1875 and Colonial Botanist in 1881, a position he maintained until his death (White 1950; Sanderson 2007). Bailey collected widely in Queensland, visiting Rockingham Bay, the upper Herbert River and Seaview Range in 1873, Cairns, nearby ranges and the Barron River in 1877, and the Bellenden Ker Range in 1889 (Johnston 1916; Dowe and Broughton 2007). In *Queensland Flora*, Bailey included 15 genera and 29 species of palms for Queensland (Bailey 1902). During his career, Bailey described 12 new Australian palm taxa and reassigned others to different genera (Dowe 2007 b).

■ KAREL DOMIN

One of the few botanists to work on palms was Karel Domin (b.1882, d.1953), a Czech botanist who visited

Australia from December 1909 to April 1910 (Chapman 1990). He collected in Queensland and New South Wales. From his collections now in herbaria in Prague, Kew and Göttingen, he compiled a floristic treatment of the areas that he visited in Australia (Domin 1915). Domin described two new palm species, *Bacularia microcarya* (= *Linospadix microcaryus*) and *B. aequisegmentosa* (= *Linospadix palmerianus*).

■ ODOARDO BECCARI

The Italian botanist Odoardo Beccari (b.1843, d.1920) (Fig. 2.4) was the most prolific palm taxonomist to work in the Asia–Pacific region (Moore 1981). Based at the University of Florence, Beccari travelled widely in south-east Asia and New Guinea (Cuccuini and Nepi 2006). He visited Australia and met Mueller in Melbourne (van Steenis-Kruseman 1950). His contribution to Australian palm taxonomy was significant. He described new taxa, many of which were New Guinea taxa also distributed in Australia, and others endemic to Australia (Beccari 1931; Beccari and Pichi-Sermolli 1955).

■ MAX BURRET

Karl Ewald Maximilian Burret (b.1883, d.1964) (Fig. 2.4), a German botanist based in the Berlin herbarium, developed an interest in tropical botany and became a specialist palm taxonomist who was most productive during the 1930s. Although Burret travelled in Europe, Brazil and south-east Asia, he was primarily a herbarium-based botanist whose many new palm taxa were described on minor characters (Potztal 1958). Burret's contribution to Australian palm taxonomy was significant, although all his Australian taxa have now been reduced to synonymy.

■ POST-1900

By the beginning of the 20th century, a broad understanding of the diversity and distribution of the Australian palm flora was established. Most genera had been recognised, but only about half of the currently recognised species had been described. Taxonomic work on Australian palms since 1900 has primarily dealt with the addition of new species to the established genera, some generic reassignments and the addition of a small number of new genera.

■ RECENT CONTRIBUTIONS TO AUSTRALIAN PALM BOTANY

After the work of Burret in the 1930s, there was a wane in the taxonomy of Australian palms for 30 years. Although accounts appeared on the distribution and origin of some *Livistona* species (Lothian 1958, 1959), little botanical effort was applied to palms. It was suggested that the apparent lack of interest in palm taxonomy was due to the nature of palms, being difficult to collect and to conserve in herbaria (Johnson 1981). However, a number of botanists took an interest in the family, albeit in a synoptical approach. A classification outline of Australian palms with keys to subfamilies and some suggestion of the taxonomic work required was provided by Rodd (1983). Rodd also commenced revising *Livistona*, publishing the results 15 years later in *Telopea* where 16 species were recognised for Australia, five of which were described as new (Rodd 1998). A.K. Irvine researched Queensland palms from the 1970s to the 1990s and produced taxonomic works that established two new genera, *Wodyetia* (Irvine 1983) and *Oraniopsis* (Dransfield *et al.* 1985). Irvine collaborated with Dowe to provide a revision of *Linospadix* (Dowe and Irvine 1997). There have been accounts of palms occurring in Australian territories and possessions, such as Christmas Island (Du Puy and Telford 1993), Lord Howe Island and Norfolk Island (Green 1994). A revision has been completed for *Archontophoenix* (Dowe and Hodel 1994) and there have been treatments and nomenclatural assessments for *Livistona* (Rodd 1998; Dowe and Barfod 2001; Dowe and Jones 2004; Dowe 2009a), *Licuala* (Barfod and Dowe 2005) and *Ptychosperma* (Dowe 2007a).

■ SUMMARY

This chapter discusses the most significant early documentation of Australian palms. The first Australian palm species were described in 1810, including *Livistona humilis*, *L. inermis*, *L. australis* (as *Corypha australis*) and *Ptychosperma elegans* (as *Seaforthia elegans*). New taxa have been described at irregular intervals, the most recent in 2005. Periods of taxonomic activity were often the result of the production of floras or the work of specialist palm botanists (Table 2.1).

Table 2.1 Chronology of the taxonomic recognition of Australian palm taxa

Year	Author(s)	No. of taxa	Source or publication
1770	Solander	3 spp.	*Endeavour* Voyage documents: unpublished informal names
1810	Brown	4 spp.	*Prodromus Florae Novae Hollandiae*
1814	Brown	6 spp.	*Addenda* in Flinders
1850	Martius	7 spp.	*Historia Naturalis Palmarum*
1875	Wendland and Drude	26 spp.	*Palmae Australasicae, Linnaea* 39
1878	Bentham	21 spp.	*Flora Australiensis,* Vol. 7
1899	Mueller	25 spp.	*2nd Systematic Census*
1983	Rodd	51 spp.	*Flowering Plants in Australia*
1984	Jones	57 spp.	*Palms in Australia*
1998	Rodd	59 spp.	*Telopea* 8
2009	Du Puy and Telford (1983), Green (1994), Dowe and Jones (*pers. comm.*)	60 spp.	*Flora of Australia* Vols. 39, 49, 50

3. Historical biogeography

This chapter introduces the historical biogeography of Australian palms and examines the development of the palm flora over time and space. Aspects of the fossil record are cautiously invoked to probe the deep history of palms in Australia. Palms are placed in the context of climate change across the Australian continent, adaptation and past and present regional distributional relationships.

■ THE FOSSIL RECORD

Compared to other areas of the world, Australia's palm fossil record is limited (Greenwood and Conran 2000) and it is difficult to assign modern affinities or establish clear historical contexts. Harley (2006), in a summary of global palm fossils, described the world's palm fossil record as 'rich and widespread' but cautioned against assigning modern affinity too freely. Overall, palm fossil records have a strong bias toward the northern hemisphere but indicate that the palm floras of some areas, such as Africa and India, were previously much more diverse than they are today (Pan *et al.* 2006). The earliest unequivocal palm fossils date from the Late Cretaceous and were recovered from sites in South Carolina and New Jersey (Berry 1914; Daghlian 1981; Dransfield *et al.* 2008).

Palms appear in the fossil record in Australia later than they do elsewhere in the world. The first indisputable records date to the Palaeocene and, with increasing diversity and distributional range, into the Eocene and Oligocene. They become less common through to the Late Miocene and Pliocene (Fig. 3.1). There are a number of possible reasons for the late first appearance of palm fossils in Australia, and for their lesser diversity compared to other areas:

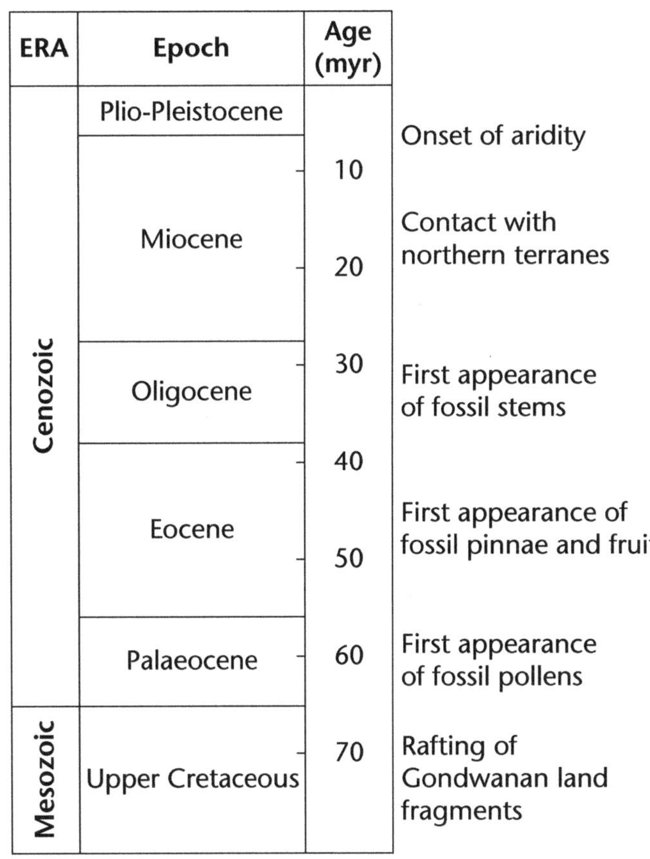

Figure 3.1 Geological time-scale with major events and estimated time of first appearance of palm fossils in Australia and New Zealand.

- palms evolved elsewhere and only reached Australian landmasses following a later radiation;
- the taphonomic potential may not have been available in the Australian environment, if palms were indeed present in the same numbers and diversity as elsewhere;
- sedimentary basins, where fossils have relatively greater potential to survive, have been eroded over much of the Australian landscape;

- the search for palm fossils has not been as thorough, or as successful, as elsewhere.

■ AUSTRALIAN AND NEW ZEALAND PALM FOSSILS

Although this work primarily deals with the Australian palm flora, the inclusion of New Zealand in the examination of the regional evolution of palms is warranted on the basis of a closely related floristic development, particularly with similarities of Eocene and Oligocene palaeofloras in each area (Martin 1994). The palm flora of New Zealand includes a single extant palm genus, *Rhopalostylis*, with two species, of which one, *R. baueri*, also occurs on Norfolk Island in the Australian geopolitical region. The palm fossil record suggests palms were more diverse in New Zealand, at least in the Eocene and Miocene, than they are now (Mildenhall 1980; Lee *et al.* 2001).

As with palm fossils elsewhere, those from Australia and New Zealand conform to two categories of preservation: macrofossils, which consist of organs such as stems, leaves, pinnae, inflorescence parts, fruit and root impressions; and microfossils, such as phytoliths or pollens. Microfossils include examples of monosulcate, disulcate and meridionally zonasulcate types that, in some cases, have been aligned to modern palm taxa (Fig. 3.2). Phytoliths (plant stones) are silica-based structures formed within and between the living cells of most monocotyledons and some dicotyledons, and are readily preserved as fossils (Kershaw *et al.* 2000; Albert *et al.* 2009). They are abundant in palms primarily as echinate spheres c. 5–20 µm in diameter (Fig. 3.2), although some other types are known (Wallis 2001, 2003).

The earliest palm fossils in Australia and New Zealand have been dated to the Palaeocene (Mildenhall 1980; Pole 1993; Greenwood and Conran 2000; McGowran *et al.* 2000) and there are records from all epochs, with varying degrees of diversity, from the Eocene to the Holocene (Couper 1952, 1953; Ballance *et al.* 1981; Greenwood and Conran 2000; Conran and Rozefelds 2003). In addition to macrofossils being assigned to modern taxa and placed in more or less irregular time sequences, palm pollen types have been found in most stratigraphic sequences through the Palaeogene and Neogene (Martin 1994; Macphail *et al.* 1994; Blackburn and Sluiter 1994; Dowe 1995; Greenwood and Conran 2000; Raine *et al.* 2008). Phytolith analysis in Australia is presently underutilised although investigations have dated previous distributions of some palm species in the present epoch. Wallis (2001) recorded two forms of palm phytoliths at sites in the Napier Range, Western Australia, and suggested palms were present there 33–40 k years bp, beyond their present distribution to the north in King Leopold Range. Climate change was suggested as a possible cause of their decline in the southern location (Wallis 2001).

The Australian and New Zealand palm fossils that have been related to extant taxa include the following.

- *Nypa*, with the extant species *Nypa fruticans* (Fig. 3.3), with macrofossils and meridionally zonasulcate pollen as the fossil genus *Spinizonocolpites* from the Palaeocene to the Pleistocene (Cookson and Eisenack 1967; Stover and Partridge 1973; Kemp and Harris 1977; Shafik 1978; Pole 1998), and the Eocene species *Nypa australis* (Pole and Macphail 1996) (Fig. 3.3). Muller (1968) and Harley *et al.* (1991) proposed that *Nypa* was formerly more specious than at present, as there is considerable diversity in fossil pollens compared to the uniformity of extant *N. fruticans* (Baker *et al.* 1998). Collinson (1993) found fossil fruits attributed to *Nypa* within the size range of extant *Nypa fruticans*.
- *Cocos nucifera*, as a Late Pliocene fruit (Rigby 1995) (Fig. 3.4).
- Cocosoid affinity, as *Cocos zeylandica* with Miocene fruit (Berry 1926; Ballance *et al.* 1981) (Fig.3.4).
- *Rhopalostylis* cf. *sapida*, with Miocene pinnae, inflorescences and fruit as *Phoenicites zeelandica* (Ettingsh.) Pole (Pole 1993) (Fig. 3.5), and as monosulcate pollen and Quaternary root systems (Ettingshausen 1887; Pole 1993, 2007; Gregory and Campbell 2003).
- Coryphoid affinity, as Oligocene stems (Conran and Rozefelds 2003) (Fig. 3.5).
- Calamoid affinity, as Palaeocene to Miocene disulcate pollens in Australia and New Zealand, and as Late Eocene pinnae, spines and fruit in New Zealand (Lee *et al.* 2009). The fossil pollens have been named *Dicolpopollis bungonensis* and *D. metroxylonoides*, and putatively related to *Calamus* and/or *Metroxylon* (Khan 1976; Truswell and Owen 1988; Greenwood and Conran 2000). Extant *Calamus* species, distributed east of Wallace's Line (to include Australian species) are possible speciation events from an eastward radiation across Wallace's Line from about the Miocene (Dransfield 1981; Baker *et al.* 1998; Baker and Dransfield 2000)

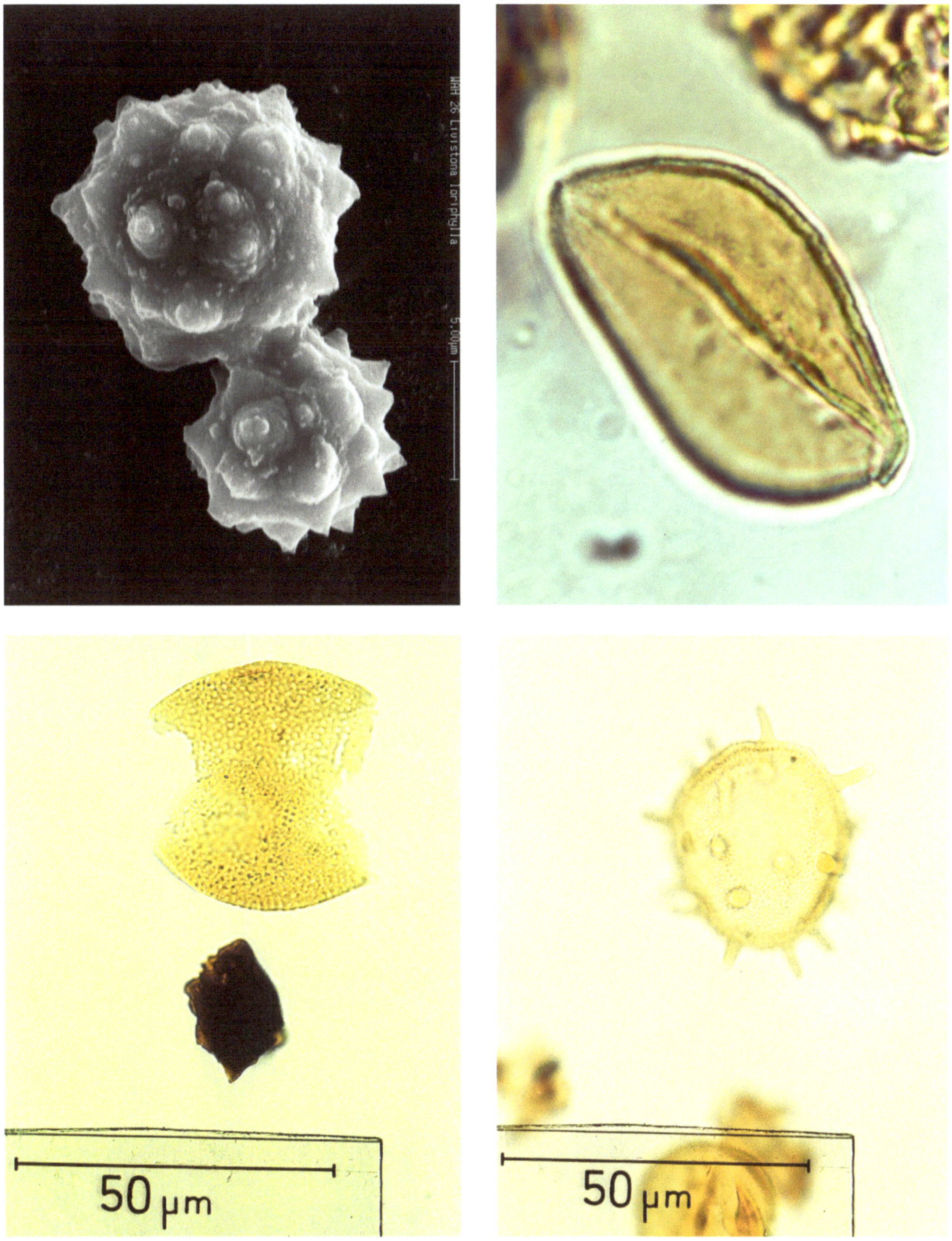

Figure 3.2 Examples of phytoliths and fossil palm pollens from Australia and New Zealand. **Top left:** Phytolith from inflorescences of *Livistona lorophylla*, Napier Range, Western Australia (PERTH 04621786). Image by Lynley Wallis, with permission. **Top right:** Monosulcate pollen type: Arecipites sp., Kawarau River, Central Otago, New Zealand, Early to Middle Miocene. Collection of GNS Science, New Zealand: photo by J.I. Raine. **Bottom left:** Disulcate pollen type: Dicolpopollis sp., Great Australian Bight. Collected by Geoscience Australia, Canberra: photo by Mike Macphail. **Bottom right:** Meridionally zonosulcate pollen type: *Spinizonocolpites prominatus*, from late Early Eocene, Macquarie Harbour Beds, Regatta Point, Tasmania. Collected by Geoscience Australia, Canberra: photo by Mike Macphail.

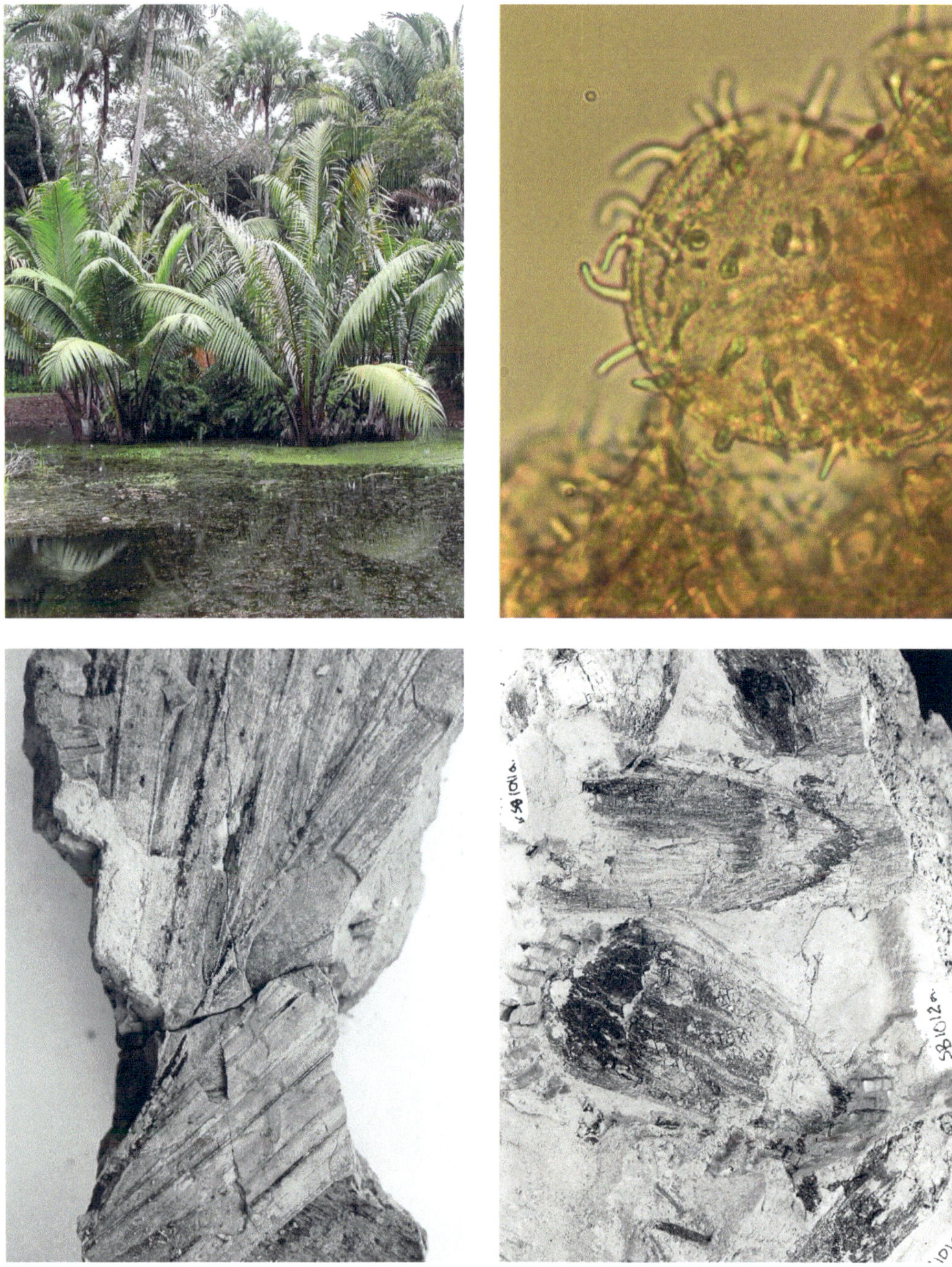

Figure 3.3 *Nypa fruticans* and fossils with affinity to *Nypa*. **Top left:** *Nypa fruticans*, Centenary Lakes, Cairns, north-east Queensland. **Top right:** Fossil pollen *Spinizonocolpites* sp., with affinity to *Nypa*, Lowana Rd, Regatta Point, Tasmania, Early Eocene. Collection and images by Greg Jordan, University of Tasmania. **Bottom left:** Fossil leaf, *Nypa australis*, Regatta Point, Tasmania, Eocene. **Bottom right:** Fossil fruit, *Nypa australis*, Regatta Point, Tasmania, Eocene. *Nypa australis* images by Mike Pole, with permission.

or possibly earlier in the Eocene (Harley and Morley 1995). This eastward radiation post-dates the fossil record in Australia of Calamoid fossil pollens from the Palaeocene and Mid Eocene (Truswell and Owen 1988), and the Late Eocene fossil record of pinnae, spines and fruit (Fig. 3.5) in New Zealand (Lee *et al.* 2009; Hartwich *et al.* in prep.). It suggests the presence of *Calamus* and/or *Metroxylon* or relatives in Australia

Figure 3.4 Fossils with Cocosoid affinity. **Left:** Fruit of *Cocos nucifera*, Chinchilla sands, southern Queensland, Late Pliocene. Image from Rigby (1995). **Right:** Endocarp of *Cocos zeylandica*, Cooper's Beach, Mangonui, Northland, New Zealand, Miocene. With permission of GNS Science, New Zealand: photo by Marianna Terezow.

and New Zealand prior to the putative Miocene radiation (Dransfield 1987). Based on the presence of disulcate fossil pollens, Baker and Dransfield (2000) speculated that a southern Calamoid line may have existed in Australia prior to Miocene contact but did not persist, and that extant Australian *Calamus* are derived from Asian stock.

Some Palaeocene to Miocene fossils have been tentatively related to the Arecoid palms (macrofossils and monosulcate pollens) and the Coryphoid palms (monosulcate pollens) as *Arecipites* spp. Researchers recommend caution with identification as this form of pollen also occurs in other plant groups, such as the Liliaceae (Truswell 1990) and is an Arecoid-like pollen of uncertain affinity (Dransfield *et al.* 2008). The fossil genus *Longapertites*, which is similar to *Dicolpopollis* but with an extended sulcus, was present on Australia's northern margin by the Late Cretaceous, but only has a rare Early Eocene occurrence in central Australia (McGowran *et al.* 2000). Harley (2006) and Dransfield *et al.* (2008) related this pollen type to a number of extant palm taxa presently occurring in Australia, including *Hydriastele* and *Licuala*.

Some well-preserved macrofossil palms from Australian sites have yet to be investigated and formally described. They include a pinnate-leafed palm of Eocene/Miocene age from near Island Lagoon, South Australia (*South Australian Museum P14209*) (Fig. 3.6) and stem portions of unknown age from a gravel deposit in the channel of Fortescue River, near Livistona Pool, Millstream, Western Australia (*Western Australian Museum P90.7a*) (Fig. 3.6). Leaves attributed to *Linospadix* are reported as Mid Miocene, from Anglesea, Victoria (Christophel and Greenwood 1988; Greenwood and Conran 2000). Eocene palm leaves are reported from Barton Range, South Australia (Greenwood and Conran 2000) and from Oxley, Queensland (Ettingshausen 1895).

The fossil record indicates palms were widespread over much of the Australian continent (Fig. 3.7) and New Zealand in previous epochs and were locally uncommon to dominant in many areas where they no longer occur (Truswell *et al.* 1987; Martin 1993; Greenwood and Conran 2000; Pole 2007). Fossil palm pollens first appear in the Palaeocene, fossil pinnae and fruit in the Eocene and stems in the Oligocene (Table 3.1). Fossil leaves are exclusively of pinnate or pinnately veined entire leaves. There are no palmate leaf fossils known from Australia. There is some evidence from the fossil record to suggest that genera such as *Archontophoenix*, *Calamus*, *Cocos*, *Linospadix* and *Rhopalostylis*, or their ancestors or close relatives, may have been present on the continent and adjacent landmasses prior to Miocene contact with Laurasian terranes, and

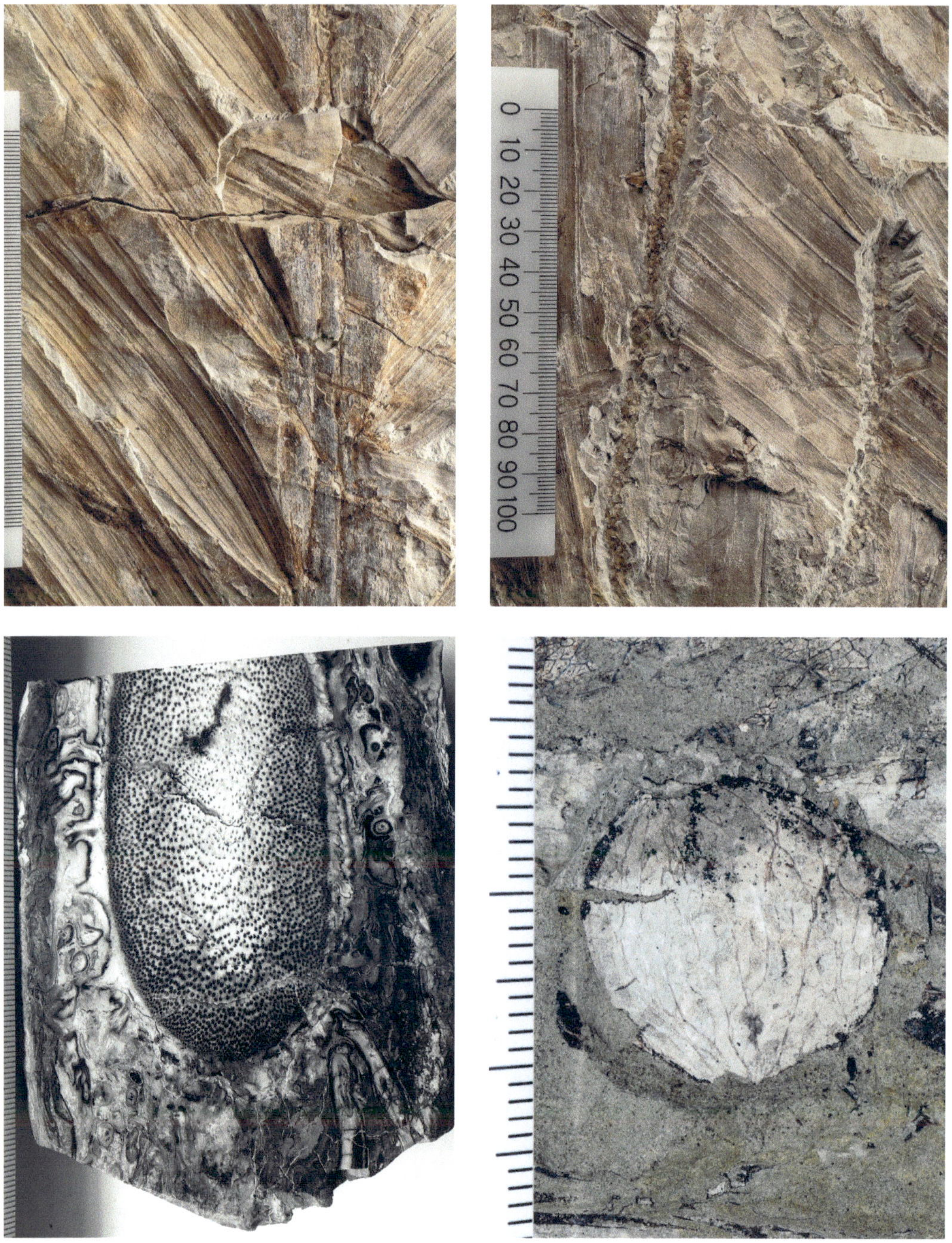

Figure 3.5 Fossils with Arecoid, Coryphoid and Calamoid affinity. **Top left:** Fossil leaf, *Phoenicites zeelandica*, Central Otago, New Zealand, Early Miocene. **Top right:** Fossil inflorescence rachillae and fruit, *Phoenicites zeelandica*. Reproduced with permission of Department of Geology, University of Otago, New Zealand: photos by R. Ewan Fordyce. **Bottom left:** Fossil periminera-lised Oligocene stem, *Palmoxylon queenslandicum*, Springsure, central Queensland. **Bottom right:** Fossil fruit, Calamoid affinity, Pikopiko, South Island, New Zealand, Late Eocene. Images of *P. queenslandicum* and Calamoid fruit by John C. Conran, University of Adelaide, with permission.

therefore may be interpreted as autochthonous elements if identifications are correct. There is also evidence, again cautiously interpreted, that some evolutionary lines have only recently become established in Australia, with their closest relatives in New Guinea and central and western Malesia: this proposal will be further examined below.

■ CLIMATE CHANGE SINCE THE LATE CRETACEOUS

Broad-scale biogeographical studies for the Australasian region mostly agree that landmasses related to present-day Australia and associated terranes drifted northward as a consequence of plate tectonics with the break-up of Gondwana (Veevers 1991; Kroenke 1996; Metcalfe 2002). Various scenarios present different arrangements of landmasses and the timing of geological and environmental events (Lange 1982; Truswell and Harris 1982; Truswell 1993; Frakes 1999; Johnson 2004). The moving landmasses may have retained their original biotic elements, and significant exchange or radiation of other biotic elements occurred only when contact was made with other terranes (Morley 2002). With eventual stabilisation or slowing-down of tectonic movement, many floras were in place by the Eocene–Miocene, possibly earlier for organisms with high vagility (Burrett *et al.* 1991). By the Miocene and into the Pliocene, the inherited flora which was to become an 'Australian flora' had been affected by

Figure 3.6 Fossil palms from Australian sites. **Top:** Pinnate-leafed palm of Eocene/Miocene age from near Island Lagoon, South Australia [South Australian Museum P14209]. Image by N. Pledge. **Bottom left:** Exterior view of a palm stem portion of unknown age from a gravel deposit in the channel of Fortescue River, near Livistona Pool, Millstream, Western Australia [Western Australian Museum P90.7a]. **Bottom right:** Interior view of same.

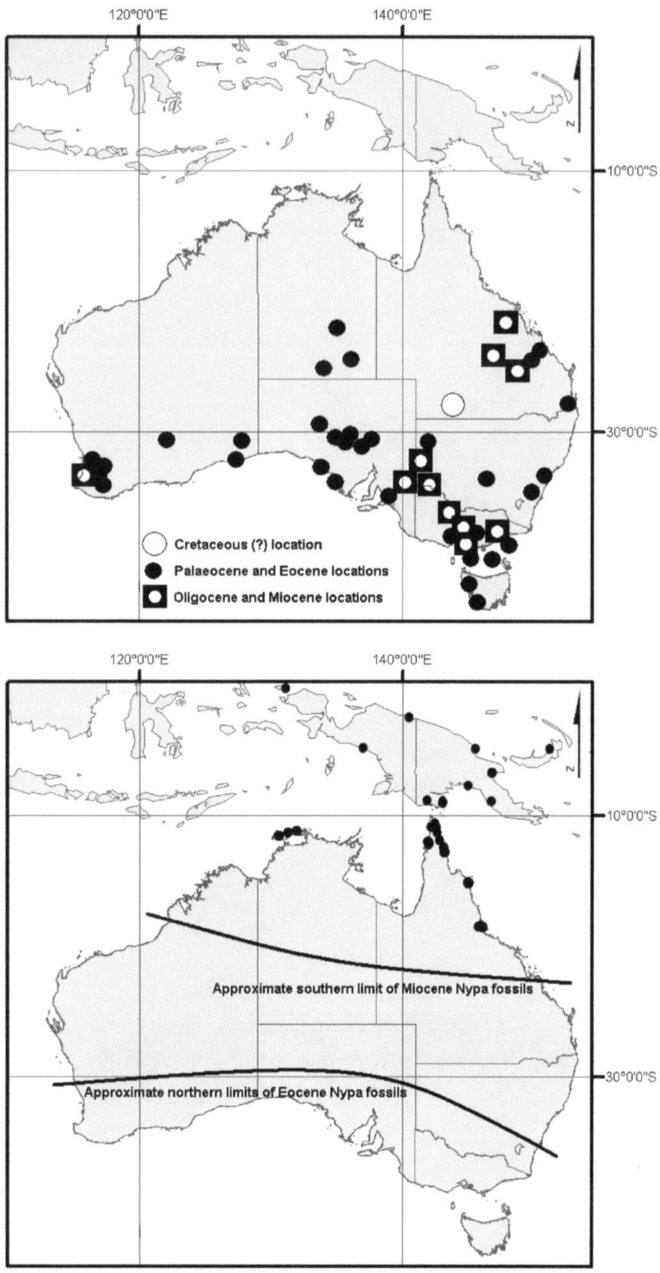

Figure 3.7 Top: Map indicating the sites where macrofossils and fossil pollens of palms have been recovered in Australia: O = Cretaceous location, leaves attributed to an early representative or relative of the Areciflorae (Pole 1999) [one site only] • = Palaeocene and Eocene locations; ◘ = Oligocene and Miocene locations (see Table 3.1 for details). **Bottom:** The approximate limits of Eocene and Miocene fossil *Nypa* records and the extant distribution of *Nypa fruticans* in Australia and New Guinea indicated by filled circles. Distribution based on herbarium specimen locations. NB: Distribution of *N. fruticans* is more extensive and continuous in coastal New Guinea than indicated by herbarium specimens.

increasing aridity and seasonality of rainfall – the continent continued to move into different climate zones and global and regional climate patterns changed (Martin 1990; Kershaw *et al.* 1994; Greenwood 1996; Crisp *et al.* 2004).

The appearance and subsequent disappearance of palms from areas in southern Australia, evidenced primarily by fossil pollens, have been invoked as indicators of climate change. *Nypa* (as *Spinizonocolpites* spp.) first appeared in southern Australia in the fossil record of the Palaeocene to Early Eocene, and has been recovered from sites in south-western Western Australia, South Australia, Victoria and Tasmania (Churchill 1973; Alley and Clarke 1992; Pole and Macphail 1996) (Fig. 3.7). By the Mid to Late Eocene, *Nypa* fossils no longer appeared in these southern areas but had begun to appear in northern areas (e.g. Capricornia Basin, Queensland), where they persisted to at least the Early Miocene (Hekel 1972; Macphail *et al.* 1994). By the end of the Miocene, there is considerable concordance between fossil records and the extant distribution of *Nypa* (Baker *et al.* 1998). The present distribution of *Nypa* in Australia is restricted and disjunct, confined to northern Queensland and the Cobourg Peninsula and Melville Island of the Northern Territory (Covacevich 1981; Duke 2006), although it is widespread and common along the coasts of New Guinea and north through south-east Asia to Sri Lanka and the Ryukyu Islands of Japan. Fossil *Nypa* pollen has been recovered from sites in Borneo for all epochs from the Late Cretaceous to the Quaternary, which suggests it has been occupying that area continuously (Muller 1972), in contrast to the relatively brief appearances in the Australian fossil record. The global fossil record of *Nypa* indicates that it has occurred in numerous places throughout the world at various times (Gee 2001). It is assumed that the megathermal climatic requirements of *Nypa* have remained constant over time, and therefore areas where it once occurred would have had similar climatic regimes and environments as *Nypa* requires today. Collinson (1993) showed that the fruiting biology of *Nypa* has remained unchanged, as the fossil fruits display no significant difference in average size or shape from modern fruits.

Similar conclusions are possible for other palm fossils. The appearance of putative Calamoid relatives, described under the pollen form genus *Dicolpopollis* and as pinnae, spines and fruit from the Pikopiko fossil forest, New Zealand (Lee *et al.* 2009; Hartwich *et al.* in prep.), commenced in the Eocene of southern Australia and in the Miocene of New Zealand, and persisted to at least the Pliocene in the former. *Dicolpopollis bungonensis*, from southern New South Wales, has been tentatively related

Table 3.1 Summary of fossil palms recognised and/or described for Australia and New Zealand

Fossil taxon	Modern affinity
Areciflorae (Pole 1999) Mid Cretaceous, Thargomindah, Erromanga Basin, Qld	Arecales (?)
Arecoid leaves, putatively recognised but undescribed Eocene, Barton Range, SA; Eocene, Oxley, Qld	Arecoid
Mid Eocene, Anglesea, Vic. (Christophel and Greenwood 1988)	*Linospadix* (?)
Eocene/Oligocene, Plantagenet Beds, WA	*Archontophoenix* (?)
Eocene/Oligocene, Tertiary Island and Stuart Creek, SA	Arecoid
Eocene/Miocene, near Island Lagoon, SA	Arecoid (?)
Arecoid pollen, putatively recognised but undescribed Palaeogene, Fowlers Gap, NSW; Palaeocene to Mid Eocene, Haggard Hill, SA; Early Eocene, Lake Eyre Basin, SA; Eocene, Darkin Swamp, WA; Mid Eocene, Rundle, Qld; Mid Eocene, Maslin Bay, SA; Eocene to Miocene, Bottle Hill, SA; Late Eocene, Muradup, WA; Late Eocene, NW of Kalgoorlie, WA; Late Eocene to Mid Miocene, Lachlan River, Murray Basin, NSW; Oligocene to Miocene, Capella, Qld; Oligocene to Miocene, Murray Basin, Vic.; Late Oligocene to Early Miocene, Barossa Basin, SA; Late Oligocene–Early Miocene, Morwell, Vic.	Arecoid
Arecipites insularis Harris (in Kemp and Harris 1977) pollen – Palaeocene, Ninetyeast Ridge, Indian Ocean	Arecoid
Arecipites minutiscabratus (McIntyre) Milne (1988) [*Monosulcites minutiscabratus* McIntyre (1968)] pollen – Palaeogene, New Zealand; Eocene, Eucla Basin (100 km east of Norseman), WA	Arecoid
Arecipites otagoensis (Couper) Mildenhall and Pochnall (1989) [*Monosulcites otagoensis* Couper (1960)] pollen – Palaeogene to Neogene, NZ	Arecoid
Arecipites subverrucatus (Pocknall) Mildenhall and Pocknall (1989) [*Monosulcites subverrucatus* Pocknall (1985)] pollen – Palaeogene, NZ	Arecoid
Arecipites waitakiensis (McIntyre) Kemp and Harris (1977) [*Monosulcites waitakiensis* McIntyre (1968)] pollen – Palaeogene to Neogene, Pareora, NZ; Eocene, Eucla Basin (100 km east of Norseman), WA; Eocene, Bungonia, NSW; Oligocene, Berwick Quarry, Melbourne, Vic.; Late Oligocene to Early Miocene, WA; Oligocene to Miocene, Oakvale, western Murray Basin, SA; Late Oligocene, Ninetyeast Ridge, Indian Ocean	Arecoid
Arecipites sp. (Martin 1994; Truswell and Marchant 1986; Macphail *et al.* 1993; Dettman and Clifford 2000) pollen – Early Eocene, Tas.; Mid to Late Eocene, Hale River (c. 100 km NE of Alice Springs), southern NT; Mid to Late Eocene to Oligocene, Yamba Basin, Rockhampton, Qld	Arecoid
Calamoideae (Lee *et al.* 2009) pinnae with spines and fruit – Late Eocene, South Island, NZ	*Calamus* (?)
Cocos nucifera L. fruit – Late Pliocene, Qld (Rigby 1995)	*Cocos nucifera*
Cocos zeylandica Berry (1926) pollen and endocarps – Miocene, Mangonui, Coopers Beach, North Island, NZ; Miocene, Waikiekie, North Island, NZ; Miocene, Matawai, North Island, NZ	*Parajubaea*(?), Endt and Hayward 1997)

(*table continued*)

Table 3.1 Continued

Fossil taxon	Modern affinity
Dicolpopollis bungonensis Truswell and Owen (1988) pollen – Mid Eocene, Bungonia, NSW; Oligocene to Miocene, Oakvale, western Murray Basin, SA	*Calamus moti*
Dicolpopollis metroxylonoides Khan (1976) pollen – Palaeocene, Little Bay, NSW; Oligocene to Miocene, Geera Clay (near Renmark), Murray–Darling Basin; Late Oligocene, Gippsland Basin, Vic.; Late Miocene, NZ	*Metroxylon* (*Calamus* ?)
Longapertites sp. pollen – Early Eocene, central Australia (McGowran *et al.* 2000)	Arecoid
Monosulcites palisadus Couper (1960) pollen – Cretaceous (?), NZ	Arecoid
Nypa australis Pole and Macphail (1996) pinnae and fruit – Lower Eocene, Regatta Point, Strahan, Tas.	*Nypa* sp.
Palmidites maximus Couper (1953) pollen – Miocene to Pliocene, NZ	Arecoid
Palmoxylon queenslandicum Conran and Rozefelds (2003) trunk – Oligocene, Minerva Creek, Springsure, Qld	Coryphoid
Phoenicites zeelandica (Ettingsh.) Pole (1993) [*Seaforthia zeelandica* Ettingsh. (1887)] pinnae, inflorescence rachillae and fruit – Miocene, NZ	Arecoid – Rhopalostylidinae (?)
Proxapertites sp. pollen – Early Eocene, central Australia, (McGowran *et al.* 2000)	Arecoid
Rhopalostylis sapida H.Wendl. & Drude (1875) pinnae – Palaeogene to Neogene, Manuherikia, South Island, NZ; pollen – Miocene, Mangonui, Coopers Beach, North Island, NZ; root impressions – Quaternary, Henderson Bay, North Island, NZ	*Rhopalostylis sapida*
Spinizonocolpites prominatus (McIntyre) Stover and Evans (in Stover and Partridge 1973) [*Spinizonocolpites baculatus* Muller (1968); *Spinizonocolpites echinatus* Muller (1968); *Spinizonocolpites perspinosus* (Couper) Hekel (1972); *Monosulcites perspinosus* Couper (1953); *Baltisphaeridium taylorii* Cookson and Eisenack (1965)] pollen – Palaeocene, Bight and Duntroon Basins (c. 1000 km west of Adelaide), SA; Palaeocene, Tas.; Palaeocene to Late Eocene, NZ; Eocene, Tas.; Eocene, Bass and Otway Basins, Bass Strait; Eocene, off-shore Gippsland Basin, Vic.; Eocene, Eucla Basin, WA; Early to Mid Eocene, Vic.; Mid Eocene, Kings Park , Perth Basin, WA; Oligocene to Miocene, near Rockhampton, Qld; Late Oligocene, Ninetyeast Ridge; Late Oligocene to Early Miocene, Barossa Basin, SA	*Nypa* sp.

to the extant *Calamus moti* F.M.Bailey (Truswell and Owen 1988), which is presently restricted to the Wet Tropics Bioregion of north-east Queensland. The appearance in and departure from the fossil record of *Dicolpopollis* in southern Australia by the Miocene may be evidence of increasing then decreasing temperatures during that time (Truswell *et al.* 1985; Truswell and Owen 1988). Pole (2003) interpreted the Early Miocene increase and Late Miocene decrease of Arecoid pollen from New Zealand as evidence of changing moisture levels rather than temperature, which he related to changing atmospheric circulation patterns then affecting the southern hemisphere. Assuming palms are indicators of meso-megathermal climatic regimes, their appearance in and disappearance from areas in the Australian region circumstantially support similar evidence suggested for the flora in general (Macphail *et al.* 1994; Frakes 1999; Pole 2003). The fossil record across many families indicates there has been an overall contraction of rainforest and an expansion of sclerophyllous vegetation in Australia since the Palaeogene (McGowran *et al.* 2000).

■ CO-OCCURRENCE OF PALM SPECIES IN AUSTRALIA AND NEW GUINEA

Comparison of the floras across Torres Strait has been the subject of considerable research and discussion (Hedley

1893; Webb and Tracey 1972; Hoogland 1972; Rowe 2007). It has revealed that both low and high genetic diversity (Moran *et al.* 1989; Prider and Christophel 2000) occur in taxa distributed in northern Australia and New Guinea.

The distribution of palms presently co-occurring in New Guinea and northern Australia can be partly explained by submergence of the Torres Strait land-bridge through rising sea-levels (Chivas *et al.* 2001; Prebble *et al.* 2005; Rowe 2007). The present landmass of New Guinea was initially formed by the collision and uplift of crustal plates in the Mid Miocene (Doutch 1972) although high mountain-building has been a more recent event. Marine transgressions into the area commenced in the Holocene, followed by alternating marine/estuarine and terrestrial conditions with rising and falling sea-levels (Reeves *et al.* 2008). The latest marine transgression in Torres Strait has been estimated as occurring 6000–9700 years bp, with the present arrangement of coastlines stabilising about 6000 years bp (Barham 1999; Johnson 2004) (Fig. 3.8). The distribution of some endemic palm species in Australia, such as *Livistona rigida*, presently occurring on rivers that flow into the Gulf of Carpentaria and formerly on palaeo-rivers into Lake Carpentaria (Fig. 3.8), is the result of rising sea-levels. The palm species that presently co-occur in Australia and New Guinea are presented in Table 3.2.

For the shared palm taxa, it is difficult to determine if their former distribution included what was to become northern Australia and New Guinea prior to marine transgression and the severing of the Australia/New Guinea land-bridge, but it must be suspected for at least three species, *Livistona benthamii, L. muelleri* and *Corypha utan*. Fossil palm pollens have been recovered from sites in Torres Strait and the present Carpentaria Basin (Prebble *et al.* 2005), but these cannot be specifically related to extant species. It indicates that palms were present on the land-bridges that connected northern Australia to New Guinea. Terrestrial environments were continuous from Cape York to southern New Guinea prior to 10 k years bp, and from east Arhnem Land to south-western New Guinea prior to 18 k years bp. The northerly or southerly movement of plants along these land-bridges was primarily driven by cyclic aridity and associated environmental impact (De Deckker 2001). A previous more or less uninterrupted north–south corridor of closed forest habitats has been proposed for the eastern portion prior to marine transgression (Nix and Kalma 1972). This corridor would have allowed for the dispersal of species, and accounts for the present distribution patterns of some rainforest species of low vagility.

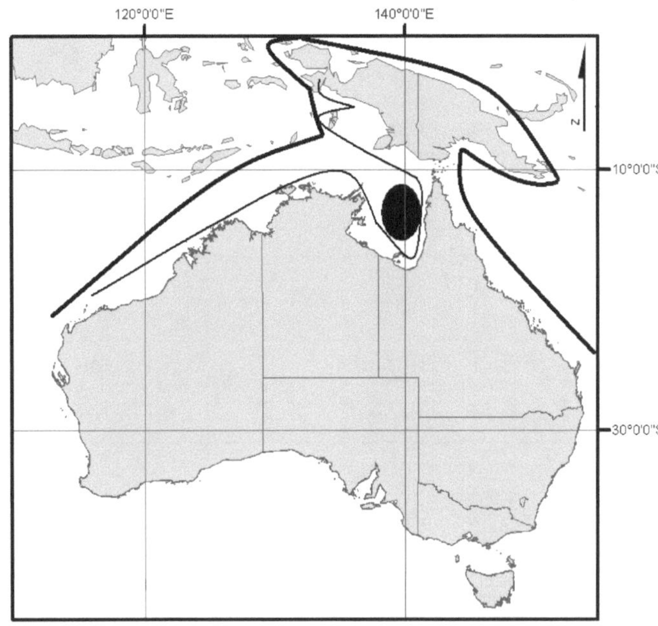

Figure 3.8 The approximate position of shorelines at 18 000 years bp (thick line) and at 10 000 years bp (thin line), and the approximate position of Lake Carpentaria (outlined oval). Adapted from Barham (1999).

There are some distributions that cannot be fully explained by these scenarios, for example, the occurrence of *Hydriastele ramsayi* in the Top End of the Northern Territory. The genus *Hydriastele*, with 48 species, extends from Sulawesi east to Fiji, is most diverse in New Guinea, has two endemic species in Australia and one shared with New Guinea. Of the endemic Australian species, *H. wendlandiana* occupies mesic closed forest similar to that required by most of the New Guinea species, and appears to be closely related to some occurring there. *Hydriastele ramsayi* occupies a unique habitat for the genus, in seasonally dry open woodland that is subject to regular fire and in which the palm forms part of the dominant canopy layer (Fig. 3.9). The adaptation of a single species to open woodland with a monsoonal rainfall/fire regime, in a large genus in which all other species occur in closed moist unburnt forest, may indicate that the genus has been present in Australia for a considerable time. The diversion time of *H. ramsayi* from other species has not been calculated.

Among other species co-occurring in Australia and New Guinea, *Arenga microcarpa* is recorded as rare in the Northern Territory, occurring in near-coastal closed forest habitats. In New Guinea it is a common species distributed throughout the lowlands. *Corypha utan* occurs in the Northern Territory and Queensland, with moderate-sized

Table 3.2 Palm species co-occurring in Australia and New Guinea

Species	Australian populations
Calamus aruensis	Qld, Cape York to Jardine River
Calamus vitiensis	Qld, northern Cape York Peninsula to Wet Tropics Bioregion
Calamus warburgii	Qld, Macrossan Range and Iron Range
Nypa fruticans	Qld, Jardine River to Herbert River; NT, Melville Island and Cobourg Peninsula
Livistona benthamii	Qld, Cape York to Iron Range; NT, Top End
Livistona muelleri	Qld, Torres Strait islands to Cairns
Arenga microcarpa	NT, Top End and western Gulf of Carpentaria
Corypha utan	Qld, Southern Cape York Peninsula; NT, Arnhem Land
Ptychosperma macarthurii	Qld, Cape York to Iron Range; NT, near Darwin
Hydriastele costata	Qld, Cape York to Iron Range

Figure 3.9 *Hydriastele ramsayi* occupies a unique habitat for the genus, in seasonally dry open woodland that is subject to regular fire and in which the palm forms part of the dominant canopy layer. East Branch, West Alligator River, Northern Territory.

populations in east Arnhem Land and large widespread populations in central and southern Cape York Peninsula, Queensland (Fig. 3.10). The species occurs in similar habitats in Australia and southern New Guinea, on flood plains or as a riparian element in seasonally dry monsoonal areas in clay alluvial soils. It is distributed as far west as Sri Lanka, more or less equally across Wallace's Line, and is possibly of recent dispersal (Dransfield 1987) (Fig. 3.10). A high dispersal potential is suggested by the monocarpic massive flowering/fruiting habit with seed disseminated by migrating frugivorous birds.

Livistona displays high diversity in Australia with 18 species, of which 16 are endemic. The two non-endemic species, *Livistona benthamii* and *L. muelleri*, occur in both northern Australian and southern New Guinea (Fig. 3.11). *Livistona benthamii* has wide distribution in Australia, most often as a riparian or seasonal swamp forest element, across the Top End of the Northern Territory, in Cape York Peninsula from Lockerbie Scrub south to Iron Range, and in southern New Guinea. *Livistona muelleri* most often occurs in open to semi-closed woodlands. It is distributed in Queensland from Cape York south to about Cairns, the Torres Strait islands and southern New Guinea. Preliminary molecular analyses (Dowe 2001; Isagi *pers. comm.*) indicate that these two species are in the Australian clade, while other *Livistona* in New Guinea belong to the *Livistona rotundifolia* clade, which has distribution in Indonesia and the Philippines. The New Guinea species in the *L. rotundifolia* clade appear to be a recent radiation from a western source (west of Wallace's Line), and support the concept of distributional and developmental bicentricity in *Livistona* (Dransfield 1981). Pollen records from Holocene sites in the Gulf of Carpentaria have been identified as Arecaceae (Prebble *et al.* 2005), but a relationship to the two *Livistona* species has not been established. It is reasonable to assume the two species were present on the Australia/New Guinea land-bridges and on the shores of Lake Carpentaria prior to the marine transgression which led to the isolation of populations on either side of Torres Strait.

It is possible that some of the shared palm species may be recent migrants to northern Australia, post marine transgression; there is evidence that the southern migration patterns of frugivorous birds favour the dispersal of propagules in a southerly direction (Tucker 1988; Dingle 2004). Some of the above-listed species, such as *Calamus aruensis*, *C. warburgii* and *Hydriastele costata* (Fig. 3.12), are among the most abundant palm species in New Guinea and appear highly vagile as well as adaptable to different types of habitats. The first two have relatively small populations in Australia, suggesting limited introduction.

■ SUMMARY

The historical events that have influenced the distribution of palms in Australia and New Zealand and associated terranes can be summarised as follows:

- Cretaceous Gondwana break-up with subsequent continental drift and the isolation and rafting of the Australian Plate;
- Palaeocene Tasman rifting;
- Eocene formation of proto-New Guinea;
- Oligocene to Miocene volcanism and island creation associated with the Lord Howe Rise;
- Miocene contact of the Australian Plate with other terranes (Wallace's Line);
- Miocene to Pliocene mountain-building in New Guinea;
- Pliocene to Pleistocene aridity, central and southern Australia;
- Holocene marine transgression of Torres Strait.

Palms first appear in the fossil record in Australia in the Palaeocene or possibly late Cretaceous, and occupied areas where they do not occur today. Appearance in and

Figure 3.10 Top: *Corypha utan*, Lakefield National Park, Cape York Peninsula, north Queensland. **Bottom:** Distribution of *Corypha utan*. Wallace's Line is indicated by a heavy line.

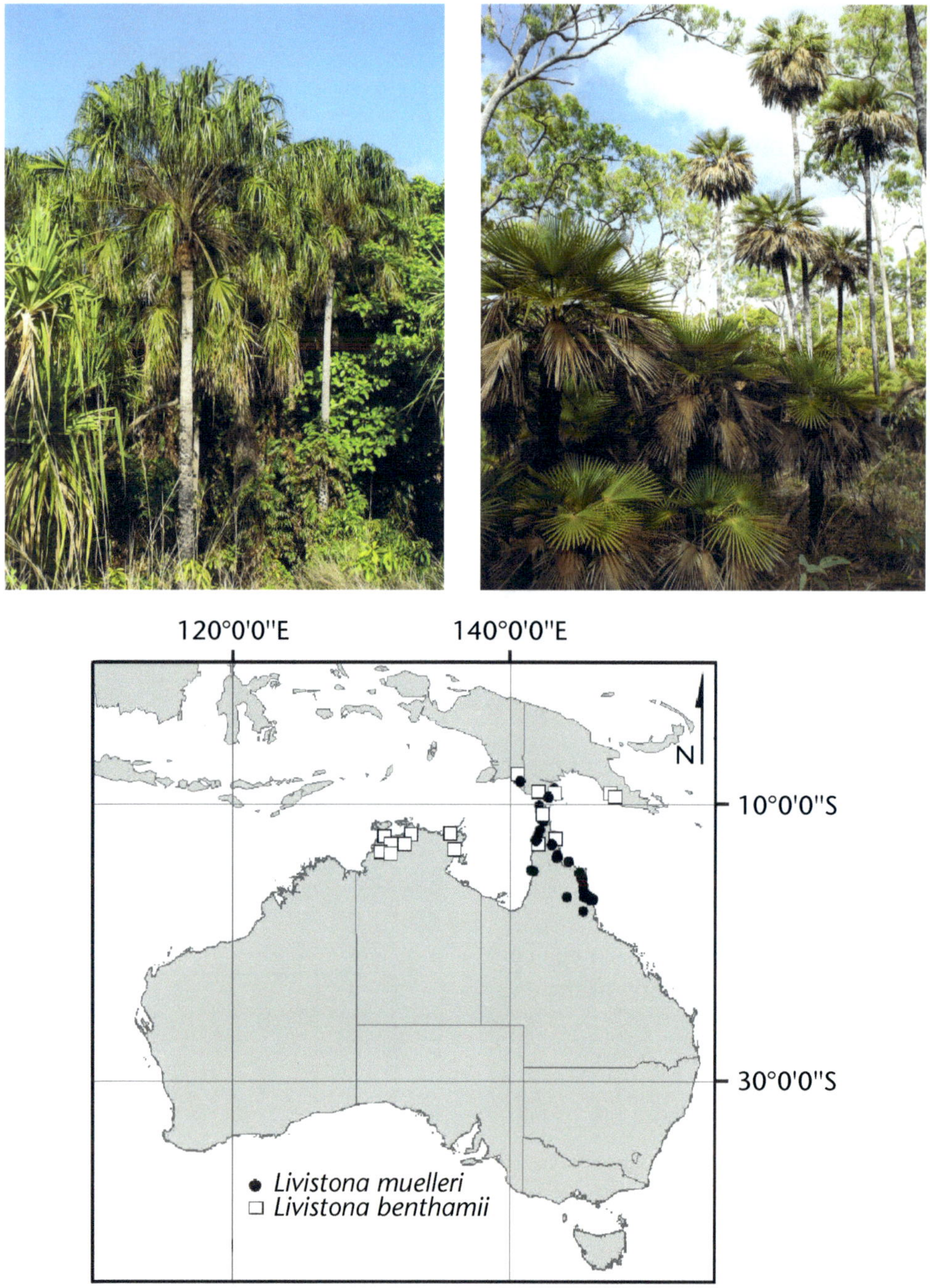

Figure 3.11 *Livistona* species co-occurring in Australia and New Guinea: **Top left:** *Livistona benthamii*, Holmes Jungle, Darwin, Northern Territory. **Top right:** *Livistona muelleri*, south of Jardine River, Cape York Peninsula, north Queensland. **Bottom:** Distribution of *Livistona muelleri* and *Livistona benthamii* in Australia and New Guinea.

disappearance from the fossil record suggests alternating changes in megathermal to mesothermal climate regimes and increasing aridity. The extant Australian palm flora consists of an autochthonous element, and a migrant outlier element established by short- and long-distance dispersal in past and recent times. Processes of vicariant and allopatric speciation have been exerted on autochthonous and migrant elements with high speciation levels in genera such as *Archontophoenix*, *Calamus*, *Linospadix* and *Livistona*.

Figure 3.12 Palms co-occurring in Australia and New Guinea. **Top left:** *Calamus aruensis*, Roma Flats, Lockerbie Scrub, Cape York Peninsula, north Queensland. **Top right:** *Calamus warburgii*, West Claudie River, Iron Range, north Queensland. **Bottom left:** *Hydriastele costata*, Lockerbie Scrub, Cape York Peninsula, north Queensland. **Bottom right:** *Ptychosperma macarthurii*, Pajinka, Cape York Peninsula, north Queensland.

■ SYSTEMATIC ARRANGEMENT OF FOSSIL PALM TAXA FROM AUSTRALIA AND NEW ZEALAND

Synonyms of some taxa applied to non-Australian/New Zealand specimens are not included in this arrangement.

Arecipites insularis Harris, in EM Kemp and WK Harris, *Spec. Pap. Palaeontol.* 19: 54, Plate 4, figs 5, 6 (1977). Type: Indian Ocean, Ninetyeast Ridge, Site 214, 11°20.2″S, 88°43.1″E, Core 43, 402 m sub-bottom, *Slide ST2498/1, 11.8 x 109.3, Py 1018* (GSSA).

Arecipites minutiscabratus (McIntyre) Milne, *Mem. Ass. Australas. Palaeontols.* 5: 295, Fig. 5D (1988); *Monosulcites minutiscabratus* McIntyre, *New Zeal. J. Bot.* 6: 201, Plate 6, figs 82–84 (1968). Holotype: New Zealand. *Slide L2879/1, 42.7 x 125.9* (GNS [NZ]).

Arecipites otagoensis (Couper) Mildenhall & Pochnall, *New Zealand Geological Survey Paleontological Bulletin* 59: 37, Plate 6, Fig. 4 (1989); *Monosulcites otagoensis* Couper, *New Zealand Geological Survey Paleontological Bulletin* 32: 70, Plate 12, figs 13–16 (1960). Holotype: New Zealand. South Island, Central Otago, S125/500 B, *Slide L886/4* (GNS [NZ]).

Arecipites subverrucatus (Pocknall) Mildenhall & Pocknall, *New Zealand Geological Survey Paleontological Bulletin* 59: 38 (1989); *Monosulcites subverrucatus* Pocknall. *New Zeal. J. Geo. Geop.* 28: 340, figs 7–11 (1985). Holotype: New Zealand. North Island, Waikato, Raglan area, Karamu, north of Limeworks Loop Road and Fillery Road junction, *Slide L10931, SM 2809, 15.0 x 101.9* (GNS [NZ]).

Arecipites waitakiensis (McIntyre) Kemp & Harris, *Spec. Pap. Palaeontol.* 19: 54 (1977); *Monosulcites waitakiensis* McIntyre, *New Zeal. J. Bot* 6: 201, Fig. 79 (1968). Holotype: New Zealand. *Slide L1789/4, 50.9 x 109.7* (GNS [NZ]).

Baltisphaeridium taylori Cookson & Eisenack, *Proc. Roy. Soc. Vic.* 79: 137, Plate 16, figs 9, 10 (1965). Holotype: Australia. Victoria, Casterton, Glenelg River, *Kenley s.n., Slide P24083* (Botany Dept, University of Melbourne).

Cocos zeylandica Berry, *Am. J. Sci.* 69: 181, figs 4–8 (1926). Holotype: New Zealand. North Island, Mangonui, *GE 9619.2–GE9619.4* (Auckland Museum).

Dicolpopollis bungonensis Truswell & Owen, *Mem. Ass. Australas. Palaeontols* 5: 270, Fig. 5C, D (1988). Holotype: Australia. New South Wales, Bungonia, *27165* (CPC).

Dicolpopollis metroxylonoides Khan, *Aust. J. Bot.* 24: 760, Fig. 13 (1976). Holotype: Papua New Guinea. Central Province Delta, Iviri No.1 Well, 07°35′54″S, 144°46′30″E, *Core 13C: S5, 84, 4* (Palynology Section, School of Biological Sciences, University of Sydney).

Monosulcites palisadus Couper, *New Zealand Geological Survey Paleontological Bulletin* 22: 65, Plate 8, Fig. 132 (1953). Type: New Zealand. South Island, Ohai, S168/571, *Slide L194/2* (GNS [NZ]).

Nypa australis Pole & Macphail, *Rev. Palaeobot. Palyno.* 92: 59, Plate II, 1–3; Plate III, 1, 3; Plate IV, 1, 3, 5 (1996). Holotype: Australia. Tasmania, Strahan, Regatta Point, 1991–1994, *M. Pole SB638* (Dept of Plant Sciences, University of Tasmania, Hobart).

Palmidites maximus Couper, *New Zealand Geological Survey Paleontological Bulletin* 22: 57, Plate 7, Fig. 103 (1953). Holotype: New Zealand. South Island, Toetoes Bay, *Slide L60/2* (GNS [NZ]).

Palmoxylon queenslandicum Conran & Rozefelds, *Alcheringa* 27: 128, fig 206 (2003). Holotype: Australia. Queensland. Springsure, Minerva Creek, 23°53′S, 148°17′E, undated, *F. and M.Newman s.n.* (QMF 18208).

Phoenicites zeelandica (Ettingsh.) Pole, *J. Roy. Soc. New Zeal.* 23: 287, figs 2, 3 (1993); *Seaforthia zeelandica* Ettingsh., *Denkschriften der Akademie der Wissenschaften, Wien* 53: 158, Plate 1, Fig. 25 (1887). Lectotype: New Zealand, South Island, Gold Fields Department District of Dunstan, clay containing fossils from Kawarau Basin, *zp169* (Canterbury Museum) (*fide* Gregg 1975).

Spinizonocolpites baculatus Muller, *Micropaleontology* 14: 11, Plate 3, Fig. 2 (1968). Holotype: Malaysia, Sarawak, near Lundu, S.S.O.L. sample My-1411, *Slide Sar. 23-1* (BM).

Spinizonocolpites echinatus Muller, *Micropaleontology* 14: 11, Plate 3, Fig. 3 (1968). Holotype: Malaysia, Sarawak, near Lundu, S.S.O.L. sample My-1411, *Slide Sar. 24-1* (BM).

Spinizonocolpites perspinosus (Couper) Hekel, *Geological Survey of Queensland Publication 355, Palaeontological Papers* 30: 8 (1972). *Monosulcites perspinosus* Couper, *New Zealand Geological Survey Paleontological Bulletin* 22: 65, Plate 8, Fig. 133 (1953). Type: New Zealand. North Island, Wanganui, *Slide L43/7* (GNS [NZ]).

Spinizonocolpites prominatus (McIntyre) Stover and Evans, *in* LE Stover and AD Partridge, *Geological Society of Australia, Special Publication* 4: 67, Plate 3, Fig. 5 (1973); *Monosulcites prominatus* McIntyre, *New Zeal. J. Bot.* 3: 214, Fig. 33 (1965). Holotype: New Zealand. *Slide L 2879/1, 37.1 x 125.4* (GNS [NZ]).

4. Distribution and ecology

The present distribution of Australian palms is the result of a complex historical interaction of geology, climate, evolution and adaptation. Chapter 3 explained something of their deep history, as interpreted from the fossil record. In summary, the fossil record indicates that palms were at various times distributed across much of the Australian landmasses, first appearing in the Palaeocene and becoming widespread through the Eocene and Oligocene in much of what is present-day southern Australia. Palms became much less common in southern Australia during the Miocene and Pliocene, and are largely absent from the fossil record in southern Australia by the Quaternary. This chapter will examine Australian palms from aspects of their present distribution and ecological amplitude. Their broader regional relationships will be investigated, and related to aspects of climate, habitat, phylogenetics and ecology.

■ AUSTRALIAN PALMS AND THEIR ENVIRONMENTS

The first reference to Australian palms in a phytogeographic and ecological context was provided by Cunningham (1827), who described the distribution limits of the known six species in Australia based on observations of broad climatic, environmental and edaphic factors (see Chapter 2). Later taxonomists provided basic habitat information as part of the descriptions of new species (Mueller 1865) or flora treatments (Bentham 1878). 'Notes' on distributional relationships, habitat preferences and associated vegetation were limited to brief inclusions in the taxonomic literature and it was not until about the mid 20th century, when autecological research and landscape ecology achieved prominence, that greater detail was given. Distribution and ecological data are now an expected and integral component of taxonomic research. Despite increased research on the ecology of the Australian flora, palms have received relatively little attention and remain significantly understudied compared to palms in the Americas and south-east Asia.

■ PALMS IN THE LANDSCAPE

In broad ecological applications, some Australian palms have been identified as keystone species in landscape and environmental ecology. Webb (1959), Tracey and Webb (1975) and Webb and Tracey (1981) recognised a number of Australian vegetation types based on the presence, either dominance or co-dominance, of palm species in rainforests and woodlands. Many of the vegetation structural classification types that they established have been modified for inclusion in ecosystem and habitat descriptors (Sattler and Williams 1999) and widely adopted in ecological research for describing those vegetation types. Examples of ecosystems with palm epithets include:

- mesophyll vine forest with dominant *Archontophoenix alexandrae;*
- *Corypha utan* open woodland on alluvial plains and old beach ridges;
- mesophyll vine forest with dominant *Licuala ramsayi* on wet lowlands;
- deciduous vine thicket with *Wodyetia bifurcata* on granite boulders (Fig. 4.1).

Figure 4.1 Ecosystems that have palms as dominant species. **Top left:** Mesophyll vine forest with dominant *Archontophoenix alexandrae*, Cairns, north-east Queensland. **Top right:** *Corypha utan* open woodland on alluvial plains and old beach ridges, Lakefield National Park, Cape York Peninsula, north-Queensland. **Bottom left:** Mesophyll vine forest with dominant *Licuala ramsayi* on wet lowlands, Licuala State Forest, Mission Beach, north Queensland. **Bottom right:** Deciduous vine thicket with *Wodyetia bifurcata* on granite boulders, Melville Range, north Queensland.

Because of their morphological and physiognomic distinctiveness, palms, whether rainforest species such as *A. alexandrae* or *L. ramsayi*, or open forest species such as *C. utan*, clearly establish the basic structure and ecological character of the particular vegetation type.

■ PHYTOGEOGRAPHICAL REGIONS IN AUSTRALIA

Phytogeographical regions have been defined in various ways in Australia (Hooker 1859; Burbidge 1960; Barlow 1981; Crisp *et al.* 1999) based on climatic parameters, species endemism, distribution of shared/non-shared taxa within discrete areas, and regional and global relationships. The Australian continent can be broadly divided into five biomes (Fig. 4.2): 'south-eastern temperate' (Bassiana), 'south-western temperate, 'monsoonal tropics' (Torresia) and 'Eremean' and 'aseasonal-wet' (Tumbuna) (Burbidge 1960; Hope 1994; Crisp *et al.* 1999, 2004). The biomes can be further subdivided into bioregions, of which c. 85 are recognised. These have increasingly become the standard designations for ecological research (Environment Australia 2000). Palms occur mainly within the monsoonal tropics and aseasonal-wet biomes, with minor representation in the south-eastern temperate and Eremean biomes. Palms are absent from the south-western temperate biome. The regional abundance of palm species in Australia is indicated in Figure 4.3.

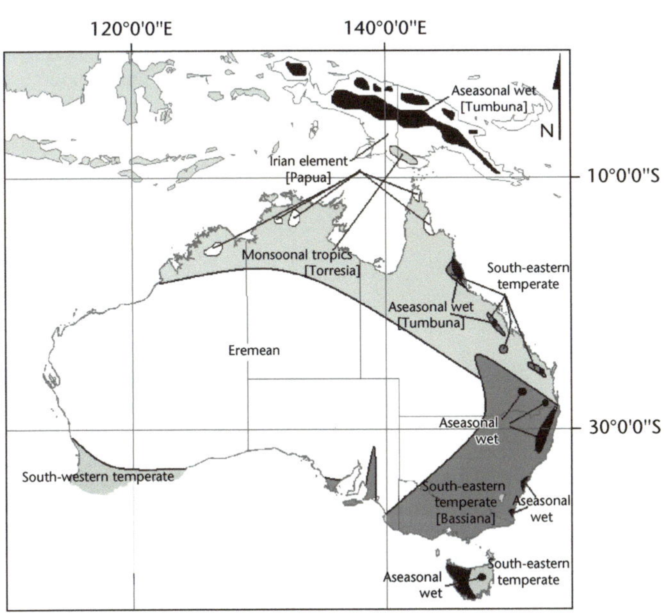

Figure 4.2 Australian biomes. Adapted from Burbidge (1960), Hope (1994) and Crisp *et al.* (1999, 2004).

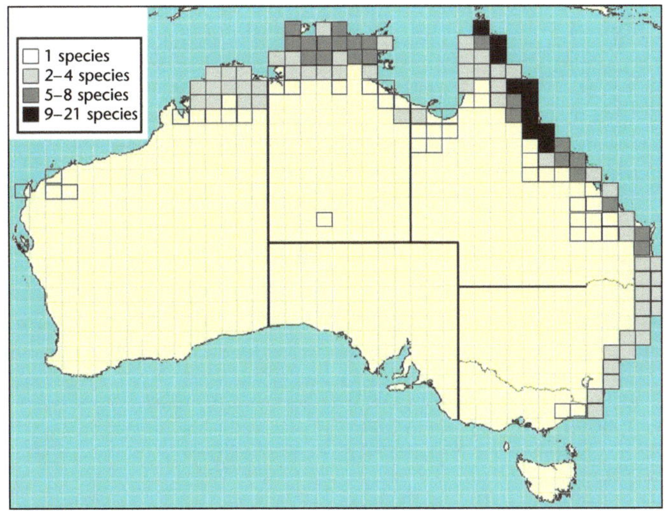

Figure 4.3 Species richness of palms in Australia, as calculated by number of species in 1°x 1° squares. Clear square = 1 sp., light grey square = 2–4 spp., dark grey square = 5–8 spp., black square = 9–21 spp.

■ DISTRIBUTION PATTERNS

Australia's palm flora is mostly an extension of the palm flora that occurs west of Wallace's Line and extends eastward and southward through east Malesia, Australia, Melanesia and New Zealand to Oceania (Table 4.1). The palm floras east and west of Wallace's Line appear to have different origins in most evolutionary lines (Dransfield 1981; Dransfield *et al.* 2008), but the distribution of some elements suggest that there has been intermingling and some bidirectional radiation across Wallace's Line (Dransfield 1987; Baker *et al.* 1998). Australian palms are most closely related to those in New Guinea, and secondarily to those in islands in the south-west Pacific including Lord Howe Island, Norfolk Island and New Zealand. No genera are shared with New Caledonia, although close supra-generic relationships exist.

Tectonic movement, the time of separation and movement of landmasses and the creation of land through volcanism helps to explain much of the context of the distribution patterns of many elements in the Australian palm flora. Australia, New Guinea, New Caledonia and New Zealand are hypothesised to have emerged from the landmass of East Gondwana (Johnson 2004). Lord Howe Island and Norfolk Island are the result of marine

Table 4.1 Distribution of genera that occur in Australia

	Endemic to Australia and off-shore territories	Shared with areas east of Wallace's Line	Shared with areas west of Wallace's Line
Archontophoenix	Yes		
Arenga		Yes	Yes
Calamus		Yes	Yes
Carpentaria	Yes		
Caryota		Yes	Yes
Cocos		Yes	Yes
Corypha		Yes	Yes
Hedyscepe	Yes		
Howea	Yes		
Hydriastele		Yes	No
Laccospadix	Yes		
Lepidorrhachis	Yes		
Licuala		Yes	Yes
Linospadix		Yes	No
Livistona		Yes	Yes
Normanbya	Yes		
Nypa		Yes	Yes
Oraniopsis	Yes		
Ptychosperma		Yes	No
Rhopalostylis		Yes	No
Wodyetia	Yes		

volcanism, the former emerging 6.9–6.4 Mya (McDougall *et al.* 1981) and the latter 3.05–2.3 Mya (Jones and McDougall 1973). It is estimated that proto-New Caledonia commenced separating from the parent Australian landmass c. 65 Mya and attained its present form and position 40–36 Mya (Rawling and Lister 1999). Proto-New Zealand attained its present location in the Early Tertiary and most of its landmasses were in their present positions at least 10 Mya (Johnson 2004). Apart from a small number of investigations (see below), molecular techniques have yet to be used for dating estimated divergence times and aligning geological events for the Australian palm flora.

■ REMOTE SPECIES OF *LIVISTONA*

Preliminary molecular investigations (Dowe 2001; Isagi *pers. comm.*) indicate that the Australian species of *Livistona* form a monophyletic clade that is sister to all

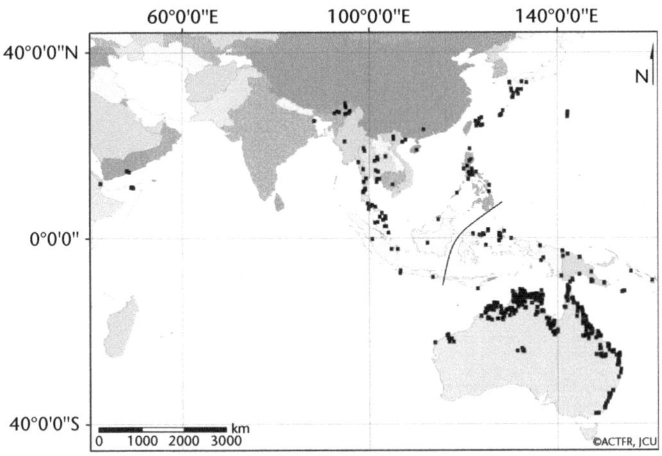

Figure 4.4 Distribution of *Livistona*, a genus distributed either side of Wallace's Line.

other *Livistona* species (see Chapter 8). This pattern is reflected in distribution of the genus, with somewhat equal distribution across Wallace's Line (Fig. 4.4). The molecular data suggest a single introduction into Australia. Although the Australian species form a monophyletic clade, there is only weak support for a hierarchical arrangement of the species and limited explanation of radiation patterns and evolutionary history within Australia. Despite the lack of molecular support at the species level, some distributional relationships between species can be inferred.

It can be assumed that the first radiation of *Livistona* into Australia was from moist habitats, and entered Australia possibly during the Mid Tertiary when the climate was able to support rainforest and other mesic habitats. The progenitors underwent adaptive radiation and speciation then fragmentation and retreat to preferred habitats in response to continental aridity. Although no pre-Oligocene macrofossils of *Livistona* have been recovered in Australia, the presence of Arecoid-type fossil pollen from Palaeocene to Miocene deposits indicates that palms, not to discount *Livistona*, were widespread across much of what is now the southern half of the Australian continent.

The presence of *Livistona* species in remote locations in Australia has been attributed to relictualism (Lothian 1958; Latz 1975; Bowman 2000) and habitat fragmentation (Webb and Tracey 1981; Webb *et al.* 1986) or long-distance dispersal (Martin 1982). The most remote species are *L. alfredii* and *L. mariae* (Fig. 4.5). The former is confined to the Fortescue River system and some adjacent smaller systems in the Pilbara, with an outlier population in Cape Range; the latter is confined to the Finke River system in central Australia (Fig. 4.5). *Livistona alfredii* and *L. mariae* occur in areas with rainfall of 500–600 mm and c. 300 mm per annum respectively. The persistence of habitat that has been able to support *L. mariae* in an arid region has been attributed to slow groundwater discharge from the Hermannsburg Sandstone into the Finke River system. Based on stable isotope analyses, Wischusen *et al.* (2004) estimated that this system has been in place since at least the Mid Pleistocene, and has provided an uninterrupted refuge for mesic plants in an arid environment that has been subjected to considerable climate change.

Livistona alfredii is more remote from other palm species than is *L. mariae*. Molecular and cladistic evidence suggests a close relationship to *L. victoriae* in the eastern Kimberley and western Northern Territory (Dowe 2001). Fossil stems attributable to palms have been recovered from the Fortescue River (where *L. alfredii* has its most dense present populations), but their silicate nature suggests recent formation (N. Pledge, *pers. comm.*, Western Australian Museum). Reconstructions of palaeoclimates suggest the Pilbara region was wetter in the Eocene through to the Miocene, and the potential for palms to disperse through a more benign environment was possible. Since the Early Pleistocene, the Pilbara region has experienced periods of drier conditions than at present (Wyrwoll 1993), which suggests that *L. alfredii* has possibly persisted as a refugial element for considerable time. Present refugia, with gorge vegetation consisting of *Callitris, Ficus* and *Brachychiton*, are recorded in the Hamersley Range to the north (Morton *et al.* 1995) but no palms are presently known there. Habitat fragmentation, caused by increasing aridity, appears the most plausible explanation for the persistence of these remote *Livistona* species as the floristic composition of their habitats and other refugial areas in central Australia are similar (Morton *et al.* 1995), which does not support long-distance dispersal as an explanation. The closest relatives of *L. mariae* include *L. rigida*, also its geographically closest species to the north and north-east, and *L. lanuginosa* in eastern Queensland, from which *L. mariae* is geographically isolated.

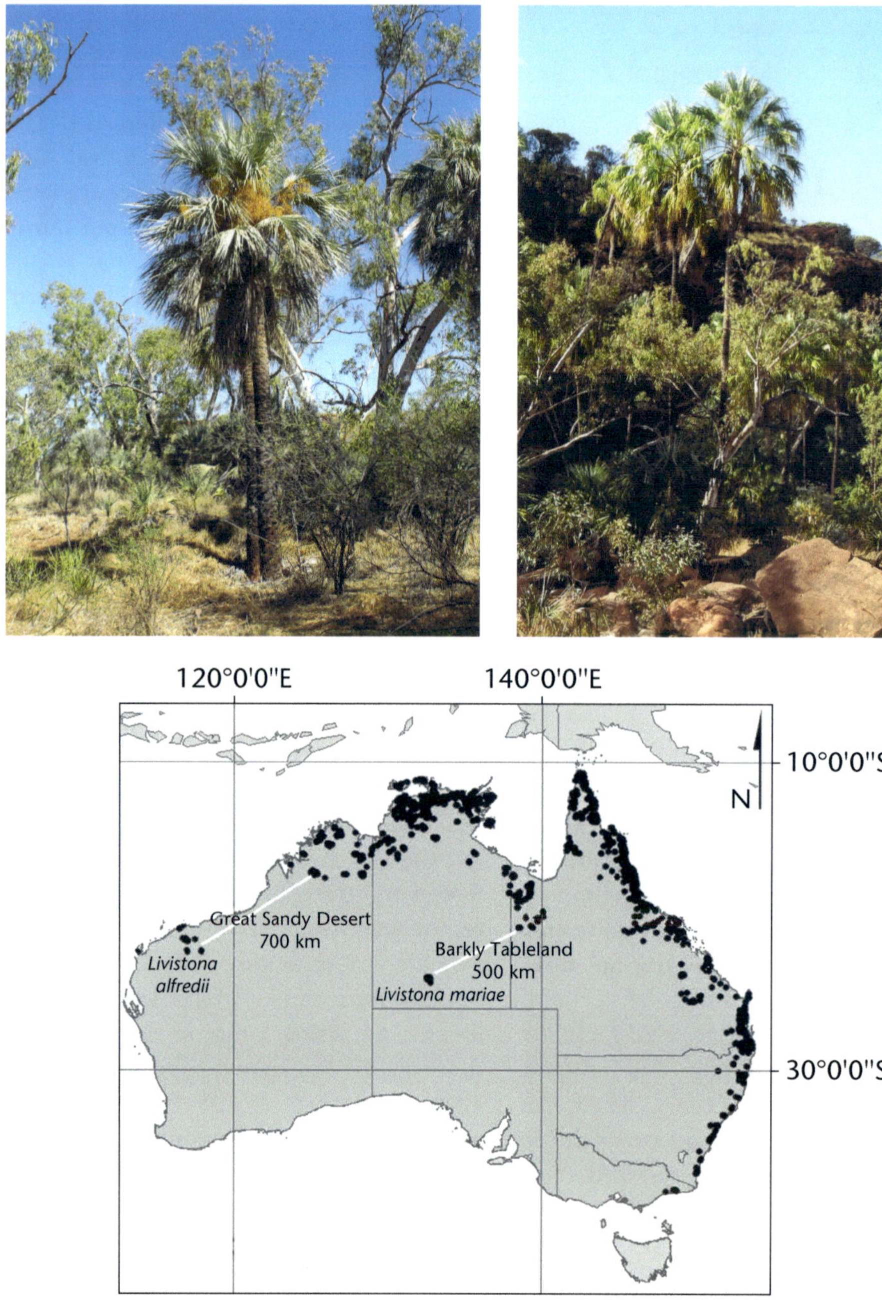

Figure 4.5 Remote palm species in Australia. **Top left:** *Livistona alfredii*, Fortescue River, Millstream, Pilbara, Western Australia. **Top right:** *Livistona mariae*, Palm Valley, Finke River, Northern Territory. **Bottom:** Locations of populations of *Livistona alfredii* and *Livistona mariae* with distances to nearest populations of other *Livistona* species, and the environmental barriers separating the populations. Filled circles indicate recorded distribution of *Livistona* in Australia.

■ REMOTE GLOBAL DISTRIBUTION: *ORANIOPSIS*

Oraniopsis, a monotypic genus with its only species *O. appendiculata* (Fig. 4.6), is endemic to the Wet Tropics Bioregion of north-east Queensland, predominantly above 800 m asl but with small and scattered populations at lower altitudes in high-rainfall areas. It is strikingly remote from closely related taxa in Madagascar and South America. Dransfield *et al.* (1985) suggested a Late Cretaceous Gondwanan rafting to explain the distribution

Figure 4.6 An example of remote global distribution of an Australian palm, *Oraniopsis appendiculata*, an endemic species with closest relatives in Madagascar and South America. Campbell Creek, Walter Hill Ranges, north-east Queensland.

of the Ceroxyloid group, and implicated the processes of vicariance rather than long-distance dispersal. Uhl and Dransfield (1987) presented a modified hypothesis in which dispersal, still from a Gondwanan origin, was via a trans-Antarctic Early Tertiary route. Contrary to these proposals, Trénel *et al.* (2007) suggested that vicariance could not have been a factor as the timing of post-Gondwanan landmass movements (120–80 Mya) is not congruent with estimated Late Eocene–Late Oligocene (42–24 Mya) divergence times of *Oraniopsis* from other taxa within the Ceroxyleae, estimated from molecular data. This indicated a possible long-distance trans-Atlantic or trans-African Mid Tertiary dispersal for *Oraniopsis*. If the estimated divergence times are correct, the persistence of *Oraniopsis* in Australia for such an extended period (minimum 24 My) is difficult to explain – much of the area in which it could have resided has been subject to extreme variations in climatic conditions since the Mid Tertiary. There is evidence that refugia have allowed the uninterrupted preservation of rainforest habitats since the Pleistocene, or possibly earlier, in much of Australia, despite intervening periods of much cooler and/or much drier conditions (Hilbert *et al.* 2007). Refugial areas have been associated with mountain tops and river valleys, the optimal habitats for *Oraniopsis*, which can remain relatively moist during dry periods and support the persistence of palaeoendemics (Webb *et al.* 1986; Prider and Christophel 2000).

■ LORD HOWE ISLAND

Although the dispersal and establishment of palms throughout the Australian continent can only be conjectured, the palms of Lord Howe Island (Fig. 4.7) have been the focus of considerable research and plausible molecular-based explanations have been provided for the timing of their arrival and speciation. Lord Howe Island, which emerged 6.9–6.3 Mya, has endemic genera and species and long-distance dispersal appears to account for initial establishment. It is suggested that the two species of *Howea* evolved by sympatric speciation, as their estimated divergence time, 1.92 Mya to <1 Mya, occurs later than the formation of Lord Howe Island (Savolainen *et al.* 2006). Phylogenetic root nodes for other palm taxa from Lord Howe Island, *Hedyscepe* and *Lepidorrhachis*, are also consistent with the geological time-frame of the island (Savolainen *et al.* 2006). Among the closest relatives of *Howea* is *Laccospadix*, a monotypic genus endemic to north-east Queensland; the estimated time of divergence of *Howea* and *Laccospadix* is 4.57–5.53 Mya, which suggests a possible dispersal event from Australia to Lord Howe Island (Savolainen *et al.* 2006). Submerged islands have been suggested as a migration route from Australia to New Caledonia (Renner *et al.* 2000; Barker *et al.* 2007); this could also be the case for Lord Howe Island. The now submerged Lord Howe Rise may have provided the intermediate habitats (Green 1994; Lee *et al.* 2001).

■ ECOLOGICAL AMPLITUDE

Globally, palms are predominant in warm to hot environments with moderate to high rainfall, and have reduced diversity and species numbers in cool and/or

Figure 4.7 Palm species on Lord Howe Island. **Top left:** *Howea belmoreana*. Photo by W.J. Baker, Royal Botanic Gardens Kew. **Top right:** *Howea forsteriana*. Photo by W.J. Baker, Royal Botanic Gardens Kew. **Bottom left:** *Hedyscepe canterburyana*. Photo by Jeanne Price. **Bottom right:** *Lepidorrhachis mooreana*, Mt Gower. Photo by Jo Wilkins.

dry environments (Moore 1973; Tomlinson 1979, 1990; Bjorholm *et al.* 2006). Rainforest environments account for about 75% of species (Dransfield 1978), but there is also significant representation in woodlands, savannas, riparian and wetland habitats, mangroves and semi-deserts. Species that occupy dry or cool/cold environments tend to be derived from certain evolutionary lines in the family, such as the Borasseae, Chuniophoeniceae, Phoeniceae and Trachycarpeae in the Coryphoideae, the Cocoseae and the Areceae in the Arecoideae and the Ceroxyleae in the Ceroxyloideae (Parsons 2007; see Bjorholm *et al.* 2005 for habitat preferences of neotropical palms). Within their environments, palms tend to occupy permanently or seasonally moist to wet niches. The ecological amplitude of Australian palms generally conforms to that expressed in palms from other areas. Physiological properties such as extreme height (*Archontophoenix* spp.), high density of individuals in some populations (*Archontophoenix cunninghamiana*, *Livistona humilis* and *Licuala ramsayi*) (Gorman 1996), prevalence of climbing palms (*Calamus* spp.) (Putz 1990) and environmental dominance (*Corypha utan*, *Livistona nasmophila* and *Wodyetia bifurcata*) (Dowe 2009a) provide the characteristic and unique structure of Australian environments where palms are present.

■ CLIMATE

Australia is one of the driest continents, with an average continent-wide rainfall of c. 400 mm per annum (Johnson 2004). High annual rainfall, 1500–3000 (5000) mm per annum, is limited to areas on the north and east coasts and western Tasmania. It is primarily controlled by seasonal air pressure movement (monsoonal influences in northern areas) and prevalent wind direction and orography (uplands intercepting the prevailing winds). Australian palms are predominant in areas where rainfall is 1200 mm or greater per annum, with c. 70% of species occurring in those areas (Fig. 4.8). An average annual minimum temperature of 15°C is a major demarcation in overall distribution of the Australian flora, but for palms this is complicated by adiabatic lapse rate temperature decrease in high-elevation areas, especially in north-east Queensland and offshore territories such as Lord Howe Island and Norfolk Island. Eight species (13%) occur exclusively within the 15°C average or lower annual minimum temperature limit. Cooler temperatures in areas where palms occur can be tempered by high rainfall, e.g. *Livistona australis* in Victoria (Ladd 1978; Cameron 1987; Helman 1987; Orscheg and Parsons 1996a, 1996b) and *Lepidorrhachis mooreana* on Lord Howe Island (Pickard 1983). The moderating effect of maritime influences may play a role in such areas.

■ EDAPHIC PREFERENCES

Globally the edaphic preferences of palms are broad, but species are recorded on unusual formations such as limestone, peat and ultramafics (Tomlinson 1990; Dransfield and Beentje 1995; Hodel and Pintaud 1998; Pintaud *et al.* 1999; Dransfield *et al.* 2008). Australian soils are, on the whole, highly weathered and generally low in nutrients, but are locally highly variable according to their origins, position in the landscape and climate (Johnson 2004). High-nutrient soils tend to support rainforest vegetation types, which often include palms (Winter and Atherton 1987). The tolerance of waterlogged or impeded drainage microhabitats is a common feature of palms in many areas. It is at the species level that palms display preferences for particular edaphic types (Table 4.2). Species may occur on a number of edaphic types; the table is a guide to the most common occurrence.

■ FIRE

Fire is among the primary forces that have shaped the Australian flora (Bowman 2000; Kershaw *et al.* 2002). The interaction of palms and fire has not been studied in detail in Australia although species can be clearly identified as fire-tolerant or not. Fire-tolerant species include the *Livistona* species, *Corypha utan* and *Hydriastele ramsayi*.

Fire and *Livistona*

Molecular evidence indicates the Australian *Livistona* species (including the two shared with New Guinea) form a monophyletic clade and most likely developed from a single introduction into Australia (Dowe 2001; Isagi *pers. comm.*). Following migration into what was possibly a moist and fire-limited environment, forces of climate change, adaptive radiation, increasing continental aridity and fragmentation have produced the disjunct and widespread distribution patterns so characteristic of *Livistona* in Australia today.

The adaptation of Australian Livistonas to fire possibly developed during the Late Tertiary (assuming that migration had taken place by that time), when fire was increasingly a factor in the Australian environment (Kershaw *et al.* 2002). The extra-Australian *Livistona* species in areas to the north in Malesia and south-east Asia invariably occupy moist habitats from which fire is largely absent (Dowe 2009a). However, a number of questions must be posed. Was the ancestral *Livistona* pre-adapted to fire prior to its arrival into Australia? Or has fire-adaptation developed post-arrival?

Livistona species possess a number of morphological, physiological and regenerational characteristics that may be related to fire adaptation. They include:

- retained leafbases in the lower portion of the stems (the portion that will be subjected to most types of low-intensity fires) in juveniles and subadults;

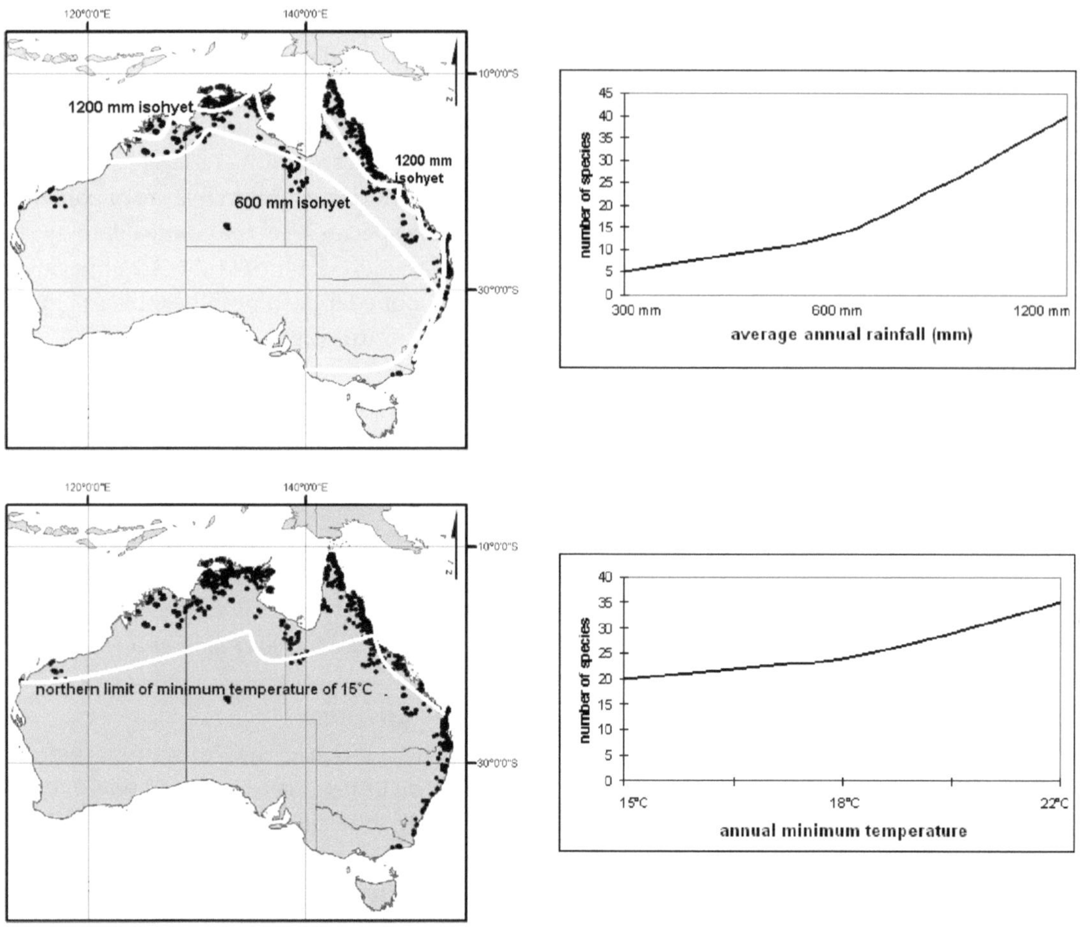

Figure 4.8 **Top left:** Relationship of rainfall to distribution and species diversity of Australian palms. Distribution based on herbarium data. Rainfall isohyets: continuous inland line is the 600 mm isohyet, discontinuous near-coast line is the 1200 mm isohyet. **Top right:** Graph showing relationship of number of species to annual rainfall. **Bottom left:** Relationship of temperature to distribution and species diversity of Australian palms. Line indicates the northern limit of average annual minimum temperature of 15°C. **Bottom right:** Graph showing relationship of number of species to temperature.

- protection of the apical meristem by retained leafbases;
- the placement of dead leaves away from the central axis, so the most intense fire is remote from the apical meristem;
- remote ligular germination mode and frequent deep burial of germinants.

An array of responses to fire has been recorded for *Livistona* species, and although only a few species have been studied some generalisations can be made. Fire intensity and season appear to be the dominant factors affecting survival and/or damage to palms in tropical savannas and woodlands (Werner 2005). Intensity is primarily dependent on fuel availability and weather conditions at the time of fire (Williams *et al.* 2003). The reported fire mortality rates for *Livistona humilis* have been variable. Mortality rates can be low in high-intensity, late dry season fires (Lonsdale and Braithwaite 1991), but can be as high as 25% when burnt in low-intensity, early dry season fires (Williams *et al.* 1999). Infrequent high-intensity late dry season fires are reported to produce mortality rates approaching 65% (Williams *et al.* 2002). Prior *et al.* (2009) reported that mortality rates for *Livistona* spp. were among the lowest for all plant forms in unburnt areas, as well as locations with early dry season and late dry season fires. It appears that *L. humilis* is most susceptible to fire immediately after the wet season growth period, when leaf tissues have yet to become lignified enough to

Table 4.2 Primary edaphic associations for Australian palms

Primary edaphic association	Species
Sandstones	*Livistona inermis, L. fulva, L. mariae, L. nitida, L. victoriae*
Laterites	*Livistona eastonii*
Deep silicious sand or sandy soils	*Cocos nucifera, Hydriastele ramsayi, Livistona drudei, L. humilis*
Estuarine sediments	*Nypa fruticans*
Basalts	*Archontophoenix maxima, Arenga listeri, Hedyscepe canterburyana, Howea belmoreana, Lepidorrhachis mooreana, Linospadix microcaryus, L. monostachyos, Rhopalostylis baueri*
Granites	*Archontophoenix purpurea, Linospadix apetiolatus, Oraniopsis appendiculata, Wodyetia bifurcata*
Calcarenites	*Howea forsteriana*
Metamorphics	*Archontophoenix myolensis*
Alluviums of various origins	*Archontophoenix tuckeri, Corypha utan, Livistona alfredii, L. benthamii, L. concinna, L. lanuginosa, L. nasmophila*
Indifferent to edaphic association, but usually with high-nutrient soils	*Archontophoenix alexandrae, A cunninghamiana, Arenga australasica, A. microcarpa, Calamus* spp., *Carpentaria acuminata, Caryota albertii, Hydriastele costata, H. wendlandiana, Laccospadix australasicus, Licuala ramsayi, Linospadix minor, L. palmerianus, Livistona australis, L. decora, L. lorophylla, L. muelleri, L. rigida, Normanbya normanbyi, Ptychosperma elegans, P. macarthurii*

withstand fire damage. A comparison of population structures of *L. humilis* in experimental burnt and unburnt plots indicated that fire appears to have little impact on levels of regeneration and recruitment (Williams *et al.* 1999).

The typical fires that affect most *Livistona* species are usually low to moderate intensity. The phenomenon of 'torching' (the rapid but low-heat ignition of the crown) is a characteristic of savanna and open forest *Livistonas*, as well as other palm-like plants such as *Pandanus* species. Preliminary research in populations of *L. decora* (Fig. 4.9) in north-east Queensland indicated that early dry season low-intensity fires fuelled by drying grass and *Melaleuca* and *Corymbia* leaf litter, as well as dried palm leaves, were not intense enough to destroy seedlings whose apical meristems were 2 cm or more below the soil surface. Orscheg and Parsons (1996a) invoked the partial burial of seeds and remote-tubular germination mode to explain the survival of seedlings of *L. australis* after fire in eastern Victoria. Latz (1975) reported on the fire deaths of individuals of *L. mariae*, in Palm Valley, with greatest mortalities occurring in the 5–8 m height class. Although Latz (1975) did not suggest the cause of this situation, it could be that fuel loads of adjacent trees and the senescent leaves of the palms accumulate at that height and provide fuel for high-temperature fires.

Livistona species in mesic though seasonally dry environments, such as *L. australis, L. drudei* and *L. benthamii*, display a lesser tolerance to repeated high-intensity fires than do savanna or woodland species (Calvert *et al.* 2005). There appears to be a limit of fire intensity and frequency at which survival of even small to moderate-sized individuals are threatened, although fire rarely kills adults (Orscheg and Parsons 1996b). Floyd (1990) noted the dominance of *L. australis* in repeatedly burnt areas of rainforest margins in New South Wales. Overall, it seems that fire has enhanced the development of *Livistona* in Australia.

There has been limited research on the other fire-adapted Australian palm species, *Corypha utan* and *Hydriastele ramsayi*. Anecdotal reports note that *C. utan* occurs in a fire-prone habitat (Crowley 1995; Cowie *et al.* 2000), but its fire ecology has not been investigated.

Figure 4.9 Fire and *Livistona* species. **Left:** *Livistona muelleri* burnt by low-intensity fire that damaged only the lower leaves, near Bamaga, north Queensland. **Right:** *Livistona decora* in an early dry season low-intensity fire fuelled by drying grass, and *Melaleuca* and *Corymbia* leaf litter, Cocoa Creek, Cape Cleveland, north-east Queensland. The fire intensity of this event did not damage seedling meristems that were 2 cm or more below soil surface level.

Fires in *C. utan* habitat on Cape York Peninsula rarely scorch vegetation above 6 m height but can consume plants within 2 m of the ground (P. Stanton *pers. comm.*). Brennan (1997) concluded that wet-season burning, rather than dry-season burning, reduced fire damage to *H. ramsayi*, but did not examine regeneration or population dynamics.

■ CYCLONES

Cyclones are severe rotational low-pressure tropical storms produced by an interaction of warm ocean waters and unstable atmospheric conditions (Puotinen *et al.* 1997). They are primary influences affecting vegetation dynamics, forest structure, community composition, structural adaptations and evolutionary patterns (Hopkins 1990; Clarke and Kerrigan 2000; Grove *et al.* 2000; Griffith *et al.* 2008; Turton 2008). Since 1907, an average of 1.63 cyclones per annum have crossed the east Australian coast between Cape York (11°41′S, 142°31′E) and Port Macquarie (31°25′S, 152°54′S) (Lourensz 1981; Puotinen *et al.* 1997). The long-term evidence for cyclones in the Australian region is circumstantially contained in the palaeoclimatic record and indicates that present climatic conditions in northern Australia were first approximated in the Late Pliocene, about 2.5 Mya (Frakes 1999). Fossil pollen evidence in off-shore cores indicates that much of the present-day vegetation was stabilised in much of northern Australia since the beginning of the Pleistocene (Crowley and Gagan 1995). It can be reasonably assumed that cyclones have been active in northern Australia long enough for them to have influenced many aspects of flora and fauna evolution (Webb 1958; Bellingham 2008) and to have influenced adaptation and selection.

Adaptation of palms to wind stress

Palms may be the predominant intact and functioning plant group following cyclonic events in forests where they occur (Loope *et al.* 1994; Vandermeer 1994; Ostertag *et al.* 2005). Many palms with tall stems can flex in forces exerted by extreme winds because of their mechanical properties (Tomlinson 1990), although there is a level of stress at which stems will fracture (Huang *et al.* 2002; Kuo-Huang *et al.* 2004). Of the plant forms in the forests of northern Australia, palms occupy areas that are among the most exposed to and frequently affected by cyclones (Puotinen *et al.* 1997). Data gathered post Cyclone Larry (March 2006) (Dowe 2009b) from populations of

Figure 4.10 Damage to palms caused by the impact of Cyclone Larry, March 2006. **Top left:** *Archontophoenix alexandrae*, with damage to the apical meristem, Russell River, north-east Queensland. **Top right:** *Arenga australasica*, a clustering palm with the dominant stem snapped in mid-stem, Kurrimine Beach, north-east Queensland. **Bottom left:** *Livistona drudei*, with crowns intact but leaves shredded on one side of the crown, Hull River, north-east Queensland. **Bottom left:** *Licuala ramsayi* var. *ramsayi*, with intact apical meristems and shredded leaves, Mission Beach, north-east Queensland.

Archontophoenix alexandrae in the severe impact zone indicated there was a high mortality rate in the subadult size classes, but the impact on the seedling/stemless juvenile class size was not significantly different from normal mortality levels. Correspondingly, there was a higher than expected level of survivorship of emergent adults. The tall canopy-emergent habit suggests selection for this habit may enhance the species' ability to maintain reproductive levels, i.e. fruit production, despite the overall impact of cyclones (Fig. 4.10).

Figure 4.11 *Livistona* species that occur in cyclone-prone environments. **Top left:** *Livistona benthamii*, Holmes Jungle, Darwin, Northern Territory. **Top right:** *Livistona concinna,* Barrett Creek, north of Cooktown, north Queensland. **Bottom left:** *Livistona decora,* Plantation Creek, Ayr, north Queensland. **Bottom right:** *Livistona drudei*, Balgal Beach, north Queensland.

Preliminary investigation of the *Livistona* species of eastern Queensland has revealed that four species, *L. benthamii*, *L. concinna*, *L. decora* and *L. drudei*, are distributed in areas of high cyclone prevalence and have leaves that are adapted to withstand high wind stress (Fig. 4.11). The species are characteristically canopy emergents occupying coastal forests (Dowe, in prep.). This appears to be an example of selection for wind-tolerant leaf morphology.

Evidence suggests that *Calamus* species in Australian rainforest benefit from cyclonic disturbances (Metcalfe *et al.* 2008; Curran *et al.* 2008), and if brought down by cyclones can attain rapid growth of 3–4 m per annum and re-emerge from the canopy in only a few years (Stork 2007). Their dominance in niches where impacts are relatively severe supports this interpretation. The development of thickets of *Calamus* species can also result from non-cyclone canopy gap creation, usually caused by falling individual trees and human-induced disturbances such as logging and road construction, all of which increase light levels to the forest floor (Laurance and Curran 2008; Pohlman *et al.* 2008).

■ ALTITUDE AND ELEVATIONAL RANGES OF PALM SPECIES

Australia is one of the lowest and flattest tracts of land in the world, with an average height of 330 m asl (Johnson 2004). Apart from the Great Dividing Range that runs fairly parallel to the east coast, with an average altitude of 300 m asl and a high point of 2228 m asl at Mt Kosciusko, and some isolated inland ranges, the continent is dominated by extensive flat drainage systems, some of which trend toward the centre of the continent and have no ocean outlets.

The altitudinal range (measured as distance above sea level) has been recorded for all Australian palms with good precision. In most respects, the elevational range (measured as distance occupied irrespective of distance above sea level, i.e. highest altitude minus lowest altitude) has not been studied with regard to preferential distribution patterns for the family globally, although results of a preliminary study are presented here (Table 4.3) (Dowe, unpublished data). Based on the palm floras of the Americas, New Caledonia, New Guinea, Thailand and Australia, greatest species richness occurs at c. 300 m asl,

Table 4.3 Species richness and altitudes, and elevational ranges for palm floras in the Americas, Madagascar, New Caledonia, New Guinea, Thailand and Australia

	Species richness[1]	Elevational range[2]
Americas	0 m asl = 34 spp 250 m asl = 219 spp 3150 m asl = 3 spp	0–200 m = 53 spp 0–2200 m = 5 spp
Madagascar	0 m asl = 32 spp 300 m asl = 67 spp 2400 m asl = 1 sp	0–200 m = 58 spp 0–2000 m = 2 spp
New Caledonia	0 m asl = 14 spp 500 m asl = 24 spp 1600 m asl = 2 spp	0–200 m = 9 spp 0–1400 m = 2 spp
New Guinea	0 m asl = 32 spp 200 m asl = 100 spp 2200 m asl = 4 spp	0–200 m = 66 spp 0–2400 m = 2 sp
Thailand	0 m asl = 18 spp 500 m asl = 96 spp 2150 m asl = 1 sp	0–200 m = 33 spp 0–1600 m = 3 spp
Australia	0 m asl = 35 spp 100 m asl = 43 spp 1600 m asl = 3 spp	0–200 m = 19 spp 0–1400 m = 3 spp

Data range from 84%–100% of species in each area. Altitude data for species based on herbarium records (see Acknowledgments for herbaria visited) and published literature.
[1]Species richness at 0 m asl; altitude of greatest richness; and highest recorded altitude of palms for the region.
[2]Numbers of species in the elevational range of 0–200 m; and greatest elevational range of species.
Primary references for elevation data: Dransfield (1979), Dransfield (1984), Jaffre and Veillon (1988), Zona (1990), Hodel (1992), Dransfield and Beentje (1995), Henderson *et al.* (1995), Moraes (1996), Borchsenius and Skov (1997), Hodel and Pintaud (1998), Hodel and Vatcharakorn (1998), Pintaud *et al.* (1999) and Comeau *et al.* (2003).

with the greatest number of species in the elevational range of 0–600 m. The greatest range in elevation for species in Dowe's study is for two species in New Guinea, which occur over a range of 2400 m. Palms reach their greatest altitude in South America, at about 3500 m asl, but those species tend to have narrow elevational ranges. The highest altitude for Australian palms is c. 1600 m, the height of the highest mountains in tropical Australia.

There have been only a few studies dealing with the effects of altitude on variations in morphology and ecology of individual palm species. Bachman *et al.* (2004) assessed species diversity relative to altitude, compensated for by estimation of available land area, and found there was a mid-elevational peak in diversity in New Guinea. Population demographics and recruitment were found to be affected by altitude, among other factors, for species of palms in Ecuador, over a gradient of c. 700 m (Svenning 2001). A number of informal reports have described morphological changes in individual species, such as stature, stem dimensions and leaf and inflorescence structure along altitudinal gradients. For example, Hodel and Pintaud (1998) noted the morphological variation in *Basselinia gracilis* on Mt Panié, New Caledonia, from sea level to 1600 m. In that case, morphological variation, relative to decreasing size in habit, leaves and inflorescences, was directly related to increasing altitude. Similarly, Dransfield and Beentje (1995) noted the altitudinal variation in leaf morphology for some Madagascan species such as *Ravenea sambiranensis* (range of 1–2000 m alt.) and *Dypsis onilahensis* (range of 50–2400 m alt.). For Australian palms, the altitudinal characteristics of *Howea belmoreana* and *H. forsteriana* on Lord Howe Island have been examined (Savolainen *et al.* 2006), primarily with regard to aspects of speciation, rather than ecology. Australian mainland species remain largely unstudied in relation to elevation gradient effects.

A summary of the altitudinal range of Australia palms indicates that 15 species (25%) have elevational ranges of 1000 m or greater (Dowe, unpublished data) (Fig. 4.12). Of these, three (5%) have elevational ranges of 1400 m or greater – they occupy the total available altitudinal range in tropical Australia, from sea level to near or on the summits of some of the highest mountains. Greatest species richness (43 species) occurs at c. 100 m asl, in contrast to 300 m asl for palms globally. At the altitudinal low and high limits, species richness at 0 m asl is 35, and at 1600 m asl is three (Fig. 4.12). Nineteen species (32%) are restricted to a relatively narrow elevational range of 0–200 m. The relatively large number of species with a narrow altitudinal range, as well as greatest species richness at low altitude, appears to be a consequence of the generally low topography in Australia. In comparison, greatest species richness is recorded at higher altitudes in all other areas (Table 4.3).

■ DEMOGRAPHY AND POPULATION DYNAMICS

The few available studies of the demography and population dynamics of Australian palms indicate they have similar characteristics to palms in other areas of the world. The population structures of six Australian palms (*Archontophoenix alexandrae, Licuala ramsayi, Livistona australis, L. eastonii, L. lanuginosa* and *Oraniopsis appendiculata*) are presented in Figure 4.13. Five of these (excluding *L. ramsayi*) display a reverse-J curve structure or close approximate, in which there are relatively greater numbers of seedlings and/or juveniles than other size classes, and reducing numbers in subsequent size classes, such as large juveniles and subadults to adults. The reverse-J curve indicates that fruit production and germination is usually consistent and abundant, and that fitness of the population is high. The progressive mortality of subsequent size classes is the result of a number of causes, such as herbivory, competition, inappropriate niche selection or catastrophic events such as floods, cyclones and fires. One factor that is consistent among species is that when subadulthood and/or adulthood are reached individuals continue to grow and, in most cases, survive until the onset of senescence and death. Globally, the life-span of palm species varies; it has been estimated that individuals may live from about 70 years (*Astrocaryum mexicanum*, Sarukhán 1978) to over 700 years (*Livistona* sp., Hnatiuk 1977). The longevity of some Australian palm species has been estimated: *Archontophoenix cunninghamiana* to about 190 years (data recalculated from Waterhouse and Quinn 1978); *Livistona mariae* up to 300 years (Latz 1975); *Livistona australis* to at least 392 years (Carlile and Priddel 2009) and *Livistona eastonii* up to 720 years (Hnatiuk 1977).

Population density, another measure of fitness, has been estimated for the following species:

- *Archontophoenix alexandrae* – 5820 palms/ha in all class sizes from three populations in the Wet Tropics bioregion (Dowe 2009b);

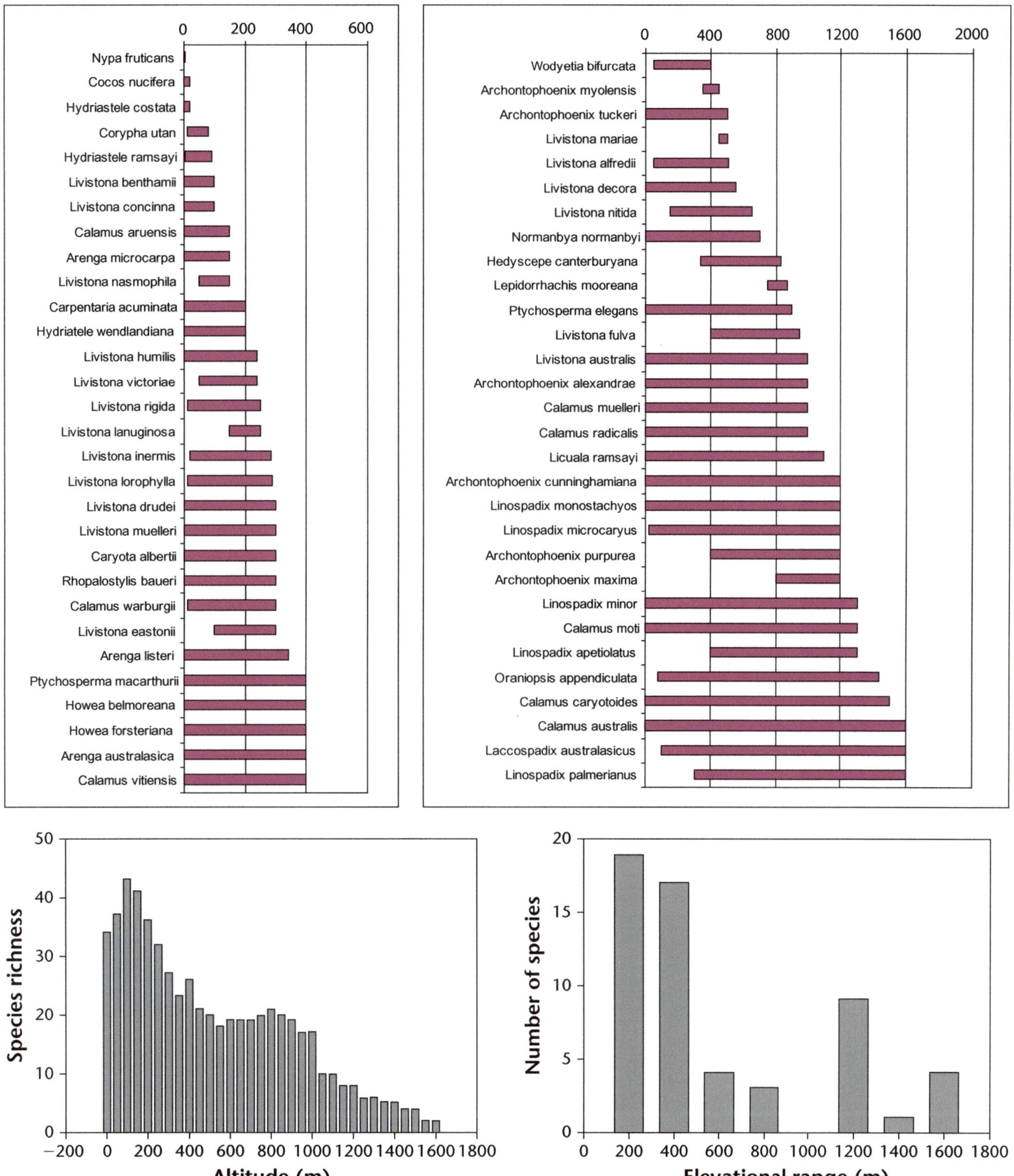

Figure 4.12 Top: Altitudinal distribution of Australian palms. Data are continued from top of left graph to bottom of right graph. Data from herbarium records. **Bottom left:** Species richness (number of species) at different altitudes for Australian palms. Greatest species richness occurs at 100 m altitude (43 spp.). **Bottom right:** Elevational range (independent of altitude) of Australian palms. The most common elevational range is 0–200 m, with 19 species in that range.

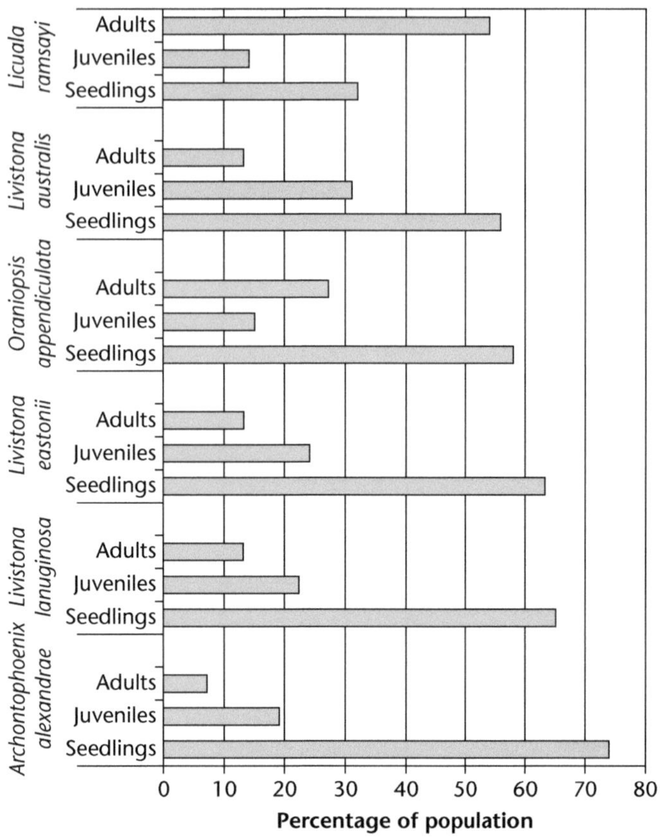

Figure 4.13 Population structure in six species of Australian palms. From top to bottom: *Licuala ramsayi* (Gorman 1996); *Livistona australis* (Orscheg and Parsons 1996a); *Oraniopsis appendiculata* (Dowe, unpublished data); *Livistona eastonii* (Hnatiuk 1977); *Livistona lanuginosa* (Pettit and Dowe 2003); *Archontophoenix alexandrae* (Dowe 2009b). Population data adapted for three size classes of seedlings, juveniles and adults.

- *Livistona eastonii* – 4135 palms/ha in all size classes from three populations on the Mitchell Plateau, Western Australia (Hnatiuk 1977);
- *Licuala ramsayi* – 3125 palms/ha in all size classes from five populations at Mission Beach, north Queensland (Gorman 1996);
- *Livistona lanuginosa* – 138 palms/ha for all size classes from eight populations in the Burdekin River catchment (Pettit and Dowe 2003);
- *Normanbya normanbyi* – 20 adult palms/ha at a single population at Oliver Creek, north-east Queensland (Lott *et al.* 1995).

Excluding *N. normanbyi*, for which the study included only adult palms, the range in population densities illustrates the variability in landscape occupation, from local dominance, such as in some populations of *A. alexandrae*, to wide dispersal in the landscape, as in *L. lanuginosa*. Despite the differences in unit area densities, the population structures of the two species are similar, with seedlings/juveniles accounting for c. 73% of the population of *A. alexandrae* and c. 60% of *L. lanuginosa*. In contrast, Gorman (1996), in a study of *Licuala ramsayi* at Mission Beach, north-east Queensland, estimated that <40% of the population consisted of seedlings/juveniles (Fig. 4.13), despite the relatively high density of the population. Gorman (1996) concluded that fruit production in *L. ramsayi* appeared to be episodic and possibly related to cyclonic or other stochastic events. To place these figures into a global perspective, the Amazon, which is recognised as one of the world's most palm-rich environments, has an estimated density of 2122 palms/ha and populations of up to 7590 palms/ha on poorly drained soils (Kahn and de Castro 1985).

■ POPULATION GENETICS

Genetic diversity at the population and community level has been investigated, utilising chemical and molecular techniques, in a small number of Australian palm species:

- *Livistona australis*, with low genetic diversity in populations in Victoria and southern New South Wales, which suggests recent dispersal from a northern population core (Orscheg and Parsons 1996a);
- *Carpentaria acuminata*, with limited gene flow between discrete populations, showed patterns of genetic diversity that were suggestive of Pleistocene fragmentation and contraction to rainforest refugia where moderate to high genetic diversity levels were documented (Shapcott 1998a, 2000);
- *Ptychosperma macarthurii*, with the Northern Territory population having limited genetic diversity and probably established from a single introduction, has been separated from Queensland populations for considerable time. A relatively high level of genetic diversity has been detected in wild populations of *P. elegans* and *P. macarthurii* in Queensland (Shapcott 1998b);
- *Lepidorrhachis mooreana*, endemic to Lord Howe Island, occurs only above 750 m on the summits of

two mountains. The genetic diversity and key demographic parameters indicate that the subpopulations on one mountain, Mt Gower, are effectively behaving as one genetic population with little evidence of genetic substructure. There is a high reproductive output with c. 40% of the population flowering synchronously in a season and 25% of plants reproductive in consecutive seasons. The population is inbred but the lack of genetic structure suggests the species has a mixed mating system with some self-pollination. The relatively large population size reduces the threats to viability and increases the relative potential for selection for adaptation (A. Shapcott, *pers. comm.*).

■ OTHER ECOLOGICAL ASPECTS

Reproductive biology, seed dispersal syndromes and other ecological aspects have been studied in some Australian palms. Summaries are included in the notes for individual species in Chapters 6 to 10.

■ SUMMARY

This chapter has examined the Australian palm flora within the broad themes of present distribution, regional relationships and ecological amplitude, interpreted within aspects of climate, habitats, phylogenetic relationships and ecology. Overall, the Australian palm flora is relatively poorly understood compared to our knowledge of other plant families such as the Myrtaceae, Mimosaceae and Proteaceae. The primary research focus for palms has been in the genera of *Calamus*, *Livistona* and *Normanbya*, and the endemic species on Lord Howe Island. The results of research into many genera, such as *Linospadix* and the monotypic endemic genera in the Ptychospermatinae, are nearly absent from the literature.

Evidence suggests the ancient Gondwanan palm lines that existed in Australia in the Early to Mid Tertiary were largely extirpated by the Miocene, and that the extant Australian palm flora is mostly of relatively recent establishment and derived from Malesian stock since the Miocene. Considering palaeoclimatic data, migrant stock of extant palms may have begun to arrive in Australia when the climate across much of the continent was more benign than at present and palms, as predominantly mesic elements, were able to colonise large areas. Environmental changes, such as increasing aridity, decreasing soil fertility and competition from the relict Gondwana stock, soon came to bear on the migrants. Habitats were affected by increasing fire occurrence as a consequence of increasing aridity, and by developing cyclonic regimes as a result of changing global air circulation patterns. Palms contracted to suitable habitats and populations, particularly in areas where aridity had a greater impact, became fragmented. The distribution pattern of *Livistona*, for example, can be explained by this process. Rainforest and other mesic habitats, preferred by most palms, eventually contracted to a few small areas that, apart from supporting relict palm species, provided habitat into which migrant palms could continue to enter from northern palm-rich areas.

5. Systematic arrangement of the Australian palm flora

This and the following chapters introduce aspects of the relationships among Australian palms and place them in a global and regional context. The most recent classification system for the Arecaceae, as presented in *Genera Palmarum: The Evolution and Classification of Palms* (Dransfield *et al.* 2008), is followed in this work. Australia has representatives of all of the five subfamilies recognised in the family. Despite limited mid-level classification diversity in the Australian palm flora, there are some instances of unique evolutionary development, primarily in the Areceae in the Arecoideae, with eight endemic genera, and the Ceroxyloideae with one endemic genus. The Calamoideae and Coryphoideae have no endemic genera but have high levels of species endemicity. Nypoideae, and its only species *N. fruticans*, occurs in outlier populations in northern Australia, to a centre of distribution in west Malesia and south-east Asia.

PHYLOGENETIC RELATIONSHIPS OF AUSTRALIAN PALMS

There are five major lines of evolution in the Arecaceae (Asmussen *et al.* 2006; Dransfield *et al.* 2008) and representatives of each occur within the Australian palm flora. The major lines of evolution are classified systematically as five subfamilies (Fig. 5.1; Tables 5.1 and 5.2). These are further divided into tribes and subtribes, and it is at these levels that the Australian palm flora becomes restricted. For example, of the 14 tribes in the largest subfamily Arecoideae, only two, Cocoseae and Areceae, occur in Australia; of the 11 subtribes within the Areceae, five have representative genera in Australia. Similarly, the Coryphoideae, with eight tribes globally, has representatives of three tribes (Trachycarpeae, Caryoteae and Corypheae) in Australia. The Calamoideae and the Ceroxyloideae are even less represented, with only a single genus in each: *Calamus* (eight species) in the former and *Oraniopsis* (one species) in the latter. The fifth subfamily, Nypoideae, is monospecific and has representation in tropical Australia with its single species *N. fruticans*.

Despite the deficient representation of mid-level classification in the Australian palm flora, the presence of some lines suggests that Australia and nearby islands have acted as a minor centre of evolution for the family. The Arecoid subtribes Archontophoenicinae, Ptychospermatinae and Linospadicinae and the Ceroxyloid tribe Ceroxyleae have endemic genera in Australia. There are no endemic genera in the Calamoideae or Coryphoideae, although both have relatively large numbers of endemic species in the genera *Calamus, Livistona, Licuala, Caryota* and *Arenga*. There is well-supported cladistic and molecular evidence to indicate that the Australian species of *Livistona* form a separate clade from all other species in the genus (Dowe 2001; Isagi *pers. comm.*).

FAMILY DESCRIPTION

Arecaceae Schultz-Sch. *Nat. Syst. Pflanzenr.* 317 (1832) *nom. cons.*; *Palmae* Juss., *Gen. pl.* 37 (1789) *nom. alt.*

Small to tall plants, monoecious or dioecious, polygamous, polygamodioecious, polygamomonoecious, and hapaxanthic or pleonanthic. **Stems** solitary or clustered, acaulescent to arborescent, erect or climbing, slender

Table 5.1 Subfamilies in the Arecaceae and the number of global and Australian taxa

	Global genera	Global species	Australian genera	Australian species
Arecoideae	107	1247	13	26
Ceroxyloideae	8	42	1	1
Coryphoideae	46	456	5	24
Nypoideae	1	1	1	1
Calamoideae	21	617	1	8
Total	183	2363	21	60

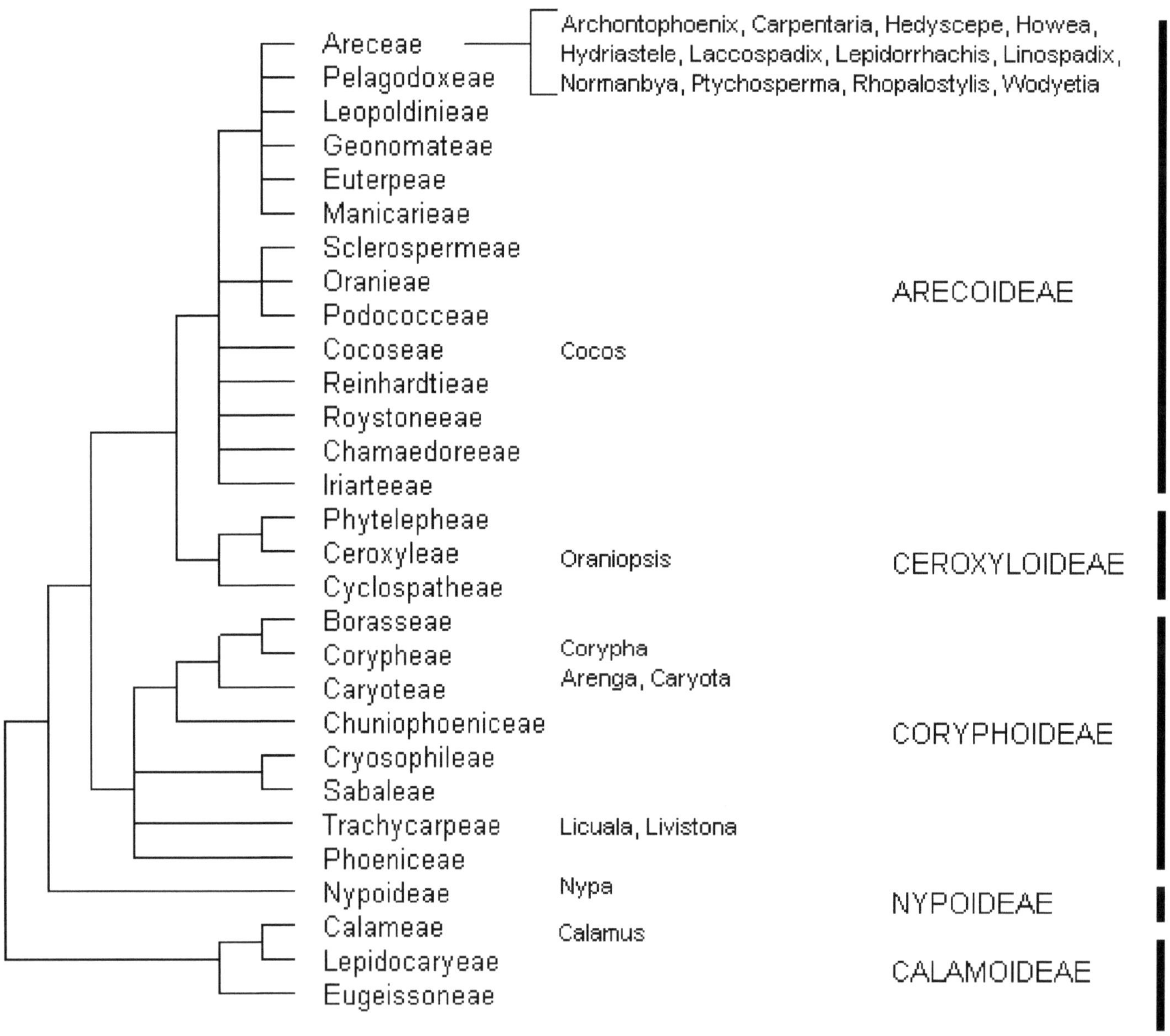

Figure 5.1 Estimated phylogenetic relationships of the palm family with indicated positions of the Australian genera. Phylogenetic tree summarised from Asmussen *et al.* (2006) and Dransfield *et al.* (2008).

Table 5.2 Summary of characteristics for subfamilies in the Arecaceae. Characters that are unique to a single subfamily are in bold

	Leaf	Sexuality	Inflorescence	Flowers	Fruit	Seed
Subfamily 1: Calamoideae Distribution: global	Pinnate, infrequently palmate, pinnae reduplicate	Monoecious, dioecious, polygamous	Highly branched, **frequently with adnation**	Solitary, **dyads** or in cincinni, unisexual	**Epicarp of imbricate scales**	1–3, usually **sarcotestal**; endosperm homogeneous or ruminate, embryo basal or lateral
Subfamily 2: Nypoideae Distribution: Old World tropics	Pinnate, pinnae reduplicate	Monoecious	**Terminated by pistillate head, staminate flowers on lateral branches**	Solitary, unisexual	**In terminal head, individual fruit compressed and irregularly angled,** epicarp smooth	1, endosperm homogeneous, rarely ruminate, embryo basal
Subfamily 3: Coryphoideae Distribution: global	Pinnate, bipinnate, palmate, **costapalmate**, pinnae and **segments induplicate**, rarely reduplicate	Monoecious, dioecious, polygamous; hermaphroditic	Usually highly branched, less frequently moderately branched, rarely spikate	Solitary, or in cincinni, or rarely in triads, unisexual or hermaphrodite	Epicarp smooth or corky-warted	1–3, endosperm homogeneous or ruminate, embryo apical to basal
Subfamily 4: Ceroxyloideae Distribution: global	Pinnate, pinnae reduplicate	Hermaphroditic or dioecious	Usually moderately branched, infrequently spikate, infrequently with pistillate head	Solitary or in **monopodial clusters**	Epicarp smooth, rough or corky-warted	1–**10**, endosperm homogeneous, embryo lateral to basal
Subfamily 5: Arecoideae Distribution: global	Pinnate, pinnae reduplicate	Monoecious or dioecious	Usually moderately branched, infrequently spikate	In triads, or in cincinni, or in **acervuli**, or rarely solitary	Epicarp usually smooth, infrequently rough or corky-warted	1–8, endosperm homogeneous or ruminate, embryo apical to basal

to stout; leafscars annular. **Leaves** forming a terminal crown, palmate, costapalmate, pinnate, imparipinnate or bipinnate, or entire. **Inflorescence** axillary, infrafoliar, interfoliar or suprafoliar, spicate or paniculate, bearing papery to woody, deciduous or persistent bracts; single prophyllar bract always present, proximal on the inflorescence, bicarinate, variously shaped; peduncular bracts 0–many, bicarinate or tubular, distal to prophyll, variously shaped. **Flowers** unisexual or bisexual; perianth of usually dissimilar sepals and petals, usually trimerous; segments fused or free, valvate or imbricate, usually thick-textured and dull-coloured; stamens 3, 6 or numerous; gynoecium apocarpous with 1–3 carpels or syncarpous with 3 or sometimes more locules, or pseudomonomerous; ovary superior, smooth or covered in imbricate scales. **Fruit** a berry or drupe, small to large, 1–10 seeded; epicarp smooth, warty, corky, scaly or spiny, usually brightly coloured; mesocarp fibrous, fleshy or dry, or obsolescent; endocarp thin, thick or absent. **Seed** adhering or not adhering to the endocarp, sometimes sarcotestal; endosperm ruminate or homogeneous, rarely with central cavity; embryo small. Eophyll simple, bifid or compound.

A family of c. 190 genera and c. 2365 species, concentrated primarily in moist equatorial, tropical and subtropical regions but with outliers in semi-arid and temperate regions. Arecaceae is represented in Australia and off-shore territories by 21 genera, of which nine are endemic, and 60 species, of which 47 are endemic. The Arecaceae is placed in its own order, the Arecales, which resolves in most recent phylogenetic studies as sister to three other orders, namely the Commelinales, the Poales and the Zingiberales, all of which are included in the broadly termed 'commelinid monocots' (Chase *et al.* 2000; Soltis *et al.* 2005).

The family name, Arecaceae Schultz-Sch., is an alternative family name to Palmae Juss. It is one of nine exceptions where alternative family names are permitted under the *International Code of Botanical Nomenclature* (Greuter *et al.* 1994). Arecaceae is the preferred named for Australian usage (Clifford 1981).

KEY TO SUBFAMILIES IN AUSTRALIA

Characters are described as they occur in Australian taxa.

1 Ovary and fruit covered in imbricate scales; seed sarcotestal; dioecious climbing palms armed with spines, bristles and/or hooks on the leaves, inflorescences and climbing organs .... **Subfam. 1. Calamoideae**

1: Ovary and fruit not covered in imbricate scales; seed not sarcotestal; non-climbing, semi-herbaceous to arborescent palms with spines and hooks confined to the petiole margins or unarmed **2**

2 Inflorescence with pistillate flowers in a terminal head; staminate flowers on spikes on lateral branches emerging from below the pistillate head; perianth segments of staminate flowers all similar, minute; stamens 3; petioles unarmed .... **Subfam. 2. Nypoideae**

2: Inflorescences not as above, but either spicate or paniculate; perianth segments in distinct sepals (calyx) and petals (corolla); stamens 6 or more; petioles with spines or hooks, or unarmed **3**

3 Leaves palmate, costapalmate, imparipinnate or bipinnate; segments induplicate; spines or teeth on the petioles of palmate or costapalmate leaves, or unarmed on imparipinnate or bipinnate leaves .... **Subfam. 3. Coryphoideae**

3: Leaves paripinnate; segments reduplicate; petioles unarmed **4**

4 Flowers solitary; peduncular bracts 3–5; plants dioecious; staminate and pistillate flowers morphologically similar .... **Subfam. 4. Ceroxyloideae**

4: Flowers in triads or arrangements derived from triads; peduncular bract 1; plants monoecious; staminate and pistillate flowers morphologically differentiated .... **Subfam. 5. Arecoideae**

6. Subfamily 1: Calamoideae

Calamoideae Griff., *Calcutta J. Nat. Hist.* 5: 4 (1844).

Bisexual, monoecious, dioecious or polygamous, hapaxanthic or pleonanthic palms; arborescent, acaulescent or climbing, usually armed with spines and hooks. **Leaves** pinnate or palmate. **Pinnae** reduplicate. **Inflorescences** paniculate or spicate. **Flowers** unisexual, in dyads or derived from dyads, or solitary; sepals connate and basally tubular; petals valvate; stamens 6 or numerous; ovary trilocular; ovules anatropous; ovary covered in imbricate scales. **Fruit** 1-seeded or 2- or 3-seeded; epicarp with spiral rows of imbricate scales; mesocarp obsolescent to fleshy; endocarp thin or obsolescent. **Seed** sarcotestal, infrequently not sarcotestal; endosperm homogeneous or ruminate; eophyll bifid or compound.

A subfamily of 21 genera and c. 620 species of world-wide distribution; in Australia with one genus and eight species (five endemic), in one tribe, Calameae, and one subtribe, Calaminae.

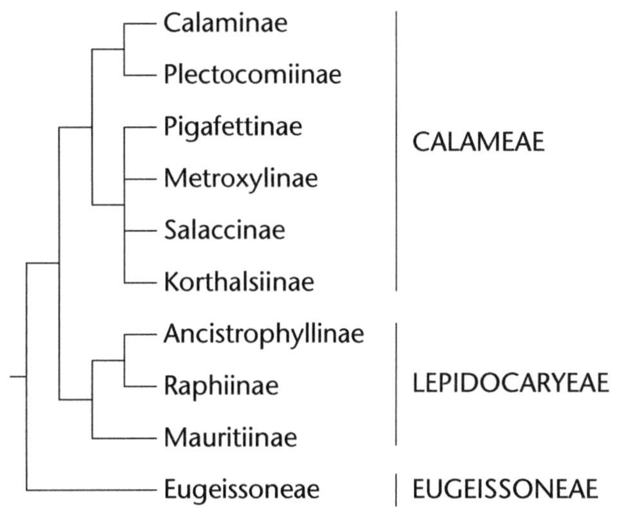

Figure 6.1 Phylogenetic tree summarising the estimated relationships of tribes and subtribes in the Calamoideae. Adapted from Baker *et al.* (2000a, 2000b), Asmussen and Chase (2001), Asmussen *et al.* (2000, 2006) and Dransfield *et al.* (2008).

Classification and relationships

The Calamoideae is monophyletic and resolves in most phylogenetic studies as sister to all other palms (Asmussen *et al.* 2006; Dransfield *et al.* 2008). The subfamily includes about one-quarter of all palms and has the largest palm genus, *Calamus*, with c. 375 species. The Calamoideae is globally distributed, and highest diversity occurs in west Malesia. Three tribes have been recognised for the subfamily (Fig. 6.1), of which one, the Calameae, has representatives in Australia. Of the other tribes, Lepidocaryeae occurs in the Americas and Africa, while Eugeissoneae has the single west Malesian genus *Eugeissona*. The fossil evidence indicates that Calamoid-like taxa were in Australia in the Palaeocene, evidenced by disulcate pollens, but early lines appear to have been extirpated by the Miocene. The extant Calamoids in Australia are most likely the result of radiation from west Malesia (see Chapter 3).

The Calameae is resolved as monophyletic in most studies (Baker *et al.* 2000a, 2000b; Asmussen *et al.* 2006; Dransfield *et al.* 2008) and consists of six subtribes, of which the Calaminae is the only one with Australian representatives (Fig. 6.2; Table 6.1). The Calaminae, with five genera, is most diverse in south-east Asia and Malesia but only one genus, *Calamus*, has representatives in Australia, with eight species (five endemic). *Calamus*, over its whole distribution, is not monophyletic and its relationship to the other genera in the Calaminae is presently unresolved (Baker *et al.* 2000b; Dransfield *et al.* 2008). Not all the Australian species have been included in molecular studies, but otherwise they are unequivocally included in *Calamus* based on morphological characteristics.

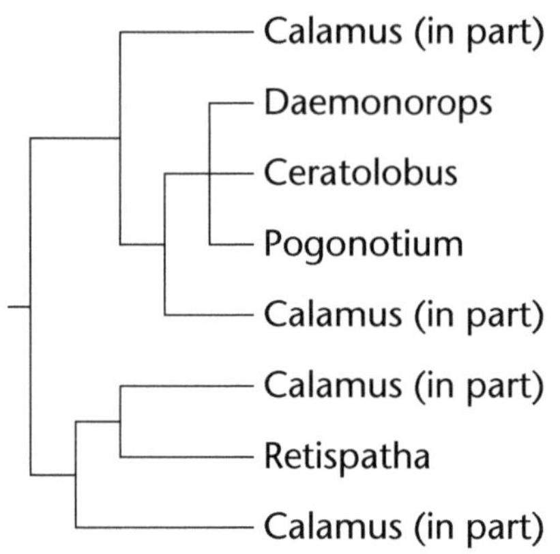

Figure 6.2 Phylogenetic tree summarising the estimated relationships of genera in the Calaminae. Adapted from Baker *et al.* (2000a).

Calamus, because of its diversity and variability, has been divided into eight sections (Furtado 1956; Dransfield 1979). The Australian species are representative of three sections:

- Sect. *Coleospathus* Becc.: Five species in Australia – *C. australis*, *C. caryotoides*, *C. moti*, *C. muelleri* and *C. radicalis*. Characterised by non-cirrate leaves; the presence of flagellae; basally tubular and apically widening inflorescence bracts; and a usually elongate inflorescence.
- Sect. *Phyllanthectus* Furtado: Two species in Australia – *C. aruensis*, *C. vitiensis*. Characterised by cirrate leaves and/or the presence of flagellae; tubular, non-splitting inflorescences bracts with non-uniform proportions; and a usually elongate inflorescence.
- Sect. *Podocephalus* Furtado: One species in Australia – *C. warburgii*. Characterised by cirrate leaves; tubular, non-splitting inflorescence bracts of uniform proportions; inflorescences not elongate; pedicellate spikelets; and sessile flower bracts.

Table 6.1 Synopsis of the Calamoideae tribes, subtribes and genera occurring in the Australian palm flora

Calameae Kunth ex Lecoq & Juillet, *Dict. Rais. Term. Bot. Fam. Nat.*: 98 (1831)
Calaminae Meisn., *Plant. Vasc. Gen.*: Tab. Diagn. 354, Comm. 265 (1842)
Calamus

Calamus

Calamus L., *Sp. Pl.*: 325 (1753). Type: *Calamus rotang* L.

Palmijuncus Rumph ex Kuntze, *Revis. Gen. Pl.* 2: 731 (1891). Type: not designated.

NB: Generic synonyms which do not have a taxonomic association with Australian taxa are not included in this treatment. A full list of generic synonyms of *Calamus* is provided in Dransfield *et al.* (2008).

Solitary or clustered, spiny, acaulescent, erect or climbing, pleonanthic, dioecious palms. **Stems** thin to thick; leafscars prominent; internodes short to very long. **Leaves** pinnate, infrequently bifid, cirrate or not, cleanly deciduous; leafbase forming a tubular leafsheath splitting opposite the petiole, usually densely armed, rarely unarmed, indumentum often abundant; ocrea often present, papery and disintegrating, or coriaceous; knee usually present, sometimes prominent; sterile flagellae present in non-cirrate species; petiole lacking or present, flat on the adaxial surface, rounded on the abaxial surface, variously armed; rachis and cirrus often armed on the abaxial surface with recurved single or multiple hooks. **Pinnae** few to numerous, single-fold, reduplicate, entire or rarely praemorse, linear to lanceolate or rhombic, terminal pair sometimes basally joined, regularly arranged or grouped or subopposite, concolorous or discolorous; variously armed or scaly, or not; bristly spines on the adaxial surface, or lacking; midribs conspicuous or not, transverse veinlets conspicuous or not; ramenta lacking. **Inflorescences** frequently flagelliform, axillary, adnate to the internode and inner surface of the leafsheath of the accompanying leaf, paniculate with few to many branches; prophyll usually inconspicuous, 2-keeled, tubular, variously armed, rarely inflated, papery or coriaceous; peduncular bracts lacking or present, variously armed; rachis bracts persistent, variously armed, usually tubular, often splitting but remaining tubular basally, rarely irregularly tattering apically, each subtending a first-order branch; first-order branch bearing a tubular bract, each subtending a second-order branch; rachillae spreading to crowded, straight to curved, bearing a basal tubular bract; rachillae bracts crowded and spiral, in the male rachillae each bract subtending 1 flower and 1 bracteole, in the female rachillae each bract subtending a dyad of 1 female flower and 2 bracteoles and 1 sterile male flower, rarely each bract subtending 2 lateral female flowers and a central sterile male flower. **Staminate flowers** symmetrical; sepals 3, basally tubular, lobed distally; petals 3, usually longer than sepals, basally tubular, with valvate lobes; stamens 6, filaments elongate, anthers medifixed, short to elongate, latrorse to introrse; sterile male flowers with empty anthers. **Pistillate flowers** larger or smaller than the staminate flowers; sepals 3, tubular, shallowly lobed; petals 3, with valvate lobes; staminodes 6, epipetalous; gynoecium tricarpellate, triovulate, covered in reflexed scales, stigmas 3, apical, reflexed, ovules basal, anatropous. **Pollen** ellipsoidal or circular; aperture equatorially disulcate; exine tectate, finely to coarsely reticulate, verrucate, sometimes with rounded excrescences. **Fruit** usually 1-seeded, rarely 2–3-seeded, stigmatic remains apical; epicarp composed of spiral rows of reflexed scales, mesocarp thin, endocarp lacking; perianth persistent or not on the fruit. **Seed** sarcotestal, rounded, grooved, angled or winged, covered with thick sarcotesta, depressed on one side; endosperm homogeneous or ruminate; embryo basal to lateral; germination adjacent-ligular; eophyll bifid or pinnate.

A genus of c. 375 species distributed in Africa, south, east and south-east Asia, Malesia, Melanesia as far east as Fiji, and Australia where eight species occur, of which five are endemic. Of the Australian species, which are commonly referred to as wait-a-whiles or lawyer vines, all occur in Queensland, with one extending into northern New South Wales. Species occur primarily in rainforest, moist sclerophyll forest, vine forest and littoral forest, from sea level to 1600 m elevation. The non-endemic species represent outlier populations that have extensive distribution in New Guinea and Melanesia. *Calamus aruensis* and *C. warburgii* are among the most common and widespread species in New Guinea but in Australia are rare species with restricted distribution. *Calamus vitiensis* occurs in New Guinea and throughout east Malesia and Melanesia as far east as Fiji, but is confined to near-coastal areas of north-east Queensland. Of the five endemic species, *C. moti* and *C. radicalis* are confined to the Wet Tropics Bioregion of Queensland, *C. australis* and *C. caryotoides* have extensive distribution from Cape York to central Queensland and north-east Queensland respectively, and *C. muelleri* from south-east Queensland to northern New South Wales.

Etymology: *'Calamus'* (Linnaeus 1753): Derived from the Latin for reed or cane, *calamus*, from the Greek *Kalamos*.

Key to species of Calamus in Australia

1 Leaf terminating in a cirrus 2
1: Leaf not terminating in a cirrus 4
2 Stem solitary .... ***7. C. vitiensis***
2: Stems clustered at the base 3
3 Leafsheath unarmed or only sparsely armed in mature plants; pinnae lanceolate to subcucullate, 13–23 each side of the rachis .... ***1. C. aruensis***
3: Leafsheath densely armed with spines; pinnae linear to narrowly lanceolate, 40–100 each side of the rachis .... ***8. C. warburgii***
4 Apex of pinnae truncate to oblique, praemorse .... ***3. C. caryotoides***
4: Apex of pinnae tapered, usually acuminate, not praemorse 5
5 Spines on the leafsheath arranged in oblique combs .... ***4. C. moti***
5: Spines on the leafsheath arranged more or less randomly 6
6 Pinnae with bristly spines on the parallel veinlets on the adaxial surface 7
6: Pinnae lacking bristly spines on the parallel veinlets on the adaxial surface .... ***2. C. australis***
7 Stems to 40 mm diam.; leaf to 3 m long with 30–55 pinnae each side of rachis; inflorescence to 6 m long, with numerous branches .... ***6. C. radicalis***
7: Stems usually 5–8 mm diam., infrequently to 16 mm diam.; leaf to 1 m long with 7–10 pinnae each side of rachis; inflorescence to 3 m long, sparsely branched .... ***5. C. muelleri***

1 Calamus aruensis

Calamus aruensis Becc., *Malesia* 3: 61 (1886); *Palmijuncus aruensis* (Becc.) Kuntze, *Revis. Gen. Pl.* 2: 733 (1891). Type: Indonesia. Papua. Aru Islands, Wokam Island, Giabu-Lengan, 1873, *O.Beccari s.n.*; holotype FI [3 sheets].

Calamus hollrungii Becc., *in* K.Schumann and M.Hollrung, *Fl. Kais. Wilh. Land*: 17 (1889). Type: Papua New Guinea. East Sepik Province. Augusta Station, *U.M.Hollrung 667*; holotype B, destroyed; isotype FI.

Calamus latisectus Burret, *Notizbl. Bot. Gart. Mus. Berlin–Dahlem* 13: 319 (1936). Type: Papua New Guinea. Morobe Province. Malahang, Aug. 1935, *Clemens 62*; holotype B, destroyed; isotype L.

Stems clustering or solitary (not recorded in Australia), climbing to 25 m, dominant stems 1–3, bare stems 10–30 mm diam.; internodes to 40 cm long, green. **Leaves** cirrate, to 5 m long, horizontal to arching, 13–23 pinnae per side; leafsheath grey bluish-green to dark green, with thin, caducous, grey to brown indumentum, spines lacking or sparse and scattered, to 26 mm long, dark green to black, solitary or in irregular partial whorls of up to 5; knee conspicuous to 125 mm long, to 37 mm wide, rarely armed; flagellae lacking; petiole 7–9 cm long, armed with long spines; rachis to 3 m long, ridged on the adaxial surface in the distal portion, spiny or with few spines on the abaxial surface, and with recurved grapnel-like hooks; cirrus 1–4 m long, with grapnel-like hooks arranged irregularly. **Pinnae** irregularly to regularly arranged, very infrequently grouped, to 70 cm long, to 12 cm wide, broadly lanceolate, subcucullate, longest pinnae near the middle of the leaf, papyraceous, semi-glossy dark green on the adaxial surface, lighter green on the abaxial surface; midrib prominent as well as 1–2 thick parallel veinlets almost as prominent; numerous thin parallel veinlets; margins with small spines; bristly spines lacking on the adaxial surface. **Staminate inflorescence** to 2.7 m long, branched to 3 orders; prophyll to 30 cm long, unarmed or with very few short spines; peduncular bract lacking, rachis bracts to 33 cm long, similar to prophyll; branches to 55 cm long, strongly recurved, bracts on primary and secondary branches funnel-shaped; rachillae to 33 cm long; rachillae bracts funnel shaped, to 2 cm long, apically broadened, obliquely truncate, prolonged into a short point on one side. **Staminate flowers** 2–3.2 mm long, 1–1.5 mm wide in bud, opening to 2–4 mm wide; sepals 1–1.2 mm long, 0.8–1 mm wide, basally connate, lobes triangular, apex acute, thin, light green, glabrous; petals 1.8–2 mm long, 0.9–1 mm wide, elongately triangular, apex acuminate, thin, light green, glabrous; stamens 1.8–2 mm long; anthers c. 0.8 mm long, yellow; pistillode c. 0.2 mm long, minutely trifid. **Pistillate inflorescence** similar to male inflorescence, but branched to 2 orders; primary branches to 55 cm long, strongly recurved; rachillae to 20 cm long, straight to curved; rachillae bracts funnel-shaped, clasping, to 2 cm long, slightly broadened at the apex, obliquely truncate, prolonged into a short point on one side. **Pistillate flowers** 2–2.7 mm long, 1–1.2 mm wide at receptivity, conical; sepals 1.8–2 mm long, 1.3–1.5 mm wide, basally connate, lobes subulate, carinate, green, glabrous; petals 1.8–2 mm long, 1–1.2 mm wide, triangular, apex acuminate, green, glabrous; ovary globose, c. 1 mm diam.; stigma lobes c. 0.3 mm long, opening to c. 0.8 mm across, recurved, glistening light brown. **Sterile staminate flowers** 2.5–3 mm long, 1–1.2 mm wide in bud, not opening widely; sepals 1–1.1 mm long, 0.5–0.7 mm wide, basally connate, lobes subulate, gibbous, green, glabrous; petals 1.8–2 mm long, 0.8–1 mm wide, basally connate, lobes triangular, apex acuminate, green, glabrous; stamens c. 1.5 mm long; anthers 0.4–0.5 mm long, versatile, empty of pollen. **Fruit** 1-seeded, globose, 8–14 mm diam., stigmatic remains c. 0.5 mm long; epicarp c. 0.2 mm thick, with 12–18 longitudinal rows of cream-white scales, shallowly channelled along the middle; margins entire, thinning, translucent; mesocarp c. 0.2 mm thick; sarcotesta c. 3 mm thick; pedicel to 2 mm long. **Seed** 8–11 mm wide, 6–7 mm high, a shallow depression on one side, the surface covered with numerous pits and irregular channels; endosperm homogeneous; embryo basal. Figures 6.3, 6.4.

Lockerbie Scrub wait-a-while

Distribution and ecology: Widespread in New Guinea and Solomon Island, but uncommon in Australia, confined to Lockerbie Scrub and Jardine River catchment, northern Cape York Peninsula, occurring in rainforest, littoral forest,

Figure 6.3 *Calamus aruensis*, lowland rainforest habitat, Jacky Jacky Creek, Lockerbie Scrub, Cape York Peninsula, north Queensland.

Figure 6.4 *Calamus aruensis*. **Top left:** Knee with lower petiole and scattered spines. **Top right:** Staminate flowers at anthesis. **Bottom left:** Pistillate and sterile staminate flowers at anthesis. **Bottom right:** Mature fruit.

swamp forest and seasonally dry monsoon forest as scattered small groups and individuals and rarely dominating a habitat (in Australia) as do other *Calamus* species, 0–150 m asl (Fig. 6.5). Flowering Jan.–Mar.; fruiting Oct.–Dec. Conservation status – *Rare* (Bostock and Holland 2007).

Typification: The holotype of *Calamus aruensis* is an unnumbered collection by Beccari [FI] collected in 1873 from the Aru Islands, Indonesia. The collection consists of sections of leaves with intact pinnae and a portion of sterile inflorescence (Fig. 6.5).

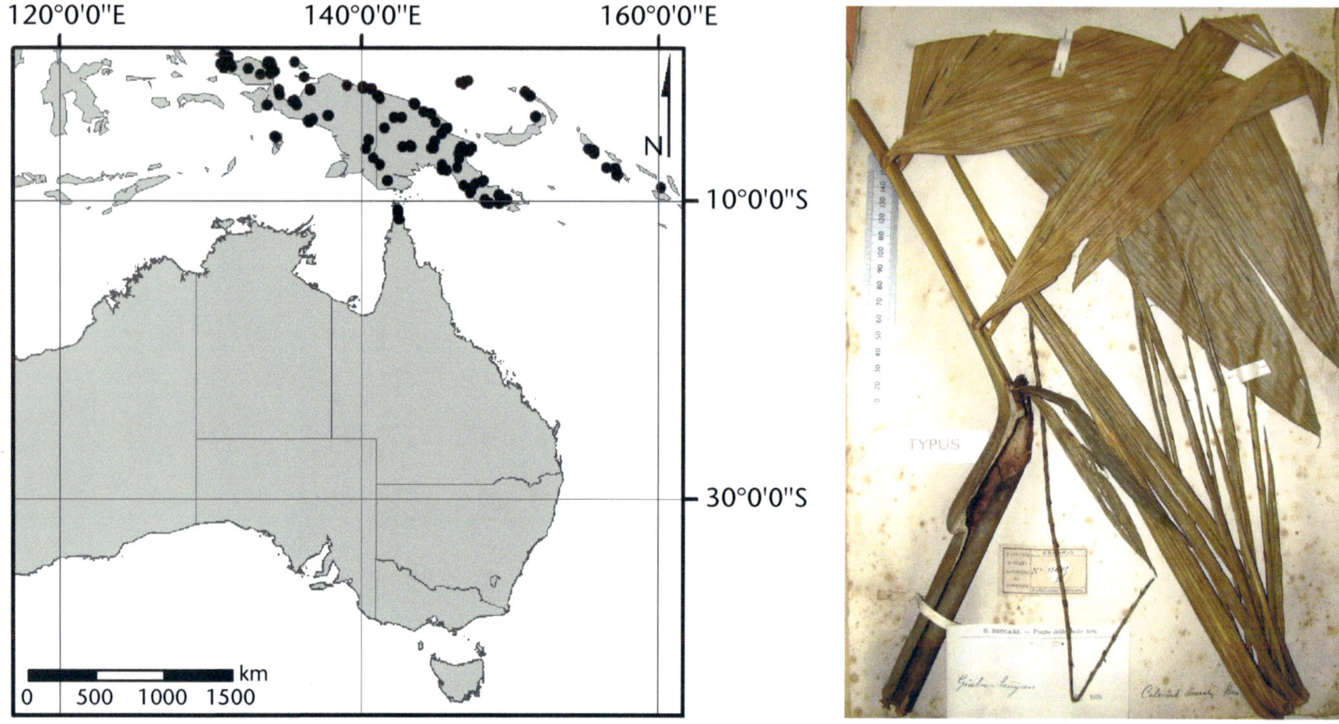

Figure 6.5 *Calamus aruensis*. **Left:** Distribution based on herbarium specimen data. **Right:** Type specimen. Indonesia. Papua. Aru Islands, Wokam Island, Giabu-Lengan, 1873, *O.Beccari s.n.*; holotype FI [1 of 3 sheets].

Etymology: *'Abita. – A Giabu–leñgan nelle Isole Aru. Maggio 1873'* (Beccari 1886, p. 61). [Distribution – On Giabua-lengan in the Aru Islands, May 1873]: With reference to the occurrence of the species in the Aru Islands, New Guinea, and from where the type specimen was collected.

Notes: Beccari (1889), in his protologue of *Calamus hollrungii*, noted the similarity of *C. hollrungii* to *C. aruensis*. Baker *et al.* (2003) found the types of these taxa to be scarcely indistinguishable and therefore synonymised *C. hollrungii* under *C. aruensis*. Burret (1936) described *C. latisectus* without any significant distinguishing characters (from *C. hollrungii*), and later (Burret 1939) synonymised it under *C. hollrungii*. In Australia, the species now known as *Calamus vitiensis* (see below) was recorded in herbarium records and other accounts under the name of *C. hollrungii* (Johnson 1981; Jones 1996).

In Australia, *Calamus aruensis* is always represented by clustering individuals, unlike in populations in New Guinea where it is predominantly represented by solitary-stemmed individuals (Baker *et al.* 2003). Usually, the leafsheaths are unarmed except for a few randomly placed small spines at the very apex near the junction of the petiole, or sometimes there are sparsely scattered long spines throughout the length of the leafsheath.

2 Calamus australis

Calamus australis Mart., *Hist. Nat. Palm. Edn 1*, 3: 213 (1838); *Palmijuncus australis* (Mart.) Kuntze, *Revis. Gen. Pl.* 2: 733 (1891). Type: Australia. Queensland. Fitzroy Is., 23 June 1819, *A.Cunningham 303*; lectotype BM [3 sheets], here designated; isolectotype K.

Calamus obstruens F.Muell., *Fragm.* 5: 48 (1865). Type: Australia. Queensland. Rockingham Bay, 17–21 Feb. 1864, *J.Dallachy 43*; holotype MEL.

Calamus jaboolum F.M.Bailey, *Queensland Bot. Bull.* 13: 14 (1896). Type: Australia. Queensland. Barron R., *E.Cowley 2 & 3*; holotype BRI.

Calamus amischus Burret, *Notizbl. Bot. Gart. Berlin–Dahlem* 15: 800 (1943). Type: Australia. Queensland. Cooktown, Cape Bedford, 1901–02, *W.Poland 203*; holotype B, destroyed.

Stems clustering, climbing to 35 m long, 1–6 dominant, bare stems 11–25 mm diam., leafscars stepped; internodes to 30 cm long, green to yellow-brown. **Leaves** non-cirrate, to 2 m long, horizontal, 12–25 pinnae per side; leafsheath glabrous, with dense single or basally united needle-like brown spines to 10 cm long, randomly arranged; knee inconspicuous; flagellae to 4 m long; petiole 5–15 cm long, spiny; rachis to 185 cm long, flat on the adaxial surface, spiny. **Pinnae** regularly arranged, linear-lanceolate, to 30 cm long, to 30 mm wide, acute to acuminate or filiform, laxly papyraceous, dark green on the adaxial surface, lighter green on the abaxial surface, with prominent midrib; margins with small spines; bristly spines on upper surface lacking. **Inflorescences,** staminate and pistillate similar, to 5 m long, branched 2–3 orders; prophyll 40–60 cm long, tubular, green with grey indumentum, moderately spiny proximally, becoming densely spiny distally; peduncular bract absent or 1, green with grey indumentum, moderately spiny proximally, becoming densely spiny distally; lateral branches c. 7; rachis bracts tubular, green with grey indumentum, moderately spiny proximally, becoming densely spiny distally; bracts subtending rachillae to 6 cm long, tightly tubular, green with scattered dark punctate scales; rachillae bracts to 2 cm long, tightly tubular, green with scattered dark punctate scales; rachillae distichous to subdistichous, to 5 cm long, green, glabrous, straight to curved; bracteoles sepal-like, red-green. **Staminate flower** 4–6 mm long, 1.5–2 mm wide in bud, opening to c. 8 mm wide; sepals 1.8–2 mm long, 1.3–1.5 mm wide, apex triangular, dark green, margin tinged dark red, glabrous; petals 5–6 mm long, 0.8–1 mm wide, triangular, acuminate, green, glabrous; stamens c. 5 mm long; anthers c. 1 mm long, yellow; pistillode c. 0.2 mm long, minutely trifid. **Pistillate flower** 2–4 mm long; sepals 0.8–1 mm long, 0.8–1 mm wide, apex broadly triangular, red, margins fringed to lacerate, glabrous; petals 1.8–2 mm long, 1.8–2 mm wide, broadly triangular, green, margins entire, glabrous; stigmas c. 1 mm long. **Sterile staminate flower** 1.8–2 mm long, 0.9–1 mm wide in bud, not widely opening, prolifically nectariferous; sepals 0.9–1 mm long, 0.8–1 mm wide, broadly triangular, dark green, apex margin tinged dark red, glabrous; petals 1.5–2 mm long, 0.9–1 mm wide, triangular, acuminate, light green with red striations; stamens c. 2 mm long; anthers c. 0.5 mm long. Fruit 1-seeded, globose to subglobose, 11–15 mm long, 9–15 mm wide; stigmatic remains extended, 3–4 mm high; epicarp c. 0.3 mm thick, with 15–18 rows of scales, cream to pale brown at maturity, scale margins reddish; mesocarp c. 0.2 mm thick; sarcotesta 1–2 mm thick; pedicel 0.5–1 mm long. **Seed** to 9 mm long, to 7 mm wide, with a depression on the long side, surface smooth or with infrequent shallow furrows; endosperm homogeneous; embryo subapical. Figures 6.6, 6.7.

Hairy Mary, lawyer vine, wait-a-while

Distribution and ecology: *Calamus australis* is the most widespread species of rattan in Australia, distributed from Pajinka, Cape York south to Byfield, central Queensland, in rainforest, moist sclerophyll forest, littoral forest, vine thickets, swamp forest and mangroves, forming dense thickets or as scattered individuals, 0–1600 m asl. (Fig. 6.8). Flowering all months; fruiting all months. Conservation status – *No present threats*.

Figure 6.6 *Calamus australis*, in moderate-altitude rainforest, Lamb Range, north-east Queensland, c. 900 m altitude.

Typification: The lectotype of *Calamus australis* is a collection by Allan Cunningham (*303*) [BM], collected in June 1819 from Fitzroy Island, off the north-east Queensland coast. The collection consists of three sheets representing a single collection: there are two sheets with a section of leaf with intact pinnae and petiole; the third sheet has a section of the leafsheath with a portion of leaf with pinnae, and attached flagellum (Fig. 6.8).

The protologue included the following: *'In insula Fitzroy dicta, inde a promotorio Cape-Grafton (16°51' lat. a., 163°33' long. or. a Ferro) versus Austrum in lat a 17° sita, absque fructificatione legit div. Allan Cunningham (1819) Junio mense, Navarchum P.P.King in nave Mermaid concomitatus.'* [Present on Fitzroy Island, from there to Cape Grafton

Figure 6.7 *Calamus australis*. **Top left:** Knee with armature. **Top right:** Staminate flowers at anthesis. **Bottom left:** Pistillate and sterile staminate flowers at anthesis. **Bottom right:** Mature fruit.

(16°51'S, 146°33') south to about 17°, collected by Allan Cunningham June (1819) without fruit, on the Voyage of the Mermaid with P.P. King].

Etymology: *'Calamus? australis'* (Martius 1838). There was no explanation for the choice of name, but possible reference to the species occurrence in Australia, from the Latin *australis*, southern.

Notes: Mueller (1865) named *C. obstruens* based on a Dallachy specimen from Rockingham Bay, and distinguished it by numerous narrower pinnae than were considered applicable for *C. australis*, but otherwise is accountable to *C. australis*. Bailey (1896) named *C. jabollum* from a Cowley collection from Barron River but the protologue provided no distinction from *C. australis*. Burret

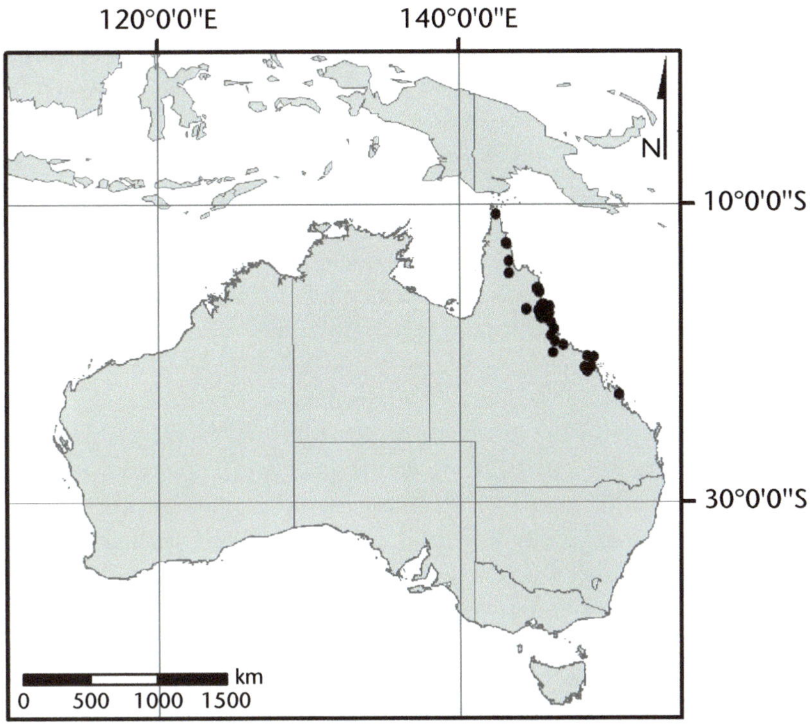

Figure 6.8 *Calamus australis*. **Left:** Distribution based on herbarium specimen data. **Right:** Type specimen. Australia. Queensland. Fitzroy Island, 23 June 1819, *A.Cunningham 303*; lectotype BM [1 of 3 sheets].

(1943) named *C. amischus* from a Wilhelm Poland collection, collected in 1901–02 from the Cooktown area, and distinguished it from *C. australis* by the density and form of the armature.

Calamus australis, although very widely distributed, is not that variable and readily distinguished from other species. In habitat, it often occurs with *C. radicalis* and *C. moti*. The pinnae of *C. radicalis* have numerous fine bristly spines, up to c. 10 mm long on the adaxial surface. These are lacking in *C. australis* (but marginal prickles occur on both species). *Calamus moti*, which also has spines on the adaxial surface, has the spines on the leafsheath arranged in oblique combs, but spines on the leafsheaths are otherwise dense and randomly arranged in both *C. australis* and *C. radicalis*. In some habitats, *C. australis* occurs with *C. caryotoides*, but that species is easily recognised by the unique praemorse pinnae, rather than linear-lanceolate pinnae in *C. australis*.

The growth habits of *C. australis* were studied by Putz (1990). The species occurs primarily in rainforest gaps or margins and vertical growth is required to maintain photosynthetic capacity but, once reaching the canopy, internode length was considerably shortened and stem slippage occurred.

3 Calamus caryotoides

Calamus caryotoides A.Cunn. ex Mart., *Hist. Nat. Palm. Edn 1*, 3: 212, tab. 116, Fig. 7, sub. 1–3 (1838); *Palmijuncus caryotoides* (A.Cunn. ex Mart.) Kuntze, *Revis. Gen. Pl.* 2: 733 (1891) [as *caryotodes*]. Type: (*New Holland Mermaid's 3rd Voyage, July–Oct. 1820*) Australia. Queensland. Endeavour River, July–Oct. 1820, *A.Cunningham s.n.*; lectotype BM, here designated; isolectotype K.

Stems clustering, climbing to 15 m long, 3–10 dominant; stems to 20 mm diam.; internodes to 20 cm long, green to brown. **Leaves** non-cirrate, to 40 cm long, horizontal, with 2–12 pinnae per side; leafsheath green, glabrous with dense, needle-like green to brown spines to 10 mm long, randomly arranged or in small horizontal rows; knee to 10 mm long, basally lipped, bare or with small spines similar to those on the leafsheath; flagellae to 4 m long; petiole 0–12 cm long, sparsely spiny, with broad-based spines; rachis grooved on the adaxial surface, sparsely spiny, rounded to angled on the abaxial surface, with single or paired recurved grapnel-like hooks. **Pinnae** regularly or irregularly arranged, apical pair broadly basally united, narrowly cuneate, to 25 cm long, to 6 cm wide, broadest toward the apex, obliquely truncate, regularly or irregularly lobed or toothed, praemorse, papyraceous, mid grey-green on the adaxial surface, only slightly lighter green on the abaxial surface, proximal pinnae sometimes much reduced; midrib inconspicuous or moderately prominent, and many moderately prominent parallel veinlets; bristly spines lacking from the adaxial surface; margins with widely spaced, distal-facing spines along whole length, larger in the basal part of the pinnae and sometimes at irregular angles. **Staminate and pistillate inflorescences** sparsely branched, to 2.5 m long; bracts tubular, tightly sheathing, sparsely armed. **Staminate flower** 3–4 mm long in bud; sepals 1.5–2 mm long, 1.2–2 mm wide, green or dull orange infused with green, glabrous; petals 3.5–4 mm long, 1.2–3 mm wide, green or cream to dull orange infused with green, glabrous; stamens 2.5–4 mm long, widely spreading at anthesis; anthers 0.5–2 mm long, versatile, yellow; pistillode minute. **Pistillate flower** to 4 mm long in bud; sepals 1–1.5 mm long, 1–1.5 mm wide, dull orange-red, glabrous; petals 3–4 mm long, to 2 mm wide, dull orange-red, glabrous; stigmas strongly recurved, to 1.5 mm long, to 3 mm wide at receptivity, lucid cream; sterile staminate flower 3–4 mm long in bud; sepals 1–1.5 mm long, dull orange-red, glabrous; petals 3–4 mm long, briefly acuminate, dull orange-red, glabrous; stamens to 3 mm long, not widely spreading; anthers sterile, 1–1.5 mm long, cream. **Fruit** 1-seeded, globose to subglobose, 10–20 mm diam., 8–15 mm wide; stigmatic remains 1–2 mm high; epicarp c. 0.2 mm thick, scales in c. 15 rows, cream-yellow to pale brown; mesocarp c. 0.2 mm thick; sarcotesta 1–2 mm thick; perianth not persistent on fruit. **Seed** c. 7 mm long, c. 5 mm wide, with a shallow depression on the long side, surface smooth; embryo basal; endosperm homogeneous. Figures 6.9, 6.10.

Fishtail wait-a-while

Distribution and ecology: *Calamus caryotoides* is widely distributed in north-east Queensland, from just north of Iron Range south to Mt Elliot, most common in drier rainforest or moist sclerophyll forest, but also in vine forests, littoral forest, and less commonly in wet rainforest. It is usually not a thicket-forming rattan but most often occurs in small groups or scattered individuals, mainly absent from disturbed areas, 0–1500 m asl (Fig. 6.11). Flowering all months; fruiting all months. Conservation status – *No present threats*.

Typification: Martius (1838), in the protologue of *Calamus caryotoides*, cited collections by Banks in 1770 and Cunningham in 1820, both being from near Endeavour River in north-east Queensland. The protologue included the following: '*In sylvaticis umbrosis, secundum oram Novae Walesiae australis, inter lat. a. 15 et 17°. Primum observatus sine fructificatione a Jos. Banks 1770; fructiferum legit all. Cunningham in expeditone speculatoria navis Mermaid dictae Navarchum P.P. King comitatus, in fluv. Endeavour-River vicinia, lat a. 15°30', Julio 1820.*' [In shady forest, coastal New South Wales, between 15 and 17°. First seen without fruit by Joseph Banks in 1770; fruit collected by Allan Cunningham in the exploratory expedition of the Mermaid commanded by P.P. King, in the vicinity of Endeavour River at 15°30', in July 1820.] Of the two specimens cited in the protologue, only the Cunningham collection has been located and has been designated as the lectotype (Fig. 6.11).

Etymology: '*pinnis parcis inaequidistantibus cuneiformibus aut (adultioribus) trapezoiden-cuneatis, antice erosis*' (Martius 1838) [pinnae irregularly arranged, cuneate (in the adult form) or trapezoidal-cuneate with the apex erose]: With reference to the resemblance of the pinnae to those in the genus *Caryota*, and the Latin *oides*, resembling.

Notes: *Calamus caryotoides* is unique among Australian rattans because the pinnae have praemorse apices and are shaped like a fishtail (hence the common name, fishtail wait-a-while). Stems may be very thin, usually about 10 mm in diameter but occasionally to 20 mm; this habit makes it only mildly obstructive in habitat compared to other rattans.

4 Calamus moti

Calamus moti F.M.Bailey, *Queensland Bot. Bull.* 13: 13 (1896). Type: Australia. Queensland. Barron R., *E.Cowley 1*; holotype BRI [7 sheets].

Stems clustering, climbing to 45 m long, 1–6 dominant, bare stems to 50 mm diam.; internodes to 65 cm long, yellow-green to light brown. **Leaves** non-cirrate, to 3 m long, strongly arched, 35–50 pinnae per side; leaf-sheath light green to yellow-green, glabrous, with needle-like green-brown spines to 20 mm long, arranged in well-spaced oblique combs; knee basally lipped, lacking spines or with scattered irregular spines mainly on the adaxial ridge; flagellae to 6 m long; petiole 0–2 cm long; rachis to 3 m long, channelled and smooth on the adaxial surface, with grapnel-like yellow hooks on the abaxial surface. **Pinnae** regularly arranged, crowded, to 50 cm long, to 3 cm wide, linear-lanceolate, acuminate,

Figure 6.9 *Calamus caryotoides*. **Top:** In moderate-altitude rainforest, Lamb Range, north-east Queensland, c. 600 m altitude. **Bottom left:** Leaf apex, pinnae with characteristic praemorse apices. **Bottom right:** Stem with leafsheath armature.

rigid, light green to yellow-green on both the adaxial and abaxial surfaces, with prominent midrib; margins with distal-facing, small spines along entire length; long thin soft bristly spines numerous on the adaxial midrib and thicker parallel veinlets. **Staminate inflorescences** to 6 m long, branched to 2 orders; prophyll to 40 cm long; peduncular bract lacking; rachis bracts lacking spines; lateral branches distichous, to 50 cm long, with stout single or 5-parted hooks; rachillae to 18 cm long; bracts subtending rachillae tightly tubular. **Staminate flowers** 5–6 mm long, 2–3 mm wide in bud, opening to c. 9 mm at anthesis; sepals 2–2.5 mm long, apex acuminate, dark green, glabrous; petals 5–5.5 mm long, 1.5–2 mm wide, apex acuminate, green, glabrous; stamens 4.5–5 mm long, filaments straight; anthers 1–1.2 mm long, yellow; pistillode minutely trifid. **Pistillate inflorescence** to 6 m long, branched to 3 orders; prophyll to 40 cm long, tightly tubular; peduncular bract 40–60 cm long, tightly tubular with strongly recurved spines in the distal portion; rachis bracts tubular, lacking spines; lateral branches distichous, to 50 cm long with short 3–5-parted grapnel-like spines; rachillae to 18 cm long; rachillae bracts funnel-shaped, clasping, to 15 mm long, slightly broadened at the apex, obliquely truncate, prolonged into a short point on one side. **Pistillate flowers** 6–9 mm long, to 3–4 mm wide at

Figure 6.10 *Calamus caryotoides*. **Top left:** Staminate flowers at anthesis. **Top right:** Pistillate and sterile staminate flowers at anthesis. **Bottom left:** Infructescence, Bicton Hill, Mission Beach, north-east Queensland. **Bottom right:** Mature fruit.

anthesis, conical; sepals 5–5.5 mm long, 3–4 mm wide, basally connate, lobes triangular, apex acuminate, green, glabrous or with scattered small rough reddish scales; petals 5–7 mm long, 3–4 mm wide, triangular, apex acuminate to acute, green, glabrous or with scattered small rough reddish scales; staminodes sagittate; ovary elongate, c. 3 mm long; stigmas 1–1.5 mm long, opening to c. 4 mm across, recurved, glistening white. **Sterile staminate flowers** 5–6 mm long, 4–5 mm wide at anthesis, not widely opening; sepals 3–4 mm long, 2.5–3 mm wide, basally connate, lobes triangular, apex acuminate, green, glabrous; petals 4–5 mm long, 2–3 mm

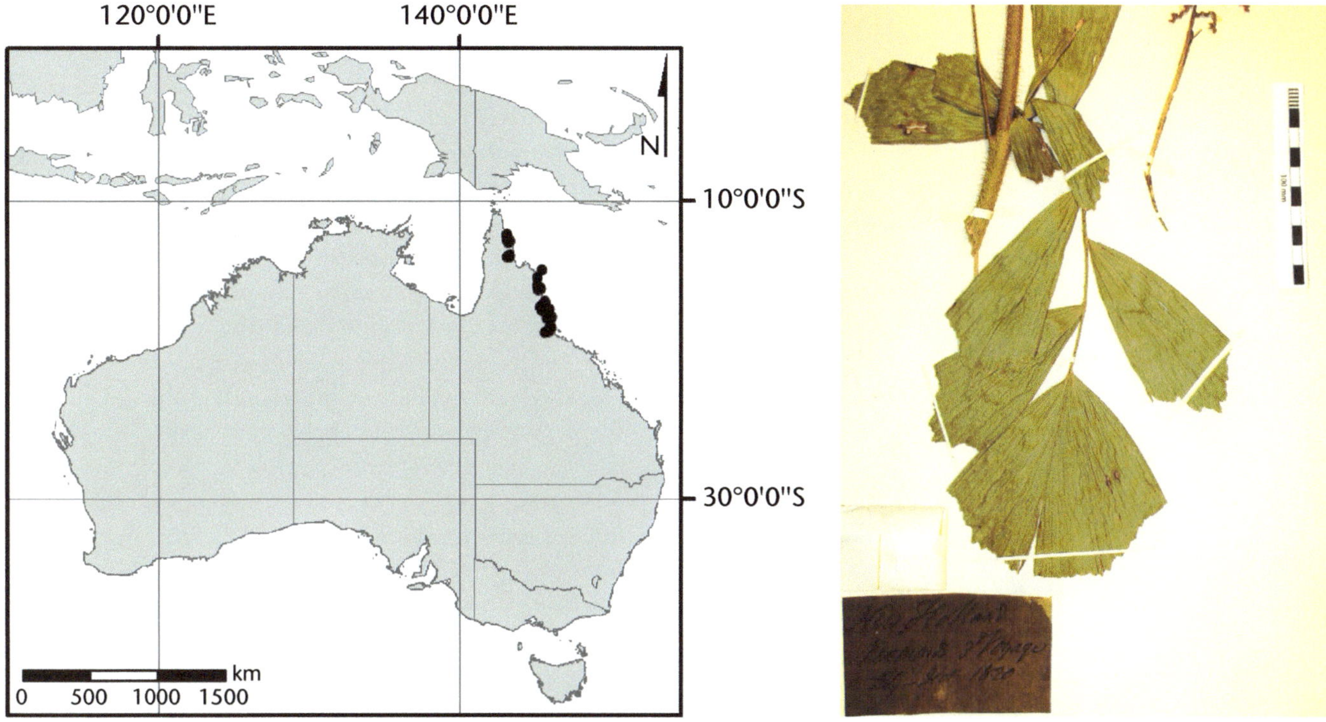

Figure 6.11 *Calamus caryotoides*. **Left:** Distribution based on herbarium specimen data. **Right:** Type specimen. Australia. Queensland. Endeavour River (New Holland Mermaid's 3rd Voyage, July–Oct. 1820), *A.Cunningham s.n.*; lectotype BM.

wide, basally connate, lobes elongately triangular, apex acuminate to acute, rigid, green, glabrous; stamens 2–3 mm long, erect, straight; anthers c. 0.5 mm long, versatile, empty of pollen. **Fruit** 1-seeded, globose to 13 mm diam.; stigmatic remains beaked, to c. 2 mm high; epicarp with 16–18 longitudinal rows of scales, cream to yellow; mesocarp to 0.2 mm thick; sarcotesta c. 3 mm thick. **Seed** to 10 mm long, to 6 mm wide, depressed on the long side, surface smooth or with few longitudinal shallow furrows and dimples; endosperm homogeneous; embryo basal. Figures 6.12, 6.13.

Yellow wait-a-while

Distribution and ecology: *Calamus moti* is distributed from near Cooktown south to about Clarke Range, north of Mackay, in rainforest, swamp forest and littoral forest, forming obstructive dense thickets especially in rainforest tree-fall gaps, rainforest margins and logged areas, 0–1300 m asl (Fig. 6.14). Flowering all months; fruiting all months. Conservation status – *No present threats*.

Typification: The holotype of *Calamus moti* is an undated collection made by E. Cowley [BRI] from Barron River in north-east Queensland. In the protologue, Bailey noted 'From Mr. Cowley I have stem and foliage specimens, and a promise of flower and fruit specimens when available'. The collection consists of a leafsheath and a leaf portion, and was sufficient to distinguish it from any previously described species with the spines arranged on the leafsheath in 'more or less false whorls' (Bailey 1896) (Fig. 6.14).

Etymology: 'Mr. Cowley tells me that the aborigines at the Barron River also distinguish three kinds by name – viz., *"Moti," "Jaboolum," and "Moologum"*' (Bailey 1896, p. 13).

Notes: *Calamus moti* is readily distinguished by the arrangement of armature on the leafsheath being in oblique combs, by the rigidly recurved leaves in mature plants, and by the leaf colour being a light apple green. The inflorescences can be very long and pendulous. The growth habits of *C. moti* were studied by Putz (1990).

Figure 6.12 *Calamus moti*, in moderate-elevation rainforest, Lamb Range, north-east Queensland, c. 800 m altitude.

Fröhlich *et al.* (1997) identified the fungus *Pestalotiopsis palmarum* (Cooke) Steyaert [Deuteromycetes] on the leaves of *C. moti*.

5 Calamus muelleri

Calamus muelleri H.Wendl., *in* H.Wendland and O.Drude, *Linnaea* 39: 193 (1875); *Palmijuncus muelleri* (H.Wendl.) Kuntze, *Revis. Gen. Pl.* 2: 733 (1891). Type: Australia. New South Wales. Clarence River, undated, *Dr. Beckler s.n.* [2 sheets] lectotype MEL, here designated.

Calamus muelleri var. *macrospermus* H.Wendl., *in* H.Wendland and O.Drude, *Linnaea* 39: 194, tab 2, Fig. 1, nos 9–11 (1875). Type: Australia. New South Wales. Richmond River, Jan. 1868, *J.A.Henderson 115*; holotype MEL 2132931, here designated.

[*Calamus australis auct. non.* Mart., F.Muell. *Fragm.* 5: 49 (1865)]

Stems clustering, climbing to 20 m long; dominant stems 1–12, 5–8 (16) mm diam.; leafscars c. 0.5 mm wide; internodes 9–30 cm long, green to yellow with age. **Leaves** non-cirrate, to 1 m long, 7–10 pinnae per side; leafsheath with dense needle-like brown spines to 10 mm long, randomly arranged or in obscure partial whorls; flagellae to 120 cm long; petiole 0–2 cm long, densely spiny; rachis flat on the adaxial surface, with retrorse, black-tipped spines. **Pinnae** regularly or irregularly arranged, lanceolate to elliptic, to 20 cm long, to 20 mm wide, acuminate, papyraceous, dark green on the adaxial surface, lighter green on the abaxial surface, with prominent midrib; bristly to thick spines on adaxial surface midrib and parallel veinlets common on juvenile plants, sparse or sometimes absent in mature plants; margins with small forward-facing spines along entire length. **Staminate and pistillate inflorescences** to 3 m long, sparsely branched, branched to 3 orders. **Staminate flowers** 5–7 mm long, 1–1.5 mm wide in bud, opening to c. 4 mm wide at anthesis; sepals 2–4 mm long, 1.8–2 mm wide, basally connate, green, glabrous; petals 4–5.5 mm long, 1.8–2 mm wide, imbricate, apically valvate, green, glabrous; stamens 1–5 mm long; filaments 1.7–3 mm long; anthers to c. 2 mm long, 0.5 mm wide, yellow; pistillode minute. **Pistillate flowers** to 5 mm long in bud; sepals 1.2–1.5 mm long, basally connate, tubular, green, glabrous; petals 2.5–3 mm long, imbricate, green, glabrous; stigmas thin, recurved. **Sterile staminate flowers** 2–3 mm long, 0.8–1 mm wide in bud, barely opening; sepals 1.2–1.5 mm long, green, glabrous; petals 2.5–3 mm long, apically pointed, green, glabrous; stamens to 2.5 mm long; anthers to 1 mm long. **Fruit** usually 1-seeded, occasionally 2-seeded, globose, 10–16 mm diam.; epicarp c. 0.2 mm thick, with 15 rows of scales, yellow at maturity; mesocarp c. 0.2 mm thick; sarcotesta 1–2 mm thick. **Seed** in 1-seeded fruit 6–10 mm long, 6–8 mm wide, depressed on one of the long sides, smooth; endosperm homogeneous; embryo lateral; seed in 2-seeded fruit hemispherical, irregular, not identical, otherwise with similar surface features to seeds in 1-seeded fruit. Figures 6.15, 6.16.

Southern lawyer cane, southern wait-a-while

Distribution and ecology: *Calamus muelleri* is disjunct from other rattans in Australia, distributed in southern Queensland, from Fraser Island, south to northern New South Wales near Dorrigo. The geographically closest rattan, *C. australis*, is c. 400 km to the north; occurs in

Figure 6.13 *Calamus moti*. **Top left:** Leafsheath with characteristic spines in oblique combs. **Top right:** Staminate flowers at anthesis. **Bottom left:** Pistillate and sterile staminate flowers at anthesis. **Bottom right:** Mature fruit.

rainforest and moist sclerophyll forest in small groups or as isolated individuals, sometimes gregarious, 0–1000 m asl (Fig. 6.17). Flowering Nov.–Mar.; fruiting all months. Conservation status – *No present threats*.

Typification: Wendland and Drude (1875) provided a number of specimens upon which the species, *Calamus muelleri*, and the subspecies, *C. muelleri* var. *macrospermus*, were described, but without differentiation: 'Clarence River! (leg. Dr. Beckler), Brisbane River!, Richmond River! (leg. Henderson), Ashers Station et Creek, Moreton Bay'. Of these, the Beckler collection from Clarence River, the Henderson collection from Richmond River and the

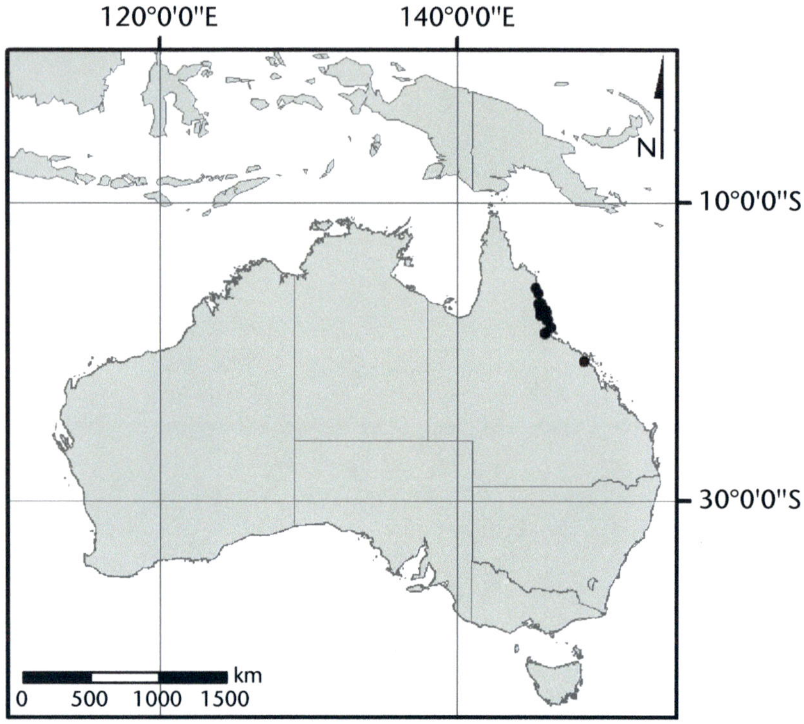

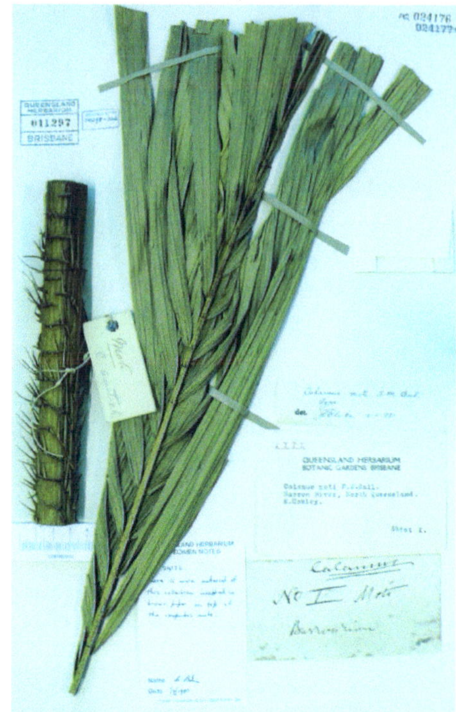

Figure 6.14 *Calamus moti*. **Left:** Distribution based on herbarium specimen data. **Right:** Type specimen. Australia. Queensland. Barron River, *E.Cowley 1*; holotype BRI [1 of 7 sheets].

Leichhardt collection from Ashers Station (sic) (= Archers Station) have been located. The collection from Brisbane River has not been located. Examination of the three known specimens indicated the Beckler and Leichhardt collections lacked fruit, while the Henderson collection included fruits that match the dimensions of those given for *C. muelleri* var. *macrospermus*. As the Beckler collection precedes that of Leichhardt in the list and is overall a more complete specimen, it is designated as the lectotype of *C. muelleri* (Fig. 6.17), and the Henderson collection as the holotype for *C. muelleri* var. *macrospermus*.

Etymology: '*in lit. ad cl. F. de Mueller*' (Wendland and Drude 1875, p. 193): Named for Baron Sir Ferdinand Jakob Heinrich von Mueller (b.1825–d.1896), Victorian Government Botanist 1853–1896, Director of Melbourne Botanic Gardens 1857–1873 and inaugural president of the Royal Society of Victoria 1860. Expeditions took him throughout Victoria and southern New South Wales, and to northern Australia with A.C. Gregory in 1855–57.

Notes: *Calamus muelleri* has a somewhat benign habit for a rattan and, similar to *C. caryotoides* in northern Queensland, does not usually form impenetrable thickets, although it can be common in some locations. However, it appears more closely related to *C. australis* and *C. radicalis*, particularly in the form of armature and inflorescence morphology.

The variety, *C. muelleri* var. *macrospermus*, was named for a collection with relatively large fruits, to 16 mm long as opposed to 12 mm long (Wendland and Drude 1875), but is otherwise attributable to *C. muelleri* as fruit size may be variable even on a single individual.

6 *Calamus radicalis*

Calamus radicalis H.Wendl. & Drude, *Linnaea* 39: 195 (1875); *Palmijuncus radicalis* (H.Wendl. & Drude) Kuntze, *Revis. Gen. Pl.* 2: 733 (1891). Type: Australia. Queensland. 35 miles [c. 56 km] N of Port Mackay (= Tully River) [Mackey in protologue], undated, *Nernst s.n*; holotype MEL [2 sheets].

Stems clustering, climbing to 40 m long, 1–3 dominant; bare stems to 40 mm diam.; internodes to 30 cm long; green to brown. **Leaves** non-cirrate, to 3 m long, horizontal to moderately arched, usually with a lateral twist, 30–55 pinnae per side; leafsheath glabrous, with

Figure 6.15 *Calamus muelleri*. **Top:** In lowland rainforest, Jowarra, Mooloolah River, Beerwah Forest Reserve, south-east Queensland. **Bottom left:** Leaf with bristles and spines. **Bottom right:** Stem with leafsheath armature, and attachment of a flagellum.

dense needle-like red-brown spines 5–15 cm long, randomly arranged; knee inconspicuous as a moderately raised portion, but lacking a definite basal lip; flagellae to 6 m long; petiole 0–2 cm long, densely spiny; rachis to 3 cm long, flat on the adaxial surface, densely spiny in the proximal portion, with spines single or clustered, straw-coloured. **Pinnae** regularly arranged, opposite to subopposite, linear-lanceolate, to 50 cm long, to 3 cm wide, acuminate, rigid, dark green on the adaxial surface, lighter green on the abaxial surface, with a prominent midrib; margins with small distal-facing spines along entire length; long soft bristly spines numerous on the adaxial midrib and parallel veinlets. **Staminate inflorescences** to 6 m long with many branches, branched to 4 orders; prophyll to 40 cm long, densely spiny; peduncular bract to 60 cm long, densely spiny toward the apex; lateral branches alternate, to 1 m long; rachillae 2–10 cm long, alternate, supporting branch narrowing between rachillae, with a single tubular bract with one side extended into a long acuminate point, green, glabrous; rachillae bracts tightly tubular, green, glabrous. **Staminate flowers** congested on rachillae, 2.5–3 mm long, 1.2–1.7 mm wide in bud, opening widely at anthesis; sepals 1.5–1.7 mm long, 0.5–0.8 mm wide, triangular, apex acuminate, green, glabrous; petals 2.5–3 mm long, 1–1.2 mm wide,

Figure 6.16 *Calamus muelleri*. **Top left:** Staminate flowers at anthesis. **Top right:** Pistillate and sterile staminate flowers at anthesis. **Bottom left:** Infructescence with maturing fruit. **Bottom right:** Mature fruit.

triangular, apex acuminate, green, glabrous; stamens 2–2.5 mm long; anthers 0.8–1 mm long, yellow; pistillode inconspicuous, sunken. **Pistillate inflorescence** to 6 m long with many branches, branched to 4 orders; primary branches to 40 cm long; rachillae 2–12 cm long, strongly curved or straight; bracts tightly tubular, green, glabrous with scattered scales toward the apex. **Pistillate flower** 1.8–2 mm long, 1.4–1.5 mm wide in bud; sepals 0.9–1 mm long, 1.4–1.5 mm wide, barely lobed, apex rounded, light green, glabrous, margin hyaline; petals 1.3–1.5 mm long, 0.8–1 mm wide, triangular, apex acuminate, light green, glabrous; stigma lobes recurved, each

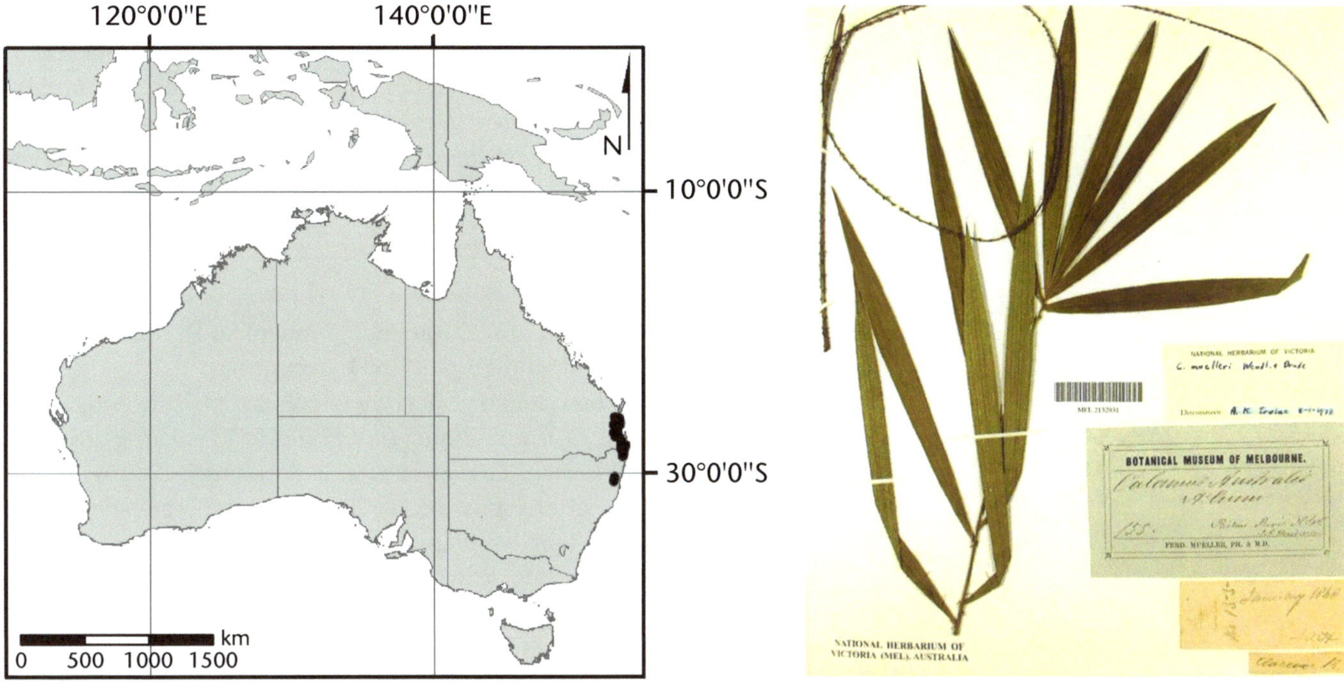

Figure 6.17 *Calamus muelleri*. **Left:** Distribution based on herbarium specimen data. **Right:** Type specimen. Australia. New South Wales. Clarence River, undated, *Dr. Beckler s.n.*, lectotype MEL [1 of 2 sheets].

0.2–0.3 mm long, glistening cream. **Sterile staminate flower** 1.8–2 mm long, 1.4–1.5 mm wide in bud, opening widely at maturity; sepals 0.8–1 mm long, deeply lobed, triangular, apex subulate, light green, glabrous; petals 1.3–1.5 mm long, 0.5–0.8 mm wide, triangular, elongate, apex acuminate, light green, glabrous; stamens 1.4–1.6 mm long; connective straight; anthers 0.4–0.5 mm long, yellow. **Fruit** 1-seeded, globose, 8–10 mm diam.; stigmatic remains c. 0.5 mm high; epicarp c. 0.2 mm thick, with 15 rows of scales, cream to pale brown at maturity, scale margins reddish; mesocarp c. 0.2 mm thick; sarcotesta 1–2 mm thick; pedicel 1–1.5 mm long. **Seed** 7–9 mm long, 5–7 mm wide, depressed on one of the long sides, irregularly furrowed; endosperm homogeneous; embryo subapical. Figures 6.18, 6.19.

Vicious hairy Mary, lawyer vine, wait-a-while

Distribution and ecology: *Calamus radicalis* is more restricted in distribution than other rattans in the Wet Tropics Bioregion, extending from the Daintree area south to about Ingham, occurring in rainforest and most common in relatively wetter locations. It mostly forms dense obstructive thickets, but may be in small groups or as isolated individuals, 0–1000 m asl (Fig. 6.20). Flowering Mar.–Sept; fruiting Oct.– Feb. Conservation status – *No present threats*.

Typification: The holotype of *Calamus radicalis* is an unnumbered and undated collection by Nernst [MEL] from north of 'Port Mackay' (spelt Mackey in the protologue, Wendland and Drude 1875, p. 196). This location is not to be confused with the present city of Mackay in central Queensland; it is an earlier name for a settlement near the mouth of the Tully River in north-east Queensland (Blake 1954). The place of collection, noted at '35 miles north', would be near the southern slopes of Bartle Frere, possibly on the banks of Russell River where some of the larger populations occur. The collection consists of two sheets with leaf portions and a flagellum. The characteristic bristly spines on the midrib and parallel veinlets of the adaxial surface of the pinnae are readily observable in the type specimen (Fig. 6.20).

Etymology: The choice of the specific epithet was not explained in the protologue, but may relate to Nernst's label notes that accompany the type specimen: 'Prickly palm. I'll try to send good specimens soon, if they can be got. So far I have seen neither flowers nor seeds. The locality is about 35 miles north of Port Mackay. The

Figure 6.18 *Calamus radicalis*, in moderate-altitude rainforest, Lamb Range, north-east Queensland, c. 800 m altitude.

flowering stems appear to come from the root and run up amongst other trees, far higher than the stems of the palm'. From the Latin *radical*, arising from a root.

Notes: *Calamus radicalis* is most similar to *C. australis*, but is readily distinguished by pinnae with bristly spines on the adaxial surface; these are lacking in *C. australis*. Additionally, it is a much larger rattan with a more robust stem, with longer spines on the leaf-sheath especially near the junction of the petiole. Flowering mode is very different in the two species, with the flowers in *C. radicalis* reaching anthesis one or two at a time on each rachillae in a basipetal progression over many weeks to months, whereas the flowering in *C. australis* is closely synchronous in single plants over a few days. Fröhlich *et al.* (1997) identified the fungus *Apiospora montagnei* [Deuteromycetes] on the leaves of *C. radicalis*.

7 Calamus vitiensis

Calamus vitiensis Warb. ex Becc., *Ann. Roy. Bot. Gard. (Calcutta)* 11: 350 (1908). Type: Fiji. Taveuni, 1200 m alt., Oct. 1881, *Weber 111*: holotype B, destroyed; isotype FI.

Calamus stipitatus Burret, *Notizbl. Bot. Gart. Mus. Berlin-Dahlem* 15: 814 (1943). Type: Solomon Islands. San Cristobal, Kira Kira, *L.J.Brass 2719*; holotype B, destroyed; isotypes BM, BO, BRI.

Calamus vanuatuensis Dowe, *Principes* 37: 206 (1993). Type: Vanuatu. Erromango, Nouankao R., *Chew Wee-Lek RSNH 118*; holotype PVNH; isotype K.

Stem solitary, slender to robust, to 15 m long, bare stems 7–22 mm diam.; internodes to 33 cm long, green to light brown. **Leaves** cirrate, to 5 m long including cirrus and petiole, pinnae 10–22 each side; leafsheath dark green, with caducous indumentum of brown to light grey, fibrous scales, spines numerous, to 40 mm long, yellow-green to brown, triangular, longer spines flexible, spine bases usually slightly swollen, spines various lengths, usually solitary or occasionally in partial whorls of up to 6, sheath mouth unarmed or lightly armed; knee to 60 mm long, to 28 mm wide, unarmed or lightly armed with short spines; ocrea to 4.5 mm, low, woody, brown, unarmed or lightly armed, collar persistent, base of ocrea extending along the petiole; flagellae lacking; petiole to 45 mm long, to 19 mm wide, channelled or flat on the adaxial surface, rounded on the abaxial surface, unarmed or with few to many short triangular spines; rachis ridged on the adaxial surface, flat with grouped grapnel-like hooks c. 2–3 cm apart on the abaxial surface, spines black-tipped; cirrus to 2 m, grapnel-like hooks arranged regularly in half-whorls on the abaxial surface. **Pinnae** regularly arranged, or in widely spaced pairs or in groups with pinnae divergent, to 43 cm long, to 7 cm wide, broadly lanceolate, cucullate, papyraceous, dark green on both surfaces, longest pinnae near middle of leaf, distal pinnae widely spaced, basal pinnae reduced, unarmed or with few bristly spines to 2.2 mm on the adaxial midrib and parallel veinlets; margins unarmed or with very few bristly spines to 5 mm long, most numerous near the apex, with indumentum sometimes scattered throughout the adaxial surface of the pinnae. **Staminate and pistillate inflorescences** similar to 2.5 m long, branched to 3 orders, c. 8 branches; prophyll to 35 cm long, to 2 cm wide, tubular, moderately 2-keeled, apex margins intact, strong recurved single spines in the distal 4–7 cm; peduncular bract lacking; rachis bracts similar to prophyll, to 25 cm long, to 1.5 cm

Figure 6.19 *Calamus radicalis*. **Top left:** Staminate inflorescence, Wongaling Beach, north-east Queensland. **Top right:** Staminate flowers at anthesis. **Bottom left:** Pistillate and sterile staminate flowers at anthesis. **Bottom right:** Mature fruit.

wide; branches to 30 cm long, straight to curved with up to 40 rachillae; rachillae to 10 cm long; rachillae bracts funnel-shaped, apically broadened, to 2 cm long. **Staminate flowers** 4–5 mm long, 1–2 mm wide in bud, opening to 7–8 mm wide at anthesis; sepals 1.5–2 mm long, apex subulate to acuminate, green, glabrous; petals 4.5–5 mm long, 1–1.5 mm wide, apex acuminate, green, glabrous; stamens 4–4.5 mm long, filaments straight to curved; anthers 1–1.2 mm long, cream-yellow; pistillode minutely trifid to sunken. **Pistillate flowers** 3.2–3.5 mm long, 1.8–2 mm wide in bud; sepals basally connate, tubular, moderately lobed, 2–2.2 mm long, 0.8–1 mm

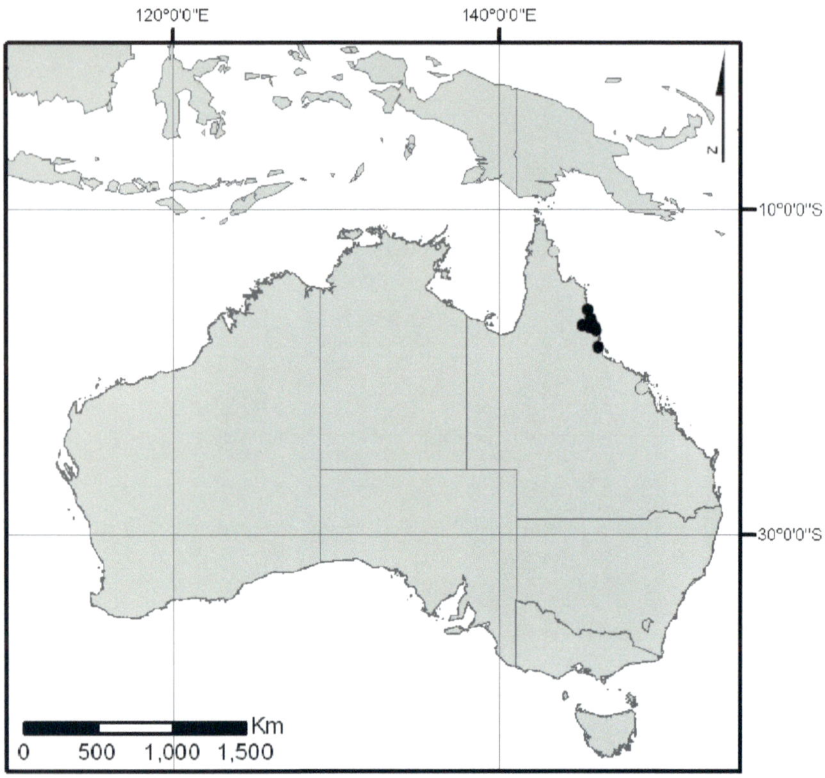

Figure 6.20 *Calamus radicalis*. **Left:** Distribution based on herbarium specimen data. **Right:** Type specimen. Australia. Queensland. 35 miles [c. 56 km] north of Port Mackay (= Tully River), undated, *Nernst s.n*; holotype MEL [1 of 2 sheets].

wide, glabrous; petals 2.5–2.7 mm long, 1.5–1.7 mm wide, basally briefly tubular, green, glabrous; staminodes 6, to 0.8 mm long, staminodal ring c. 1 mm high; ovary to 1.5 mm long, to 1.4 mm wide, globose; style to 0.8 mm long; stigmas to 0.8 mm long. **Sterile staminate flowers** not seen. **Fruit** 1-seeded, globose, 9–13 mm diam.; stigmatic remains 1–1.5 mm high; epicarp c. 0.3 mm thick, with 18 longitudinal rows of white, shallowly channelled scales with entire but uneven margins, apex brownish; mesocarp c. 0.2 mm thick; sarcotesta 1–2 mm thick; pedicel 1.5–2 mm long. **Seed** 6.5–9 mm long, 7–8 mm wide, depressed on one of the long sides, the surface deeply and irregularly furrowed; endosperm homogeneous; embryo lateral to basal. Figures 6.21, 6.22.

Dunk Island lawyer vine, solitary lawyer vine, Mission Beach wait-a-while

Distribution and ecology: *Calamus vitiensis* is distributed from New Guinea east through the Solomon Islands and Vanuatu to Fiji, and in north-east Queensland from Heathlands, Cape York Peninsula, south to Mission Beach, where it is uncommon occurring in near-coastal rainforest in small groups or as isolated individuals. In Queensland it is restricted to 0–400 m asl, but otherwise occurs up to 1200 m altitude in other locations (Fig. 6.23). Flowering Aug.–Sept.; fruiting Nov.– Jan. Conservation status – *No present threats*.

Typification: The holotype of *Calamus vitiensis, Weber 111* [B], was collected in 1881 from Taveuni, Fiji, but was destroyed during the burning of the Berlin Herbarium in 1942. However, a fragment consisting of a portion of rachilla and flowers was maintained in Florence by Beccari, and thus represents an isotype. In the protologue, Beccari indicated the name had been suggested by Warburg and was appended to the Weber collection in Berlin. Some parts of the protologue are attributed to Weber, possibly based on his field collection notes (Fig. 6.23).

Etymology: 'Habitat. – In the small Island of Taviuni of the Fiji group . . .', (Beccari 1908, p. 351): With reference to the occurrence of the species in Fiji, and from where the type specimen was collected.

Figure 6.21 *Calamus vitiensis*, in lowland rainforest, Etty Bay, north-east Queensland.

Notes: This rattan was formerly named in Queensland species lists and other treatments as *Calamus hollrungii*, but is now attributed to *C. vitiensis*, based on a revision of the *C. aruensis* complex by Baker *et al.* (2003). In the revision it was found that the type of *C. hollrungii* could not be distinguished from *C. aruensis*, and thus *C. hollrungii* was placed as a synonym of that species. Accordingly, the rattan formerly listed as *C. hollrungii* had to be attributed to another species. With examination of herbarium specimens, it was identified as *C. vitiensis*. In this process other species were placed in synonymy, but do not relate to Australian collections.

In Australia, *C. vitiensis* is readily distinguished from other rattans, being solitary-stemmed and with a cirrus extending from the leaf rachis. The only other rattan species in Australia that possess cirrate leaves are *C. aruensis* and *C. warburgii*, but both are clustering species occurring in Cape York Peninsula. *Calamus vitiensis* does occasionally occur in the vicinity of these two species, but is distinguished from *C. aruensis*, in which the leafsheath is unarmed or nearly so, by its densely armed leafsheath, and from *C. warbugii*, which has more numerous and linear acuminate pinnae, by the relatively lesser number of broad cucullate pinnae.

8 Calamus warburgii

Calamus warburgii K.Schum., *in* K.Schumann and K.Lauterbach, *Fl. Schutzgeb. Südsee*: 203 (1901). Type: Papua New Guinea. Madang Province. Schumann R., June 1896, *O.Kersting 437*; holotype LZ, destroyed; isotype FI [1 sheet].

Stems clustering, to 40 m tall long, 1–5 dominant, bare stems to 30 mm diam.; internodes to 25 cm long, green to light brown. **Leaves** cirrate, to 6.5 m long including cirrus and petiole; pinnae 40–100 each side of rachis; leafsheath dark green, brown to grey indumentum, densely armed with needle-like yellow-green spines to 60 mm long, usually solitary or infrequently in partial whorls, randomly arranged, mouth very densely spiny; knee 3–5 cm long, 1–3 cm wide, basally lipped, moderately spiny; flagellae lacking; petiole 1.5–7 cm long, to 2 cm wide, densely spiny; rachis to 2.5 m long, flat on the adaxial surface in the proximal portion, becoming sharply ridged in the distal portion, rounded on the abaxial surface, with grapnel-like hooks, 3–6 in a group, regularly placed c. 3 cm apart in half-whorls along most of distal portion; cirrus to 4 m long, grapnel-like hooks, 3–6 in a group, regularly placed c. 3 cm apart in the basal portion becoming congested toward the apex; spines 3–6 in a group, recurved, black-tipped. **Pinnae** regularly arranged, closely positioned, horizontal proximally, becoming semi-pendulous distally, linear to narrowly lanceolate, to 45 cm long, to 22 mm wide, acuminate, papyraceous, dull midgreen on the adaxial surface, only slightly lighter green on the abaxial surface, midrib prominent and a slightly less prominent parallel vein each side of midrib; erect bristly spines to 15 mm long on the midrib and parallel veins on both surfaces; apex with short bristly spines; proximal margins slightly thickened and with short adpressed spines. **Staminate and pistillate inflorescences** similar, to 2 m long, axis rigid and ± straight; c. 10 branches, distichously to subdistichously arranged; staminate inflorescence branched to 3 orders, pistillate inflorescence branched to 2 orders; prophyll to 10 cm long, to 2 cm wide, tubular, papyraceous, marcescent, 2-keeled, apical margin entire, prolonged on one side into a triangular point, unarmed or slightly spiny,

Figure 6.22 *Calamus vitiensis.* **Top left:** Leaf with characteristic grouped pinnae. **Top right:** Staminate flower at anthesis. **Bottom left:** Pistillate flowers at anthesis. **Bottom right:** Mature fruit.

or densely spiny toward the apex; peduncular bract absent; rachis bracts similar to prophyll, to 8 cm long, basally tightly tubular widening to 2 cm wide at the apex; lateral branches to 30 cm long, straight to curved with up to 24 rachillae; rachillae subdistichous, to 7 cm long, basally narrowed near attachment, flattened; rachillae bracts basally tubular, c. 2 cm long, widening at the apex, obliquely truncate, prolonged into a short point on one side; bracteoles sepal-like, inflated toward the apex, margins entire. **Staminate flower** 2–3 mm long, 1.5–2 mm wide in bud; sepals 0.8–1 mm long, triangular, acuminate, yellow-green, glabrous; petals 2.5–3 mm long, triangular, acuminate, yellow-green, glabrous; stamens 2–3 mm long; anthers 0.3–0.5 mm long, yellow, orchidaceous perfume;

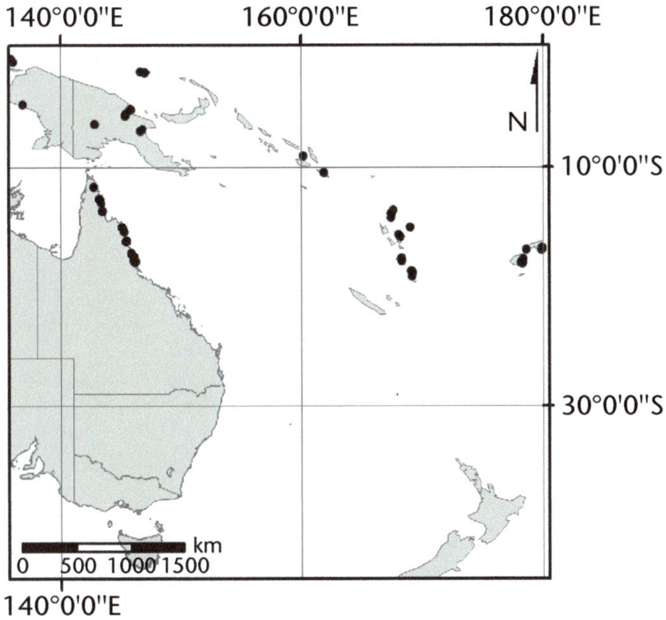

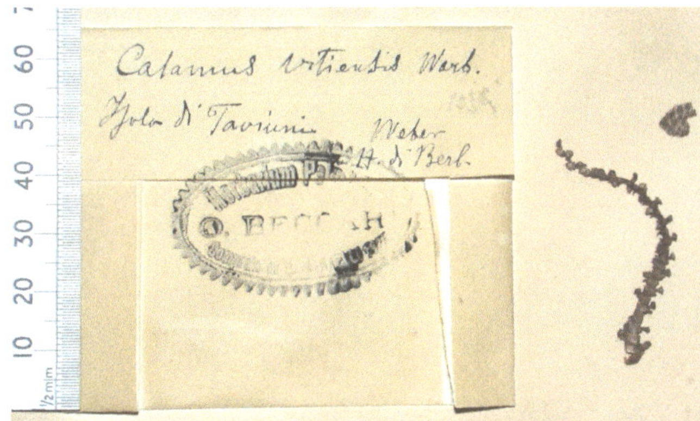

Figure 6.23 *Calamus vitiensis*. **Left:** Distribution based on herbarium specimen data. **Right:** Type specimen. Fiji. Taveuni, 1200 m alt., Oct. 1881, *Weber 111*; isotype FI [fragment].

pistillode minutely trifid. **Pistillate flower** 3–4 mm long, 2–3 mm wide in bud; sepals 1.2–1.5 mm long, 1.3–1.5 mm wide, lobes triangular, gibbous, margins minutely lacerate, green, glabrous; petals 2.2–2.5 mm long, 1.8–2 mm wide, triangular, margins entire, light green, glabrous; stigmas strongly recurved, c. 1.5 mm long, glistening white; ovary globose. **Sterile staminate flower** 1.5–2 mm long, 1.2–1.5 mm in bud, not widely opening; sepals 0.8–1 mm long, 1.3–1.5 mm wide, lobes triangular, green, glabrous; petals 1.8–2 mm long, 1.6–1.8 mm wide, triangular, green, glabrous; stamens c. 2 mm long; anthers c. 0.5 mm long. **Fruit** 1-seeded, globose, 10–13 mm diam.; stigmatic remains c. 2 mm high; epicarp c. 0.2 mm thick, with c. 15 rows of scales, white-cream at maturity, scale margins reddish; sarcotesta c. 2 mm thick; pedicel 0.5–1.5 mm long. **Seed** c. 8 mm long, c. 6 mm wide, depressed on one of the long sides, smooth; endosperm homogeneous; embryo basal. Figures 6.24, 6.25.

Iron Range wait-a-while, Claudie River lawyer vine

Distribution and ecology: *Calamus warburgii* is distributed throughout New Guinea where it is a common rattan but is uncommon in Australia, restricted to eastern Cape York Peninsula between Macrossan Range and Iron Range, in rainforest and swamp forest habitats (Fig. 6.26). It occurs as large obstructive thickets primarily as a riparian element or as small thickets and scattered individuals on hill slopes, 10–300 m asl. Flowering Feb.–Mar.; fruiting Apr.–June.

Figure 6.24 *Calamus warburgii* in lowland rainforest, West Claudie River, Iron Range, Cape York Peninsula, north Queensland.

Figure 6.25 *Calamus warburgii*. **Top left:** Pistillate inflorescence. **Top right:** Staminate flowers at anthesis. **Bottom left:** Pistillate flowers at anthesis and sterile staminate flowers in bud. **Bottom right:** Single mature fruit. Photos by John G. Pritchard.

Conservation status – *Vulnerable* (Bostock and Holland 2007).

Typification: The holotype of *Calamus warburgii* is *Kersting 437* [LZ], collected in June 1896 from Schumann River in Papua New Guinea. The collection was destroyed with the bombing of Leipzig herbarium in 1943. However, a fragment consisting of three pinnae, a portion of rachilla and fruits was maintained in Florence by Beccari, and thus represents an isotype. The identity of Schumann River is not clear, but it is most probably a tributary of Ramu River in Madang Province (Fig. 6.26).

Etymology: The choice of the specific epithet was not explained in the protologue, but most probably named for the German botanist Otto Warburg (1859–1938), who

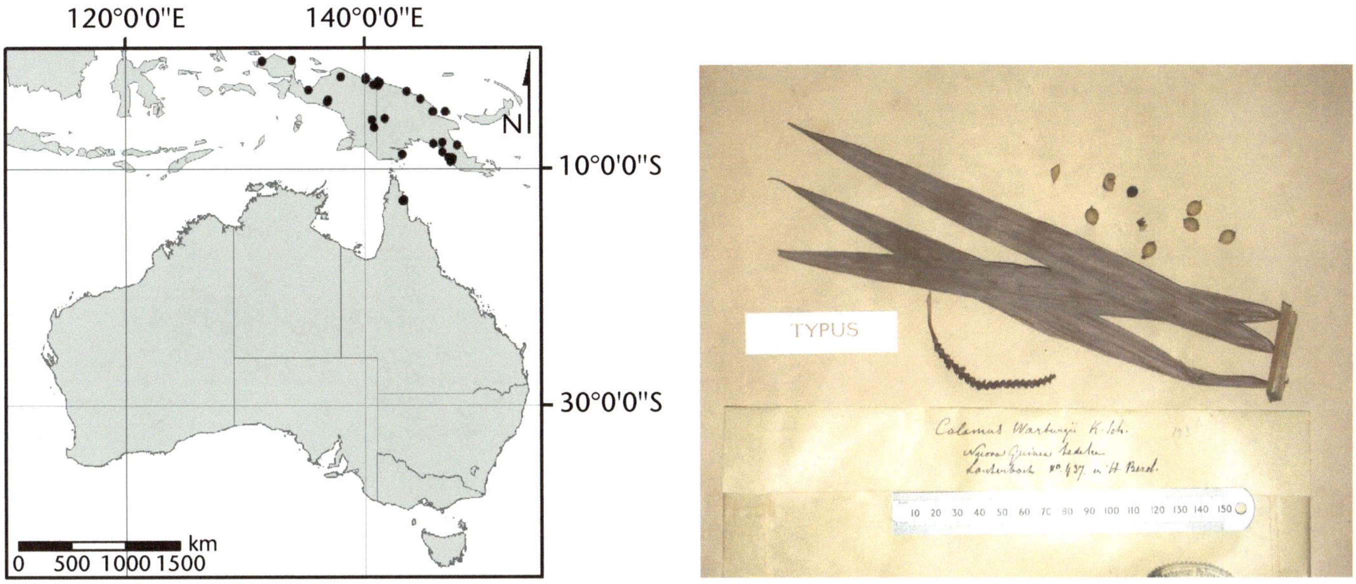

Figure 6.26 *Calamus warburgii*. **Left:** Distribution based on herbarium specimen data. **Right:** Type specimen. Papua New Guinea. Madang Province, Schumann River, June 1896, *O.Kersting 437*; isotype FI.

was an associate of Karl Schumann. Warburg studied at Bonn, Berlin and Hamburg and travelled in south-east Asia, Oceania and Australia 1885–89. He occupied various lecturing positions and professorships 1891–1930 in Germany and Israel, and was President of the World Zionist Organisation 1911–20.

Notes: *Calamus warburgii* is distinguished by long cirrate leaves with numerous closely spaced narrow-linear pinnae of a soft texture, and a short rigid inflorescence.

7. Subfamily 2: Nypoideae

Nypoideae Griff., *Palms Brit. Ind.*: 7 (1850), as *Nipinae*.

Monoecious, unarmed, pleonanthic palms. **Leaves** pinnate. **Pinnae** reduplicate. **Inflorescence** with a terminal pistillate head and staminate spikes on lateral catkin-like branches; peduncular bracts 7–9, tubular. **Flowers** dimorphic, usually solitary, infrequently in pairs or groups of 3. **Staminate flowers** with sepals and petals similar, minute; stamens 3; filaments united; pistillode lacking. **Pistillate flower** with smooth ovary; carpels 3, free. **Fruit** borne in a globose head, composed of irregularly enlarged carpels, 1-seeded, epicarp smooth, deeply grooved; mesocarp fibrous; endocarp thickened. **Seed** adhering to the endocarp, not sarcotestal; endosperm homogeneous; eophyll bifid or compound.

A monogeneric subfamily distributed throughout South-East Asia, east Asia, into the western Pacific, Melanesia and northern Australia. Feral populations occur in Central America (Duke 1991).

Classification and relationships

The Nypoideae resolves in most phylogenetic studies as sister to all other palms (Asmussen *et al.* 2006; Dransfield *et al.* 2008). The subfamily includes a single genus, *Nypa*, which has a single species, *N. fruticans*. Fossils of *Nypa* have been recovered from numerous sites across the globe, but present distribution is confined to Asia, Malesia, the western Pacific and northern Australia. Based on morphological variation of fossil pollens, other species of *Nypa* may have existed, but are now extinct. Because of its unique characteristics, the classification of *Nypa* has been inconsistent; it has been placed in its own family, Nypaceae (Mueller 1889), and in the Pandanaceae (Mueller 1881). Within the Arecaceae, the Nypoideae has no clear closest relatives (Dransfield *et al.* 2008).

Nypa

Nypa Steck, *Sagu*: 15 (1757). Type: *Nypa fruticans* Wurmb.

Nipa Thunb., *Kongl. Vetensk. Acad. Nya Handl.* 3: 231 (1782).

[*Nipa* Wurmb (*sub* Pandanaceae R.Br.), *in* F.Mueller, *Fragm.* 11: 127 (1881); *sub* Nipaceae Brongn., *in* F.Mueller, *Second Syst. Cens. Aust. Pl.*: 202 (1889)].

Prostrate, colonial, pleonanthic, monoecious palms. **Stem** stout, subterranean or decumbent on the ground surface, dorsally flattened, dichotomously branched, leafscars curved, internodes compressed. **Leaves** paripinnate; leafbase broadly clasping the stem; petiole long, stout, channelled on the adaxial surface, rounded on the abaxial surface, distally terete; rachis basally terete, distally angled, pruinose to glabrous. **Pinnae** in a single plane on the rachis, reduplicate, subopposite to opposite, acute, coriaceous, midrib prominent; ramenta on the midrib on the abaxial surface, membranous. **Inflorescence** interfoliar, solitary, erect, branched 5–6 orders, protogynous; prophyll tubular, 2-keeled; peduncular bract tubular, inflated, beaked, splitting lengthwise; rachis bracts tubular, inflated, subtending 7–9 first-order branches; peduncle erect, terete in cross-section; rachis shorter than peduncle, distally with an aggregated head of pistillate flowers; first-order branches proximal to the head of pistillate flowers, erect, terminating in catkin-like rachillae of staminate flowers overhanging the pistillate head. **Staminate flowers** solitary, crowded, sessile; 3 narrow, oblanceolate sepals; 3 imbricate, small narrow petals; stamens 3; filaments connate in a solid stalk; anthers basifixed, elongate, extrorse; pistillode lacking. **Pistillate flowers** in an aggregated head; 3 oblanceolate irregular sepals; 3 similar petals; staminodes lacking; carpels 3–4, obovoid, asymmetrical, angular-sided from pressure of adjacent carpel, distally acute; stigmatic opening funnel-shaped; ovule anatropous, dorsally attached. **Pollen** ellipsoidal to circular; aperture meridionally zonasulcate; exine tectate, finely reticulate, with supratectal spines. **Fruit** an aggregated globose head, individual fruits irregularly angular with longitudinal grooves, apex pyramidal; stigmatic remains apical; epicarp smooth; mesocarp fibrous; endocarp thick with interwoven fibres; perianth not persistent. **Seed** ovoid, grooved; hilum basal; raphe branches ascending from the base; endosperm homogeneous; embryo basal; pressure of the plumules disengages the seed from the head; eophyll bifid or compound.

Etymology: '*Nypa. Herb. Amboin. 1: 69. Tab. XVI*': Derived from the Malay name, *nipa*, for the species.

Nypa fruticans

Nypa fruticans Wurmb, *Verh. Batav. Genootsch. Kunsten* 1: 349 (1779); *Nipa fruticans* (Wurmb) Thunb., *Kongl. Vetensk. Acad. Nya Handl.* 3: 231 (1782), *orth. var*. Type: lectotype, illustration *in* K.L. von Blume, *Rumphia* 2: pls 164, 165 (1847), *fide* H.Tralau, *Kongl. Svenska Vetenskapsakad. Handl.* ser. 4, 10: 8 (1964).

Cocos nypa Lour., *Fl. Cochinch.* 2: 567 (1790). Type: not designated.

Nipa litoralis Blanco, *Fl. Filip.* 662 (1837). Type: not designated.

Nipa arborescens Wurmb ex H.Wendl., in Kerchove, *Palmiers* 252 (1878).

Nipa fruticans var. *neameana* F.M.Bailey, *Pro. Roy. Soc. Queensland* 5: 147 (1888). Type: Australia. Queensland. Herbert R., 1880, *A.Neame*; holotype BRI.

[*Nipa fruticans* Wurmb (*sub* Pandanaceae), *in* F.Mueller, *Fragm.* 11: 128 (1881)].

[*Nipa fruticans* Wurmb (*sub* Nipaceae), in F.Mueller, *Second Syst. Cens. Aust. Pl.*: 202 (1889)].

Stem to 30 cm wide, to 15 cm thick, submerged in mud or on the surface, roots on the lower surface; leafscars very prominent, 3–5 cm high. **Leaves** 4–12, 4–9 m long, erect to arcuate, 50–70 pinnae each side of rachis; leafbase to 200 cm long; petiole 60–400 cm long; rachis to 7 m long, yellow to brown. **Pinnae** 50–130 cm long, 5–8 cm wide, rigid, narrowly lanceolate, acuminate, coriaceous, dark green and glossy on the adaxial surface, pruinose on the abaxial surface, with peltate, lanceolate dark brown scales; midrib prominent on the adaxial surface; ramenta to 12 mm long. **Inflorescence** 30–150 cm long; prophyll 40–50 cm long, 20–25 cm wide, obtuse, yellow-orange-light brown; peduncular bract to 50 cm long, to 25 cm wide, acute, orangish; additional bracts to 50 cm long, to 25 cm wide, acute, orangish; first-order branches 10–15 cm long; staminate rachillae 6–9 cm long, glabrous, yellow-orange; pistillate head 4–8 cm diam., light brown. **Staminate flowers** 4–5 mm long in bud; sepals 3–4 mm long, to 0.8–1 mm wide, yellowish; petals 3.5–4 mm long, 1.8–2 mm wide, yellowish; stamens to c. 5 mm long; anthers c. 2 mm long, yellow. **Pistillate flowers** 10–15 mm long; sepals 4–5 mm long, 1.5–2 mm wide, cream-green; petals 4–5 mm long, 1–2 mm wide, cream-green; stigmatic opening funnel-shaped, c. 2 mm wide. **Fruiting head** 20–50 cm diam. at maturity. **Fruit** obovoid to wedge-shaped, 80–150 mm long, 20–90 mm wide, smooth, deeply grooved; epicarp to c. 1 mm thick, dark brown; mesocarp to 20 mm thick, fibrous; endocarp to 2 mm thick, hard. **Seed** broadly ovoid, 40–70 mm long, 40–50 mm wide. Figures 7.1, 7.2.

Figure 7.1 *Nypa fruticans*, Centenary Lakes, Cairns, north-east Queensland.

Figure 7.2 *Nypa fruticans*. **Top left:** Inflorescence at anthesis. **Top right:** Staminate flowers at anthesis. **Bottom left:** Pistillate flowers at anthesis. **Bottom right:** Mature fruit.

Mangrove palm, nipah, Nypa palm

Distribution and ecology: In Australia uncommonly on the Northern Territory coast and in north Queensland, but otherwise with widespread distribution from southern India and the Bay of Bengal coasts, throughout south-east Asia north to the Ryukyu Islands (Japan), the Philippines, Malesia and east to Solomon Islands, and with feral populations in Central America. In the Northern Territory restricted to populations on Melville Island and Cobourg Peninsula, and in Queensland with populations on Cowal Creek and Jardine River on north-west Cape York Peninsula; Harmer Creek, Olive River, Pascoe River, Lockhart

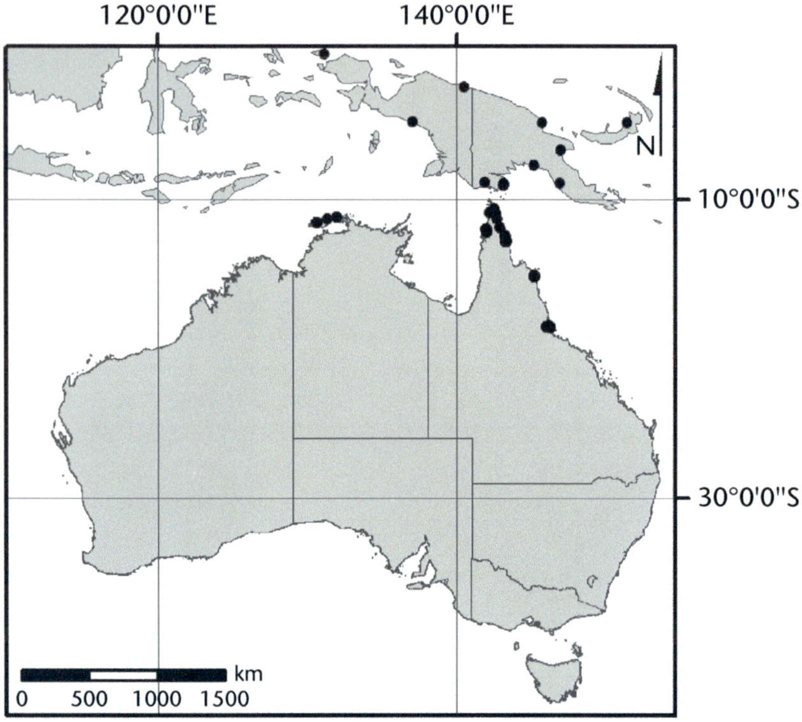

Figure 7.3 *Nypa fruticans*. **Left:** Distribution based on herbarium specimen data. **Right:** Type specimen. Lectotype, illustration *in* K.L. von Blume, *Rumphia* 2: pls. 164 [illustrated here], 165 (1847), *fide* Tralau (1964).

River on eastern Cape York Peninsula; Morgan River and McIvor River in southern Cape York Peninsula; then south with a disjunction to Herbert River (Fig. 7.3). Occurring in dense monospecific colonial populations in estuarine locations with moderate tidal parameters in anaerobic alluvial mud, at 0–5 m asl. Fossil evidence indicates that *Nypa* has at various times been widespread on the coasts of southern and eastern Australia, with macrofossils, as *Nypa australis* Pole & Macphail, including leaves and fruit located in Eocene deposits from western Tasmania (Pole and Macphail 1996), and fossil pollen, as *Spinizonocolpites* spp., from Victoria, South Australia, southern Western Australia, Tasmania and Queensland (see Chapters 3 and 4). Flowering all months; fruiting all months. Conservation status – *No present threats*.

Typification: In the absence of specimens, the illustrations in *Rumphia* 2: pls. 164, 165 (1847) were chosen as the type (Tralau 1964) (Fig. 7.3).

Etymology: *Nypa fruticans* Wurmb, *Verh. Batav. Genootsch. Kunsten* 1: 349 (1779). From the Latin *frutex*, shrub.

Notes: The variety *N. fruticans* var. *neameana* was named for the Herbert River population in Queensland, based on a collection by Arthur Neame [BRI], but there was nothing in the protologue that distinguished it from *Nypa fruticans*.

Nypa fruticans is distinguished by the creeping prostrate habit with dichotomous branching of the stem, large erect pinnate leaves, axillary inflorescences in which the staminate flowers are produced on terminal catkin-like rachillae and the pistillate flowers in globose heads positioned below the staminate flowers, and an infructescence in which the fruit are tightly aggregated into a large globose head. The adaptation of the species to tidal estuarine habitats is unique in the family. See Chapter 3 for detailed discussion of fossil record and historical biogeography.

8. Subfamily 3: Coryphoideae

Coryphoideae Burnett, *Outl. Bot.*: 398 (1835), as *Coryphidae*.

Dioecious, monoecious, polygamodioecious or polygamomonoecious, hapaxanthic or pleonanthic palms. **Leaves** palmate, costapalmate, imparipinnate or bipinnate, segments and pinnae induplicate or infrequently reduplicate. **Inflorescences** interfoliar, intrafoliar or apical, paniculate or rarely spicate; peduncular bracts lacking or few to numerous. **Flowers** bisexual or unisexual, or functionally unisexual though morphologically bisexual, solitary, or in sympodial clusters, or in cincinni, or in triads; perianth segments imbricate at the base, apically free; ovary smooth; carpels free or connate. **Fruit** 1–3-seeded, epicarp smooth or rough; mesocarp fibrous, or fleshy, or dry; endocarp thin or thick. **Seed** not adhering to endocarp, not sarcotestal; endosperm homogeneous or ruminate; eophyll simple or bifid.

A subfamily of 46 genera and c. 450 species, of worldwide distribution; in Australia with five genera with 24 species (19 endemic), in three tribes and one subtribe.

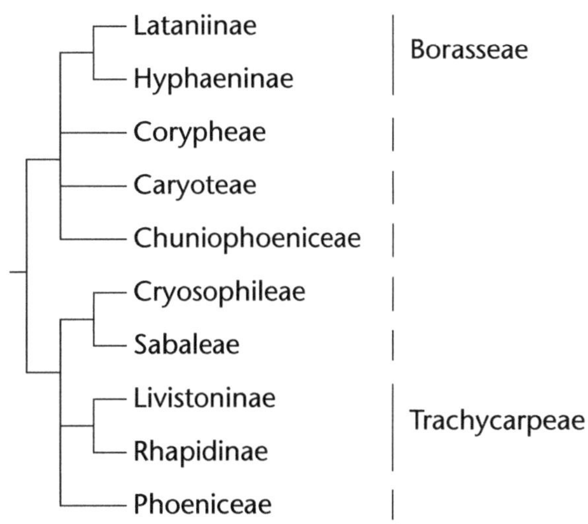

Figure 8.1 Phylogenetic tree summarising the estimated relationships of tribes and subtribes in the Coryphoideae. Adapted from Dransfield *et al.* (1990), Asmussen *et al.* (2000, 2006) and Dransfield et *al.* (2008).

Classification and relationships

The Coryphoideae is monophyletic and resolves in most phylogenetic studies as sister to a clade including the Arecoideae and the Ceroxyloideae (Hahn 2002b; Asmussen *et al.* 2006; Dransfield *et al.* 2008). The subfamily includes all palms with induplicately folded segments and pinnae. Of the Australian representatives, *Corypha*, *Licuala* and *Livistona* have palmate or costapalmate leaves, while *Arenga* and *Caryota* have pinnate or bipinnate leaves. The Coryphoideae is globally distributed, with centres of diversity in the New World and Asia/Malesia. Eight tribes have been recognised for the subfamily (Fig. 8.1), of which three, the Trachycarpeae, Caryoteae and Corypheae, have representatives in Australia (Table 8.1).

The Trachycarpeae is the most diverse and widespread tribe, with 17 genera in two subtribes, the Rhapidinae and the Livistoninae. The Rhapidinae occurs in south-east Asia and in the Mediterranean region. The Livistoninae, with five genera, has two genera, *Livistona* and *Licuala*, in Australia (Fig. 8.2). The tribe is resolved as monophyletic in most studies (Asmussen *et al.* 2006; Dransfield *et al.* 2008). *Livistona*, with 36 species (18 in Australia), and *Licuala* with c. 135 species (one in Australia), are among the most speciose genera in the Coryphoideae.

The Caryoteae is monophyletic and resolves in most phylogenetic studies as sister to a clade including the Corypheae and Borasseae (Bayton 2007; Asmussen *et al.* 2006; Dransfield *et al.* 2008). The tribe includes pinnate and bipinnate-leaved taxa, in three genera, *Caryota*,

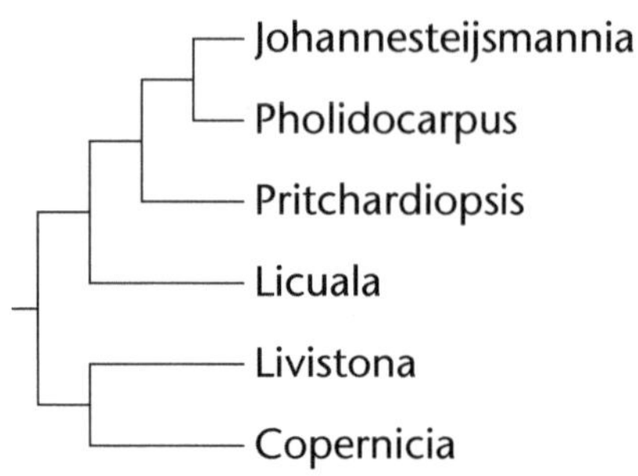

Figure 8.2 Phylogenetic tree summarising the estimated relationships of genera in the Livistoninae. Adapted from Asmussen *et al.* (2000, 2006) and Dransfield *et al.* (2008).

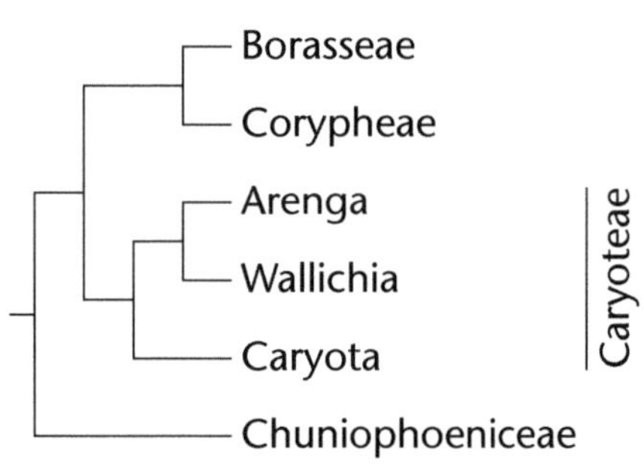

Figure 8.3 Phylogenetic tree summarising the estimated relationships of tribes and genera in the 'syncarpous clade' within the Coryphoideae. Adapted from Asmussen *et al.* (2000, 2006) and Dransfield *et al.* (2008).

Table 8.1 Synopsis of the Coryphoideae tribes, subtribes and genera in the Australian palm flora

Trachycarpeae Satake, *Hikobia* 3: 121 (1962)
Livistoninae Saakov, *Palms and their culture in the USSR* 193 (1954)
Livistona
Licuala
Caryoteae Scheff., *Ann. Jard. Bot. Buitenzorg* 1: 142 (1876)
Caryota
Arenga
Corypheae Mart., *in* S.Endlicher., *Gen. pl.*: 252 (1837)
Corypha

Arenga and *Wallichia* (Fig. 8.3); the first two genera have representatives in Australia and the species are outliers of a centre of diversity in south-east Asia and Malesia.

The Corypheae, with the single genus *Corypha* with six species (one in Australia), is resolved as monophyletic in most phylogenetic studies, and as sister to the Borasseae (Bayton 2007; Asmussen *et al.* 2006; Dransfield *et al.* 2008) (Fig. 8.3).

■ LIVISTONINAE

Livistona

Livistona R.Br., *Prodr.*: 267 (1810). Type: *Livistona humilis* R.Br. *Saribus* Blume, *Rumphia* 2: 48, tab. 95, 96 ('1836', published 1838). Type: *Saribus rotundifolius* (Lam.) Blume.

NB: Generic synonyms which do not have a taxonomic association with Australian taxa are not included in this treatment. A full list of generic synonyms is provided in Dransfield *et al.* (2008).

Solitary, short to tall, erect, pleonanthic hermaphroditic, dioecious or functionally unisexual or polygamomonoecious palms. **Stems** slender to robust, cylindrical, broader at the base, ringed with leafscars; internodes narrow to broad. **Leaves** costapalmate or palmate, when senescent sometimes forming a persistent skirt; leafbases narrowly clasping the stem, margins forming a fibrous network around the stem, frequently ligulate; petioles long, concave, flat or ridged on the adaxial surface, rounded on the abaxial surface, margins smooth or spiny, or with small prickles; hastula present; lamina regularly or irregularly segmented. **Segments** narrow, lanceolate, single-fold, induplicate, tapered to the apex, apices shallowly or deeply cleft, rigid or pendulous; divisions between segments almost to the leafblade to less than half of the length of the leafblade; segments with a prominent midrib, few to many parallel veinlets at each side of the midrib; transverse veinlets equal in thickness or thinner than the parallel veinlets. **Inflorescences** interfoliar, either unbranched at the base and first-order branches (partial inflorescences) subtended in its entire length, or basally bi- or trifurcate with collateral axes of approximately the same size contained within a common prophyll, and with first-order branches subtended along the entire length of each axis (not in Australia); partial inflorescences 1 (*L. humilis*) or numerous per axis, branched 3–6 orders when regarding the primary branch as number 0; prophyll 2-keeled; peduncular bracts lacking or if present then loosely or tightly tubular; rachis bracts tubular, each subtending a partial inflorescence; bracts on second-order branches lacking or present; rachillae straight or flexuous, glabrous or tomentose, papillose, pubescent or pruinose. **Flowers** hermaphroditic or functionally unisexual, solitary or in sympodial cluster of 2–many, spirally arranged, variously coloured; sepals triangular, fused at the base, lobed apically, glabrous, or hairy (*L. carinensis* – not in Australia); petals free, longer than the sepals, triangular, valvate, glabrous, or hairy (*L. carinensis*); stamens 6, epipetalous, shorter than the petals; anthers didymous, globose to ovoid, medifixed, versatile, latrorse; carpels wedge-shaped, basally separated; styles united, slender, shorter than the stamens; stigma 3-lobed, ovule anatropous. **Pollen** ellipsoidal; aperture monosulcate; exine tectate, finely reticulate. **Fruit** globose, subglobose, ellipsoid, obloid, obovoid, reniform, pyriform, or obpyriform, 1-seeded; stigmatic remains apical to subapical; epicarp thin or thick, smooth or with scattered pores or mildly rugose, variously coloured; suture line extending longitudinally from stigmatic remains toward the base or not readily visible; mesocarp dry or fleshy, with or without embedded fibres; endocarp bony to woody; perianth persistent or not. **Seed** globose, subglobose, ovoid, ellipsoid or reniform; endosperm homogeneous, intruded by the testa from one side with spongy-crystalline tissue to displace part or most of the endosperm; hilum broad, circular to elongate; raphe branches few or lacking; embryo small, supra- to sublateral; germination remote-tubular; eophyll simple, with longitudinal ribs.

Livistona is a genus of 36 species distributed in north-east Africa, southern Arabian Peninsula, south and east Asia, Malesia and Australia. Populations may be monospecific, in small to large groups or as scattered individuals in rainforest, montane forest, swamp forest, monsoon forest, coastal and littoral forest, semi-closed to open woodlands and savanna, semi-arid woodlands, and riparian and riverine forest, on various soils, including limestone and peat, in permanently or seasonally wet situations, or in semi-arid to arid situations where usually associated with permanent ground water or at the base of cliffs where subsurface moisture is available. Several *Livistona* species are important in horticulture and are used in urban and domestic landscapes.

Etymology: '*Dixi in memoriam viri nobilis Patricii Murray Baronis de Livistone: Balfourii amici; qui Hortum Botanicum in Praedio suo supra mille Plantis instruxerat, Edinensem tunc conditum ditavit: Galiam universam Botanices causa peragravit, ubi febre correprus succubuit. Plura de viro sui temporis meritissimo vide in Sibbaldi Memor. Balfour. p. 69, et seq.*' (Brown 1810, p. 268). [I have called it in memory of that noble man Patrick Murray, Baron de Livistone, a friend of Balfour, who arranged a botanical garden in his own estate with over a thousand plants, and enriched that of Edinburgh, which was then founded. He travelled through the whole of France for botanical reasons, when, attacked by fever, he died. For more facts concerning the most meritorious man of his time see Sibald's Memoir of Balfour, p. 69, et. seq.]: Named for Patrick Murray, Lord of Livingston (b.1634, d.1671), Scottish botanist and horticulturist whose plant collection formed the basis of the Edinburgh Botanic Garden, founded in 1670.

Relationships of Australian species

Dowe (2001) performed cladistic analyses of all *Livistona* species based on morphological characters. In summary, three moderately supported lineages within the genus were identified, of which one included all the Australia species plus *L. carinensis* from Africa/Arabia:

- *L. exigua/L. saribus* group;
- Asian/Malesian group (with two sublineages):
  - *L. chinensis* group (*L. boninensis, L. chinensis, L. endauensis, L. halongensis, L. jenkinsiana, L. tahanensis*);
  - *L. rotundifolia* group (*L. chocolatina, L. merrillii, L. papuana, L. rotundifolia, L. surru, L. tothur, L. woodfordii*);
- African/Australian group (with two sublineages):
  - *L. mariae* group (*L. alfredii, L. australis, L. carinensis, L. decora, L. lanuginosa, L. mariae, L. nasmophila, L. nitida, L. rigida, L. victoriae*);
  - *L. humilis* group (*L. benthamii, L. concinna, L. drudei, L. eastonii, L. fulva, L. inermis, L. humilis, L. lorophylla, L. muelleri*).

More recently, a preliminary molecular phylogeny of the Australian species (Fig. 8.4) and a small number of extra-Australian species using AFLPs, cpDNA sequences and nuclear DNA sequences has been provided (Dowe 2001; Isagi *pers. comm.*). Results indicate that the Australian species form a monophyletic group in all analyses, and the extra-Australian species a less well-supported group. Some of the arrangements based on phenetic and morphological analyses as noted above were in most part confirmed by molecular studies, such as Rodd's (1998) Mariae and Northwestern groups, and Dowe's (2001) *L. mariae* and *L. humilis* groups. The closest relationship of the African/Arabian *L. carinensis* was with extra-Australian species rather than the *L. mariae* group as predicted by Dowe (2001).

Livistona has recently been revised (Dowe 2009a), and aspects dealing with taxonomy and typification are examined in that work. In this current work, these aspects will be summarised rather than repeated.

Key to species of Livistona in Australia

1 Divisions between leaf segments extend almost to the base of the leaf blade 2
1: Divisions between leaf segments extend to slightly more or less than half the length of the leafblade 6
2 Petioles lacking sharp spines on the petiole margins, but reduced to small prickles .... **10. *L. inermis***
2: Petioles with sharp spines on the petiole margins 3
3 Moderate palms to 15 m tall; fruit obovoid-pyriform .... **12. *L. lorophylla***
3: Tall robust palms to 18 m tall; fruit globose 4
4 Inflorescence robust; rachis to 60–80 mm diameter at the base; bracts loose .... **5. *L. decora***
4: Inflorescence not robust; rachis to 25 m diameter at the base; bracts tightly sheathing 5
5 Leafbase and lower petiole purple; inflorescences not sexually dimorphic .... **6. *L. drudei***
5: Leafbase and lower petiole green; inflorescences sexually dimorphic with inflorescences on 'male' plants branched to 4 orders and those on'female' plants branched to 5 orders .... **4. *L. concinna***
6 Apices of leaf segments rigid; apex acute to acuminate 7
6: Apices of leaf segments drooping or pendulous; apex long attenuate 10
7 Free part of leaf segments bifurcate for at least half the length 8
7: Free part of leaf segments bifurcate for much less than half the length 9
8 Stem to 7 m tall, 5–8 cm diameter; inflorescences sexually dimorphic on separate plants; female plants with a single distal partial inflorescence (infrequently with 1–2 lower partial inflorescences); male plants with 4–7 partial inflorescences .... **9. *L. humilis***
8: Stems to 12 (21) m tall, 10–15 cm diameter; inflorescences not sexually dimorphic .... **7. *L. eastonii***
9 Abaxial surface of leaf blade with coppery brown floccose tomentum .... **8. *L. fulva***
9: Abaxial surface of lamina glabrous except for scales on the ribs .... **14. *L. muelleri***
10 Adaxial surface of lamina grey-green to glaucous, waxy 11
10: Abaxial surface of lamina dark glossy green, not waxy 16
11 Lamina not glossy, dull grey-green 12
11: Lamina semi-glossy to glossy 14
12 Inflorescence bracts densely lanuginose .... **11. *L. lanuginosa***
12: Inflorescence bracts glabrous or sparsely scaly, or densely tomentose 13
13 Stem 15–30 cm diameter; rachillae glabrous .... **18. *L. victoriae***
13: Stem 20–50 cm diameter; rachillae minutely scabrous .... **1. *L. alfredii***
14 Divisions between leaf segments extend to about one-third the length of the lamina .... **17. *L. rigida***
14: Divisions between leaf segments extend to about half the length of the lamina 15
15 Lamina ± circular in outline; distal segments usually overlapping; inflorescence bracts with dense white-grey scales .... **13. *L. mariae***
15: Lamina ± oval in outline; distal segments usually not overlapping; inflorescence bracts with sparse, scattered long scales .... **15. *L. nasmophila***
16 Remnant petioles with a bulbous base, about half the length of the original petiole .... **3. *L. benthamii***
16: Remnant petiole not bulbous, with short retained petiole 17
17 Divisions between leaf segments extend for about half the length of the lamina; fruit reddish-brown to dull black .... **2. *L. australis***
17: Divisions between leaf segments extend for about two-thirds the length of the lamina; fruit glossy jet-black .... **16. *L. nitida***

1 Livistona alfredii

Livistona alfredii F.Muell., *Victorian Naturalist* 9: 112 (1892) [as *L. Alfredi*]. Type: Australia. Western Australia. Hamersley Range, Millstream, June 1878, *J.Forrest s.n.*; lectotype MEL [4 sheets].

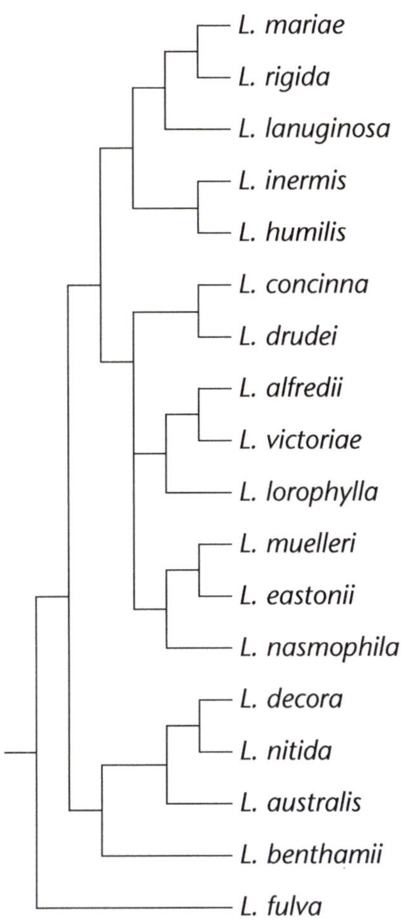

Figure 8.4 Phylogenetic tree summarising the estimated relationship of *Livistona* species in Australia. Adapted from Dowe (2001, 2009a) and Isagi (*pers. comm.*).

[*Livistona mariae auct. non* F.Muell., F.Muell., *Fragm.* 11: 54 (1878); *Livistona mariae auct. non.* F.Muell., C.A.Gardner, *Forest Depart. Bull. W.Australia* 32: 36 (1923)].

Functionally dioecious palm. **Stem** to 15 m tall, 20–50 cm dbh; leafscars prominent; internodes narrow, pale grey; petiole remnants persistent in basal 1 m or so. **Leaves** 25–30 in a ± globose crown, sometimes persistent to form a 'skirt' of senescent leaves; petiole 90–130 cm long, 25–30 mm wide, flat on the adaxial surface, rounded on the abaxial surface, margins with curved single black spines congested in the proximal portion; leafbase fibres prominent, coarse, persistent; lamina regularly segmented, ± subcircular, 90–140 cm long, rigid, pale green-grey to glaucous, waxy, dull on the adaxial surface; light green-grey and waxy on the abaxial surface. **Segments** 50–66, rigid, free for 60–70% of their length, depth of apical cleft 60–75% of the length of the free portion, apices rigid, lobes attenuate; parallel veinlets 6–8 each side of the midrib; transverse veinlets thinner than parallel veinlets. **Inflorescences** 180–270 cm long; partial inflorescences 5–7, branched to 3 orders; prophyll to 30 cm long, 7–9 cm wide, densely cream tomentose, basally pink-red, apex lacerate-fibrous; peduncular bracts 2, moderately to densely tomentose, cream; rachis bracts densely tomentose, cream; axes glabrous, cream; rachillae to 13 cm long, to 2 mm thick, glabrous, pinkish to yellow, minutely scabrous. **Flowers** solitary or in pairs, cylindrical in bud; sepals triangular, 0.8–1 mm long, acute, orange-brown; petals triangular, 2–3 mm long, acute to mucronate, orange-brown, with longitudinal brown striations; stamens 1–2 mm long. **Fruit** globose to subglobose, 25–40 mm long, 30–40 mm wide; epicarp to 0.5 mm thick, dark brown to brown-black with scattered greyish lenticellular pores; suture line extends to about half-way to base; mesocarp 5–8 mm thick; endocarp thin, brittle, crustaceous; pedicel to 3 mm long. **Seed** globose, 14–20 mm wide; embryo sub-lateral, c. 3 mm long; eophyll 3-ribbed. Figures 8.5, 8.6.

Figure 8.5 *Livistona alfredii* in floodplain habitat, Fortescue River, Millstream-Chichester National Park, Pilbara, Western Australia.

Figure 8.6 *Livistona alfredii*. **Top left:** Prophyll with indumentum. **Top right:** Crown with inflorescences. **Bottom left:** Flowers at anthesis. **Bottom right:** Mature fruit.

Millstream fan palm

Distribution and ecology: Endemic to Western Australia, in the Pilbara Region mainly in the upper reaches of the Fortescue, Robe and Ashburton Rivers. A small disjunct and senescent population occurs in the Cape Range, North West Cape Peninsula, on the western side of Exmouth Gulf (Humphreys *et al.* 1990); occurs in small populations or as scattered individuals as a canopy element often in association with *Eucalyptus leucophloia* Brooker (snappy gum), primarily as a riparian element in the flood zone of watercourses, and along drainage lines in well-drained sites, 50–510 m asl (Fig. 8.7). Flowering Sept.–Oct.; fruiting Nov.–Jan. Conservation status – *Rare* (Atkins 2008).

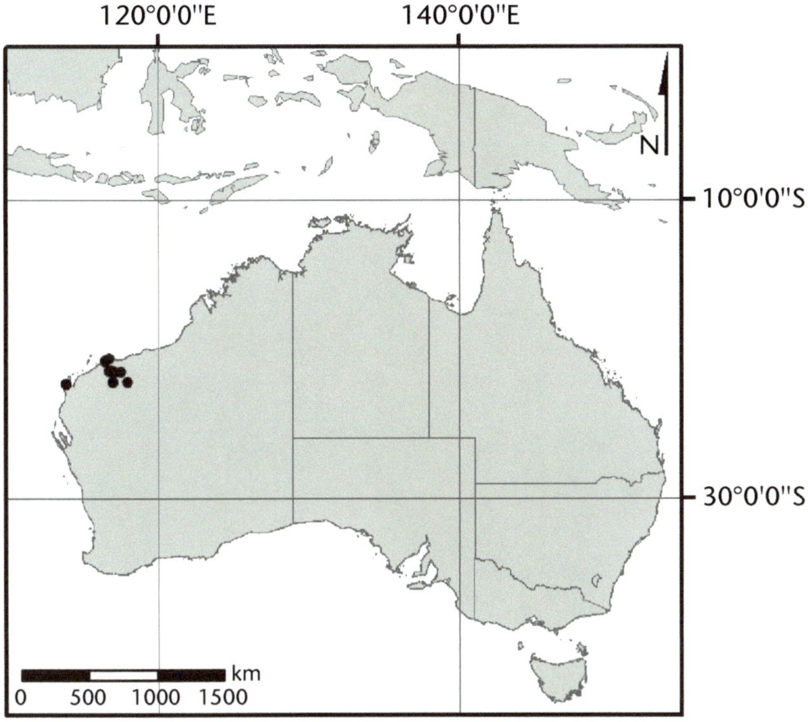

Figure 8.7 *Livistona alfredii*. **Left:** Distribution based on herbarium specimen data. **Right:** Type specimen. Australia. Western Australia. Hamersley Range, Millstream, June 1878, *J.Forrest s.n.*; lectotype MEL [1 of 4 sheets].

Typification: The protologue of *Livistona alfredii* in Victorian Naturalist did not mention any specimens (Mueller 1892). However, specimens of *L. alfredii*, collected by Forrest [MEL] in June 1878 from 'on the Millstream of the Fortescue River', were cited by Mueller (1878) as part of the distribution of *L. mariae*, and may be interpreted as the lectotype, and not the holotype as suggested by Rodd (1998) (Fig. 8.7).

Etymology: 'The West-Australian Fan-Palm has therefore now been named *L. Alfredi*, in honour of H.R.H. the Duke of Edinburgh, at whose nuptial festival the Central Australian Palm became dedicated to the Princess Marie of Russia' (Mueller 1892, p. 112): Named for Prince Alfred Ernest Albert (b.1844, d.1900), Duke of Edinburgh, second son of Queen Victoria and Prince Albert.

Notes: Based on the large fruit and seed of *L. alfredii*, Drude (1893) proposed a new monospecific section within *Livistona*, sect. *Gregorya*. Drude did not add to the original description provided by Mueller, and the species remained imperfectly understood. It was not until Beccari's (1931) treatment of *Livistona* that a full description of *L. alfredii* was provided, and the identity of the species was firmly established.

Livistona alfredii is the most geographically isolated species of *Livistona* in Australia, occurring at least 700 km from any other palm species (see Chapter 4). Molecular investigation resolves it as sister to *L. victoriae*. The geographically closest species is *L. lorophylla* in the Kimberley Region. *Livistona alfredii* is distinguished from *L. victoriae* by its much more robust stem, a dense rather than open crown; much longer inflorescence; partial inflorescences branching to three rather than four orders; and its much larger fruit – 40 mm diam. rather than 15 mm diam.

2 Livistona australis

Livistona australis (R.Br.) Mart., *Hist. Nat. Palm. Edn 1*, 3: 242 (1838); *Corypha australis* R.Br., *Prodr.*: 267 (1810); *Saribus australis* (R.Br.) Kuntze, *Revis. Gen. Pl.* 2: 736 (1891). Type: Australia. New South Wales. Port Jackson, June 1802, *R.Brown s.n.*; holotype BM [3 sheets]; isotype FI.

[*Livistona inermis auct. non* R.Br., *in* H.Wendland and Drude, *Linnaea* 39: 229 (1875)].

Functionally dioecious palm. **Stem** to 25 m tall, 25–40 cm dbh; leafscars raised; internodes brown, longitudinal fissures prominent; petiole remnants retained in the lower 2 m or so. **Leaves** 35–60 in a globose crown; petiole 150–250 cm long, 10–20 mm wide, moderately ridged on the adaxial surface, margins with short single black curved spines congested in the proximal portion, a thin light green longitudinal stripe present on the abaxial surface; leafbase fibres prominent, finely woven, sheet-like, persistent; lamina regularly segmented, ± circular, 100–130 cm long, chartaceous, dark green and glossy on the adaxial surface, slightly lighter green and non-waxy on the abaxial surface. **Segments** 80–100, distally pendulous, segments free for 49–59% of their length, depth of apical cleft 50–63% of the length of the free portion, apical lobes acute; parallel veinlets 8 each side of the midrib; transverse veinlets thinner than parallel veinlets. **Inflorescences** 140–250 cm long; partial inflorescences 6–9, branched to 5 orders; peduncular bract lacking; rachis bracts loosely tubular, densely floccose; rachillae 5–25 mm long, glabrous. **Flowers** solitary or in clusters of 2–4, funnel-shaped in bud; sepals triangular, 1.2–2 mm long, fleshy, acute, white-cream to pale yellow; petals triangular to ovate, 2.0–3.5 mm long, fleshy, subacute to obtuse, white-cream to pale yellow; stamens c. 2.5 mm long. **Fruit** globose, 12–22 mm diam.; epicarp reddish-brown to dull black, occasionally glaucous, with scattered lenticellular pores; suture line extends almost the length of the fruit; pedicel 2–3 mm long. **Seed** globose, 10–16 mm wide; eophyll 5-ribbed. Figures 8.8, 8.9.

Figure 8.8 *Livistona australis* in lowland coastal forest, Bribie Island, south-east Queensland.

Cabbage palm, cabbage tree

Distribution and ecology: Endemic to eastern Australia, in Queensland, New South Wales and Victoria. In the northern part of its range in Queensland, it is confined to high elevation above 400 m on Paluma Range, Seaview Range, Mt Elliot, Cape Upstart, Mt Abbott and Eungella Range with a major disjunction to Fraser Island, and then at low and moderate elevation south through south-east Queensland, through New South Wales and with a minor disjunction to the Brodribb River in eastern Victoria, making it the most southerly naturally occurring palm in Australia. Grows in moist areas of open forest, swamp forest, rainforest, moist sclerophyll forests as a subcanopy element or canopy emergent, occurring in large and small colonies, small groups or scattered individuals, often locally common, 0–1000 m asl (Fig. 8.10). Flowering all months; fruiting all months. Conservation status – *No present threats.*

Typification: The holotype of *Livistona australis* (basionym *Corypha australis*) is an unnumbered collection made by Robert Brown [BM] in June 1802 from Port Jackson, New South Wales. The collection consists of a smallish, possibly juvenile leaf, a length of petiole and portions of inflorescence (Fig. 8.10). Labels related to the original collection include 'New Holland', *'Corypha nuda'* (a name used by Daniel Solander 1768–71 in his unpublished *Plantae Novae Hollandiae*) and *'Corypha australis'*. Two other specimens in BM, also appended with a type label, are collections made by George Caley in 1808 and are not associated with typification.

Etymology: *'C. australis'*: Though not explicitly explained in the protologue, with reference to the occurrence of the species in Australia, from the Latin *australis*, southern.

Notes: Brown (1810) distinguished *C. australis* (= *L. australis*) from his new *Livistona* species by a few minor characters, including the length of separation of the

Figure 8.9 *Livistona australis*. **Top left:** Stem with remnant leafbases and vertical fissures. **Top right:** Crown with inflorescences, Broken River, Eungella Range, central Queensland. **Bottom left:** Flowers at anthesis. **Bottom right:** Mature fruit.

filaments, basally separated in *Corypha*, separated for most of their length in *Livistona*; endosperm hollow in *Corypha*, intruded ventrally in *Livistona*; and the position of embryo being basal in *Corypha* and dorsal in *Livistona*.

Although *L. australis* has a widespread distribution it is morphologically uniform and readily distinguished from other species of *Livistona*. In the field it most closely resembles *L. nitida* and *L. decora*. From the former it is distinguished by its less pendulous segment apices and dull rather than glossy fruits. *Livistona decora* is distinguished from *L. australis* by its deeply segmented leaves and glossy fruits.

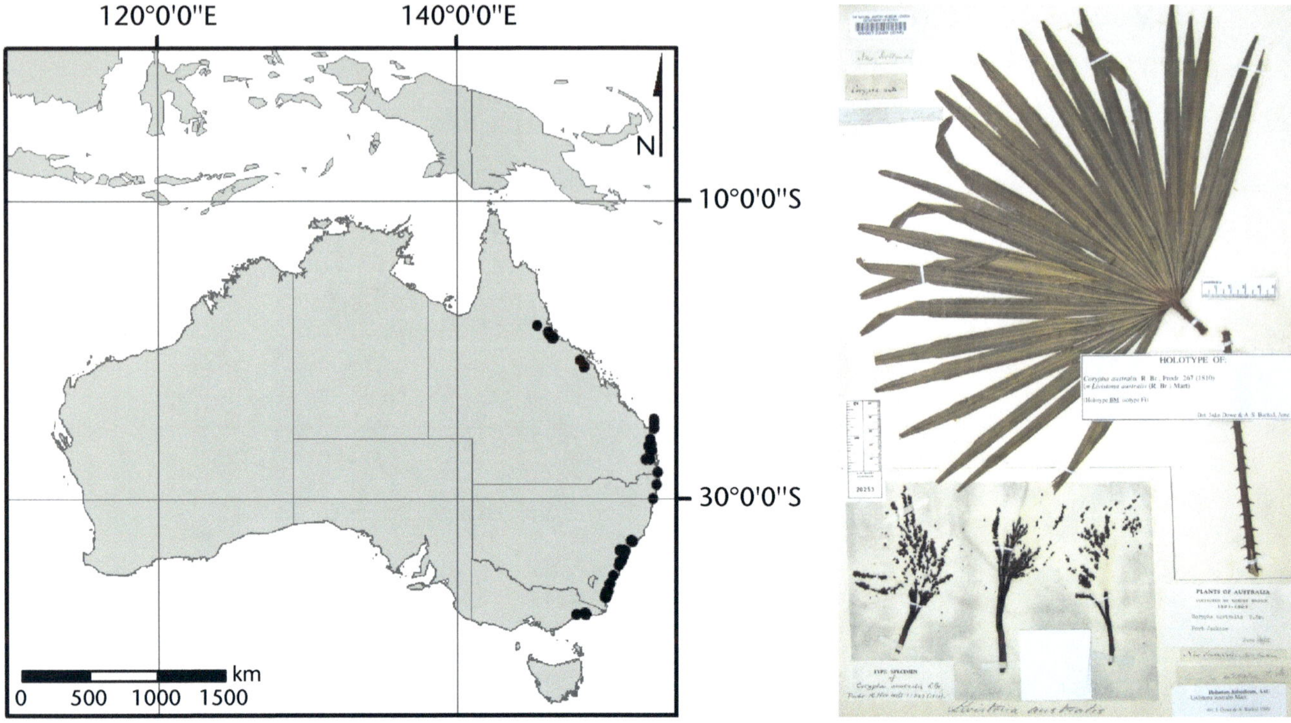

Figure 8.10 *Livistona australis*. **Left:** Distribution based on herbarium specimen data. **Right:** Type specimen. Australia. New South Wales. Port Jackson, June 1802, *R.Brown s.n.*; holotype BM [1 of 3 sheets].

3 Livistona benthamii

Livistona benthamii F.M.Bailey, *Queensl. fl.* 5: 1683 (1902) [as *L. Benthami*]; '*Livistona humilis* R.Br., var.', F.M.Bailey, *Queensland Agric. J.* 2: 130 (1898), *nom. illeg.* Type: Australia. Queensland. Cape York Peninsula, Somerset, Dec. 1897, *F.L.Jardine s.n.*; holotype BRI [1 parcel].

Livistona holtzei Becc., *Webbia* 5: 18 (1921). Type: Australia. Northern Territory. Port Darwin, undated, *Holtze s.n.*; holotype FI.

Livistona melanocarpa Burret, *J. Arnold Arbor.* 20: 190 (1939). Type: Papua New Guinea. Western Province, Daru Is., 9 Mar. 1936, *L.J.Brass 6310*; holotype BRI; isotypes A, BM, BO, L.

Functionally dioecious palm. **Stem** to 18 m tall, 13–20 cm dbh; leafscars raised; internodes grey; petiole remnants long, abscising about half-way to hastula at a low ridge on the abaxial surface, persistent. **Leaves** 30–50 in a globose crown; petiole 120–200 cm long, 10–18 mm wide, moderately ridged on the adaxial surface, margins with single curved black spines congested in the proximal portion; c. 50% of petiole remains intact on stem until deteriorating with age; petiole base bulbous at attachment; leafbase fibres prominent, coarse, persistent; lamina regularly segmented, circular, 90–160 cm long, chartaceous, dark green on the adaxial surface, lighter green, glossy and non-waxy on the abaxial surface. **Segments** 50–80, free for 60–75% of their length, depth of apical cleft 60–75% of the length of the free portion; segment apices pendulous, lobes acute; parallel veinlets 8 either side of the midrib; transverse veinlets thinner than parallel veinlets. **Inflorescences** 120–210 cm long; 7–9 partial inflorescences, branched to 2 orders; prophyll 12–22 cm long, 4–5 cm wide, 1–2 cm thick, dorsiventrally compressed, with sparse silver scales; peduncular bract lacking; rachis bracts tightly sheathing, light brown, with sparse scurfy silver scales; rachillae 5–12 mm long, patchily pubescent. **Flowers** solitary or in clusters of 2–3, tubular; sepals triangular, 0.8–2 mm long, membranous, acute, whitish-pale yellow to green; petals broadly triangular, 1.0–1.3 mm long, fleshy, acute, whitish-pale yellow to green; stamens c. 1 mm long. **Fruit** globose to obovoid or pyriform, 9–13 mm long, 9–11 mm diam.; epicarp smooth, purple-black, pruinose, suture line extends for the length of the fruit; pedicel 2–3 mm long. **Seed** ovoid, 8–9 mm long; eophyll not seen. Figures 8.11, 8.12.

Figure 8.11 *Livistona benthamii* in lowland swamp forest, Paterson Creek, Lockerbie Scrub, Cape York Peninsula, north Queensland.

Bentham's fan palm

Distribution and ecology: In Australia, in the Northern Territory from Daly River across the Top End to east Arnhem Land, and in Queensland from Moa Is. in Torres Strait through Cape York Peninsula south to McIlwraith Range, also in Papua New Guinea in the Western and Central Provinces and in Indonesia in Papua, Merauke Division. Grows in swamp forests, on alluvial flats, on the leeward side of mangroves and in seasonally inundated areas, in monsoonal thickets, swamp forest and riparian forests at low elevations, in large or small colonies and as scattered individuals, as a canopy emergent, 0–100 m asl (Fig. 8.13). Flowering Jul.–Nov.; fruiting Oct.– Mar. Conservation status – *No present threats*.

Typification: The holotype of *L. benthamii* is an unnumbered Frank Jardine [BRI] collection dated Dec. 1897 from Somerset, Cape York, Queensland. The collection consists of a single leaf and petiole sections, and an inflorescence (Fig. 8.13).

Etymology: 'after Geo. Bentham' (Bailey 1902, p. 1683): Named for George Bentham (b.1800, d.1884), botanist and systematist, author of *Flora Hongkongensis* 1861, *Flora Australiensis* 1862–78 and *Genera Plantarum* (with J.D. Hooker) 1862–83.

Notes: For the population occurring in the Northern Territory, Beccari (1921) provided the name *L. holtzei* based on *Holtze s.n.* [FI] and for New Guinea Burret (1939) provided the name *L. melanocarpa* based on *Brass 6310* [BRI] collected from Western Province, Papua New Guinea.

Livistona benthamii is distinguished by its relatively small leaves, the abscission of the petiole at about its midpoint, thus leaving persistent on the stem a section of erect protruding petiole which in time may deteriorate proximally to a bulbous-rounded base; a thinnish inflorescence in which the rachis bracts are tightly tubular, smooth and non-fibrous, and small fruit, to 13 mm diam. The leaf has a relatively large undivided portion, and the segments are pendulous or at least drooping from the disjunction of the segments.

4 *Livistona concinna*

Livistona concinna Dowe & Barfod, *Austrobaileya* 6: 166, Fig. 1 (2001). Type: Australia. Queensland. c. 5 km N of Cooktown Airport, Barrett Creek, 15°25'S, 145°11'E, 5 m alt., 17 Oct. 2000, *J.L.Dowe 607*; holotype BRI; isotypes AAU, CNS, K.

Dioecious palm. **Stem** to 30 m tall, 24–35 cm dbh; leaf-scars raised; internodes narrow; grey; petiole remnants not persistent. **Leaves** 50–65 in a globose to conical crown; petioles 120–300 cm long, glabrous, green throughout, 5–11 cm wide proximally, 2.2–3.5 cm wide in mid area, c. 2.8 mm wide distally, triangular in cross-section, moderately ridged on the adaxial surface, margins with solitary symmetric or retrorsely or latrorsely curved black spines 3–5 mm long congested in the proximal portion, distal margin unarmed, sharp, slightly winged; leafbase fibres not prominent, coarse, persistent; lamina regularly segmented, subcircular in outline, glabrous, mid-green on the adaxial surface, slightly lighter green on the abaxial surface, glossy on both surfaces, non-waxy, 155–165 cm long, c. 200 cm wide. **Segments** 60–78, apices deeply forked, distal portion pendulous; segments free for about 60% of their length, depth of apical cleft about 41% of the length of the free portion, mid-lamina segments 2.6–4 cm

Figure 8.12 *Livistona benthamii*. **Top left:** Stem with retained petioles, Fogg Dam, Northern Territory. **Top right:** Flowers at anthesis. **Bottom left:** Infructescence with mature fruit. **Bottom right:** Fruit, whole and in longitudinal section.

wide where the segments diverge, apical lobes acuminate, filamentous; hastula raised, sharp, papery on the margins; parallel veinlets 9–10 each side of the midrib; transverse veinlets thinner than parallel veinlets. **Inflorescences** sexually dimorphic; non-fruit-bearing inflorescences 120–180 cm long, slightly curved with 8–9 partial inflorescences, branched to 4 orders; peduncle 25 mm wide, 10 mm thick, glabrous; rachis to 25 mm diam.; fruit-bearing inflorescences 160–250 cm long, slightly curved, held horizontal to semi-pendulous with 8–9 partial inflorescences, branched to 5 orders; peduncle c. 30 mm wide; prophyll 27–35 cm long, glabrous; rachis to 25 mm diam.;

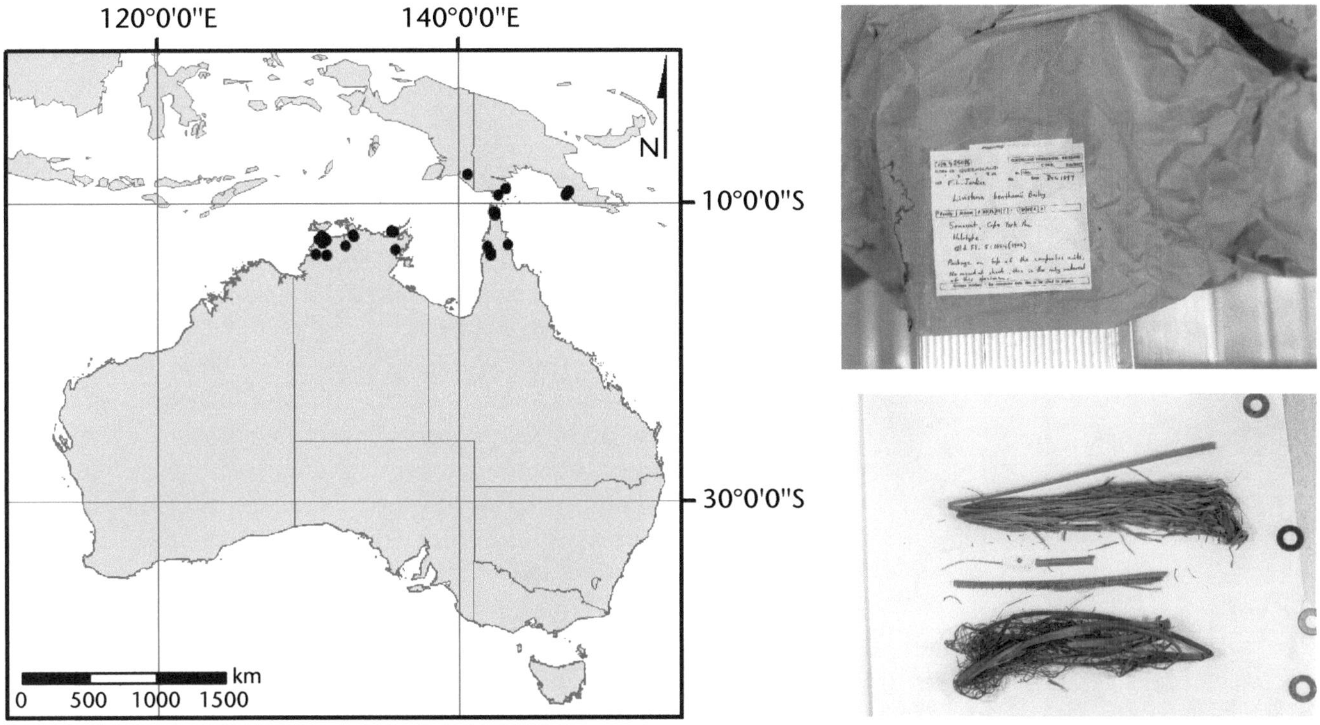

Figure 8.13 *Livistona benthamii*. **Left:** Distribution based on herbarium specimen data. **Right:** Type specimen. Australia. Queensland. Cape York Peninsula, Somerset, Dec. 1897, *F.L.Jardine s.n.*; holotype BRI.

peduncular bract lacking; rachis bracts 30–50 cm long, tubular, tightly sheathing, not disintegrating or splitting with age, glabrous, apically acute, margins entire, lateral splits uneven, one about twice as deep as the other; rachillae 5–20 cm long, glabrous. **Flowers** solitary or in clusters of 2–4, 1.6–2 mm high and 1.5–2 mm wide in bud; sepals basally fused, cupular, 3-lobed, lobes triangular, margins hyaline, 0.5–1.5 mm high, apices acute, white to cream-yellow; petals triangular, slightly asymmetric, 1–2.2 mm long, 0.8–2 mm wide at the base, apex acute, white to cream-yellow; stamen 0.5–1 mm high, filament subulate, connective 0.2 mm long; carpels c. 1 mm high. **Fruit** globose, 9–12 mm diam.; stigmatic remains subapical; epicarp smooth, shiny black, with scattered lenticellular pores, drying slightly rugose; suture line extends for about three-quarters the length of the fruit; mesocarp c. 1 mm thick, moist, oily, gritty; endocarp 0.1–0.2 mm thick, crustaceous, light brown, tessellate; pedicel to 2 mm long. **Seed** globose to subglobose; intrusion extends to half or less the width of the seed, contorted; embryo lateral to sublateral, c. 2 mm long; eophyll 5-ribbed. Figures 8.14, 8.15.

Cooktown Livistona, Cooktown fan palm, Kennedy River Livistona

Distribution and ecology: Endemic to north-east Queensland, on Flinders Island, Kennedy River and tributaries, Barrett Creek north of Cooktown, and Archer Point, occurring in small populations, small groups or isolated individuals in seasonally moist open forest, seasonally inundated *Melaleuca* swamp forest, along creek and river banks, and mangrove margins, 0–100 m asl (Fig. 8.16). Soils are usually alluvial and fire regularly occurs in most populations. Grows with *Corypha utan* in the Kennedy River area and with *L. muelleri* in the vicinity of Cooktown. Flowering Nov.–Mar.; fruiting Apr.–Oct. Conservation status – *Rare* (Bostock and Holland 2007).

Typification: The holotype of *L. concinna* is *Dowe 607* [BRI] collected 17 Oct. 2000 at Barrett Creek, north of Cooktown, Queensland. The collection consists of a leaf, petiole portions, a petiole base and the lower portion of an inflorescence (Fig. 8.16).

Etymology: 'From the Latin *concinnus*, neat, well-arranged, in reference to the organised and neat appearance

Figure 8.14 *Livistona concinna* in mangrove margin habitat, Barrett Creek, north of Cooktown, north-east Queensland.

of the petioles in the crown compared to those in other Livistona species' (Dowe and Barfod 2001, p. 167).

Notes: *Livistona concinna* is an occasionally tall palm to 30 m with a conical crown. The leaves have a large intact lamina and the segments are pendulous from the segment disjunction. Inflorescences are sexually dimorphic on different plants with the male inflorescence being less robust than the female, and globose shiny black fruit to 12 mm diam. *Livistona concinna* is sister to *L. drudei* in molecular investigations.

5 Livistona decora

Livistona decora (W.Bull) Dowe, *Austrobaileya* 6: 979 (2004); *Corypha decora* W.Bull, *Catal.*: 10 (1887). Type: Neotype. Australia. Queensland. Yeppoon-Emu Park Rd, c. 11 km S of Yeppoon, 12 May 1976, *A.N.Rodd 3069 with S.Jacobs*; neotype BRI; isoneotypes BH, K, NSW, *fide* Dowe and Jones (2004).

Livistona decipiens Becc., *Webbia* 3: 301, tab. 2 (1910). Type: Cultivation. France. Nice, May 1908, *A.Robertson-Proschowsky*; holotype FI.

Livistona decipiens var. *polyantha* Becc., *Webbia* 5: 15 [18] (1921). Type: Cultivation. Indonesia. Bogor Botanic Gardens, undated, *Beccari s.n.*; holotype FI; isotype BO.

[*Livistona inermis auct. non* R.Br., *in* H.Wendland and O.Drude, *Linnaea* 39: 229 (1875) *pro parte; Livistona australis auct. non* (R.Br.) Mart., *in* G.Bentham, *Fl. austral.* 7: 146 (1878) *pro parte*].

Functionally dioecious palm. **Stem** to 18 m tall, 25–30 cm dbh; leafscars narrow to broad; internodes smooth, brown in newly exposed parts, grey with age, with shallow vertical fissures; petiole remnants frequently retained at the base of the stem. **Leaves** 30–60 in an obloid crown; petiole 150–280 cm long, 18–24 mm wide, flat or slightly concave on the adaxial surface, green throughout, margins with single curved black spines to 20 mm long in the proximal portion; leafbase fibres moderately prominent, coarse, disintegrating; lamina regularly segmented, ± circular, 120–185 cm long, coriaceous, bright green on the adaxial surface, lighter green on the abaxial surface. **Segments** 70–84, free for 82–88% of their length, apical spilt 44–54% of the length of the free portion, lobes pendulous; parallel veinlets 6–7 each side of the midrib; transverse veinlets thinner than parallel veinlets. **Inflorescences** 100–350 cm long, with 8–13 partial inflorescences, branched to 4 orders; longest partial inflorescence to 110 cm; prophyll 37 cm long, 11–16 cm wide, woody; peduncle more or less terete, 60 mm wide at the base, glabrous; peduncular bract lacking; rachis bracts loose, glabrous to sparsely papillate; rachis to 25 mm diam., at the base; primary rachis bract tubular, loose, apically fibrous-laterate, bracts subtending each partial inflorescence basally tubular, becoming winged distally, apex pointed; in the most distal partial inflorescence there are 2 vestigial bracts at the base; rachillae 5–20 mm long, papillose. **Flowers** solitary or in clusters of 2–6, funnel-shaped, 1.5–2 mm long in bud; sepals ovate-triangular, c. 1.3 mm long, membranous, obtuse, yellow; petals broadly ovate, 1.5–1.8 mm long, fleshy, obtuse, yellow; stamens c. 1.5 mm long. **Fruit** globose, 12–18 mm diam.; epicarp shiny black, with scattered lenticellular pores; suture line extends about half the length of fruit; mesocarp 1–3 mm thick, spongy, moist, fibres aggregated toward the endocarp; endocarp thin, brittle; pedicel to 2 mm long. **Seed** globose, c. 10 mm wide; intrusion half to

Figure 8.15 *Livistona concinna*. **Top left:** Crown with petioles and leafbases. **Top right:** Leaf with segments and apical clefts. **Bottom left:** Flowers at anthesis. **Bottom right:** Fruit, whole and in longitudinal section.

three-quarters across endosperm; eophyll 5-ribbed. Figures 8.17, 8.18.

Weeping cabbage palm, ribbon fan palm

Distribution and ecology: Endemic to Queensland, from Black River, north of Townsville, to Fraser Island and Rainbow Beach, occurring most commonly in coastal and near-coastal lowlands in large dense colonies or in scattered groups, or as a riparian element, in open forest, littoral rainforest and dry rainforest at low to moderate elevations, and with isolated inland populations at moderate elevations associated with permanent springs,

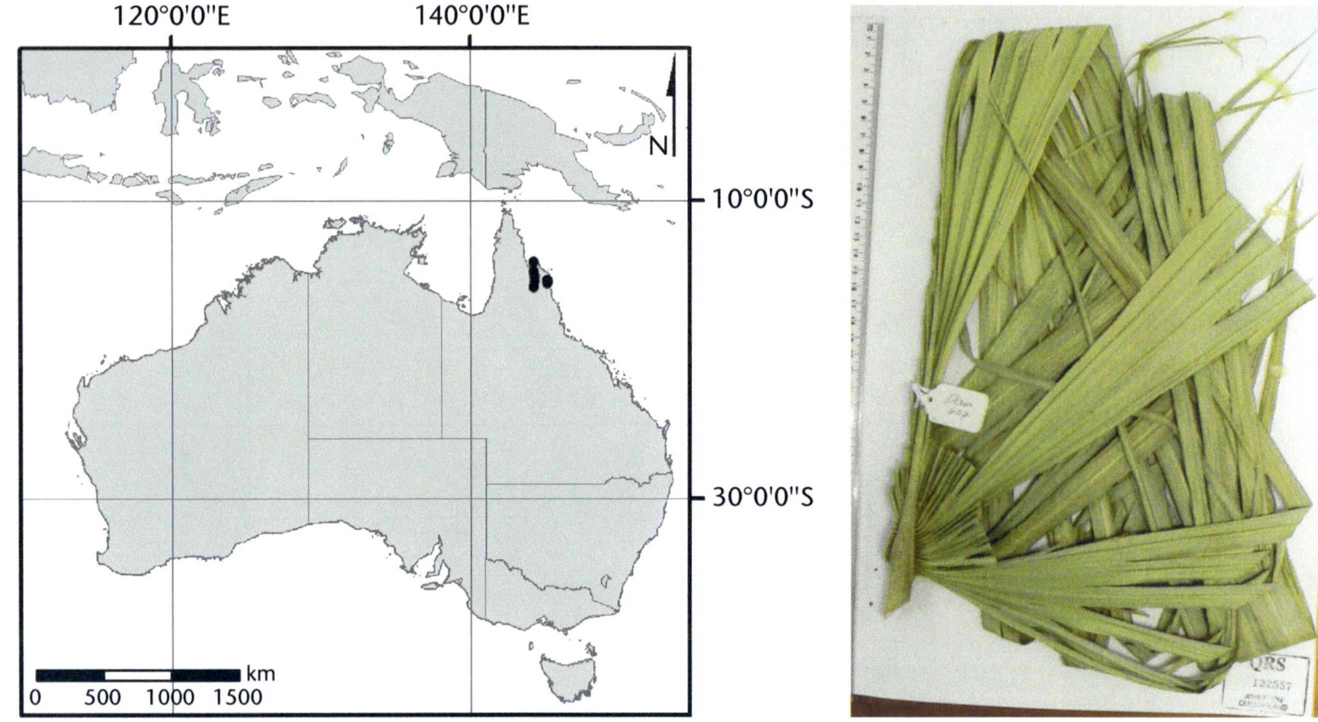

Figure 8.16 *Livistona concinna.* **Left:** Distribution based on herbarium specimen data. **Right:** Type specimen. Australia. Queensland. c. 5 km north of Cooktown Airport, Barrett Creek, 15°25′S, 145°11′E, 5 m alt., 17 Oct. 2000, *J.L.Dowe 607*; holotype BRI.

Figure 8.17 *Livistona decora* in riparian habitat, West Barratta Creek, north-east Queensland.

0–550 m asl (Fig. 8.19). Flowering Jul.–Dec.; fruiting Oct.–Apr. Conservation status – *No present threats.*

Typification: In the absence of specimens associated with the protologue, a neotype *Rodd 3069* [BRI] was proposed (Dowe and Jones 2004). The neotype consists of a complete leaf, a section of leafbase fibre, a prophyll, portions of petiole, a partial inflorescence and fruit (Fig. 8.19).

Etymology: 'elegant and ornamental species' (Bull 1887, p. 10): With reference to the attractive appearance of the species, from the Latin *decoris*, ornamental.

Notes: *Livistona decora* is most closely related to *L. nitida*, and is readily distinguished by the deeply segmented leaves and pendulous segment apices.

6 *Livistona drudei*

Livistona drudei F.Muell. ex Drude, *Bot. Jahrb. Syst.* 39: 12 (1893); Mueller, *Fragm.* 11: 55 (1878) *nom. prov.*; *Fragm.* 11: 147 (1881) *nom. in index*; Type: Australia. Queensland. Rockingham Bay, Sept. 1865, *Dallachy s.n.*; holotype MEL [2 sheets].

[*Livistona inermis auct. non.* R.Br., H.Wendland and O.Drude, *Linnaea* 39: 229 (1875) *pro parte*].

Functionally dioecious palm. **Stem** to 28 m tall, 15–20 cm dbh; leafscars narrow, pale grey; internodes broad; petiole

Figure 8.18 *Livistona decora*. **Top left:** Leaf with pendulous segments. **Top right:** Segment apices. **Bottom left:** Flowers at anthesis. **Bottom right:** Mature fruit.

remnants retained in the basal portion. **Leaves** 30–60 in a globose or conical crown; petiole 150–230 cm long, 15–25 mm wide, ridged on the adaxial surface, purple in basal portion, green distally, margins with single, curved reddish spines largest and closer in the proximal portion; leafbase fibres moderately prominent, coarse, persistent; lamina regularly segmented, ± circular, 100–150 cm long, coriaceous, dark green on the adaxial surface, lighter green and glossy on the abaxial surface. **Segments** 60–84, pendulous distally, free for 60–70% of length, depth of apical cleft about 60% of the length of the free portion, apical lobes attenuate; parallel veinlets 7 each side of the midrib;

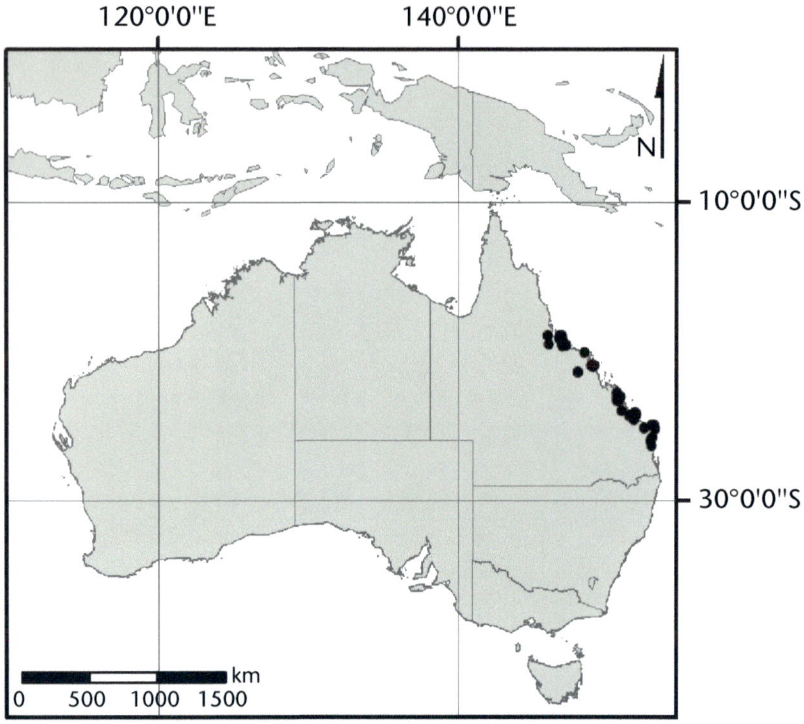

Figure 8.19 *Livistona decora*. **Left:** Distribution based on herbarium specimen data. **Right:** Type specimen. Australia. Queensland. Yeppoon–Emu Park Rd, c. 11 km south of Yeppoon, 12 May 1976, *A.N.Rodd 3069 with S.Jacobs*; isoneotype NSW [1 of 4 sheets].

transverse veinlets thinner than parallel veinlets. **Inflorescences** 150–300 cm long; 7–8 partial inflorescences, branched to 3 orders; peduncle subterete, 15–25 mm diam.; prophyll glabrous; peduncular bract 1; rachis bracts tightly tubular, glabrous, apices with acuminate lobes; rachillae 1–8 cm long, pubescent. **Flowers** solitary or in clusters of 2–5, irregularly funnel-shaped, 1.2–2.2 mm long, 1.1–1.5 mm wide in bud, opening to c. 2 mm wide at maturity; sepals basally connate, lobes narrowly triangular, 0.7–1.5 mm long, c. 0.5 mm wide, fleshy, subacute, light green-cream to yellow, glabrous; petals broadly ovate, 1.2–2 mm long, c. 1 mm wide, fleshy, apex thickened and curved inward, cream to yellow, glabrous; stamens 1–1.5 mm long, cream; anthers subglobose to ovate, c. 0.2 mm long, light yellow; carpels c. 0.8 mm long; style c. 0.2 mm long. **Fruit** globose, 10–12 mm long, 10–11 mm diam.; epicarp semi-glossy purple-black with scattered lenticellular pores; suture line extends for about half the length of fruit, marked with lip-like structures; mesocarp thin, fibrous; endocarp thin, brittle; pedicel 1–2 mm long. **Seed** ellipsoid, 8–9 mm long; eophyll 5-ribbed. Figures 8.20, 8.21.

Figure 8.20 *Livistona drudei* in coastal open forest, Clemant State Forest, Boar Creek outflow, north-east Queensland.

Figure 8.21 *Livistona drudei*. **Top left:** Crown. **Top right:** Petiole base with characteristic purple coloration. **Bottom left:** Flowers at anthesis. **Bottom right:** Mature fruit.

Northern weeping fan palm

Distribution and ecology: Endemic to north-east Queensland in coastal and near-coastal areas between Kurrimine Beach and Black River, and with disjunctions to Conway Range and Don River, occurring in small populations and groups or scattered individuals on the landward side of mangroves, open forest and swamp forest on coastal plains and along the margins of streams and estuaries, and infrequently at moderate elevation, 0–300 m asl (Fig. 8.22). Flowering Nov.–Mar.; fruiting Jan.–Jun. Conservation status – *Vulnerable* (Bostock and Holland 2007).

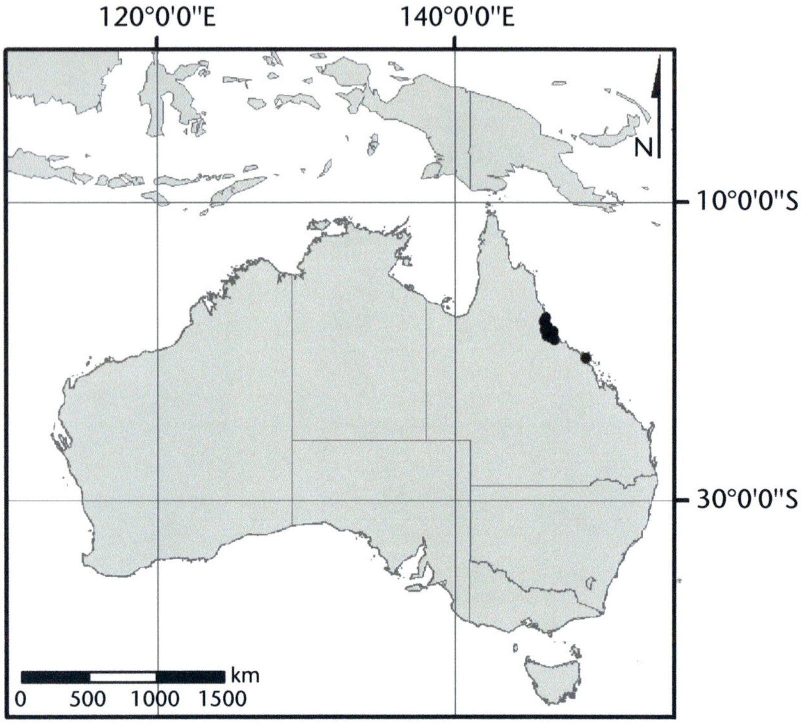

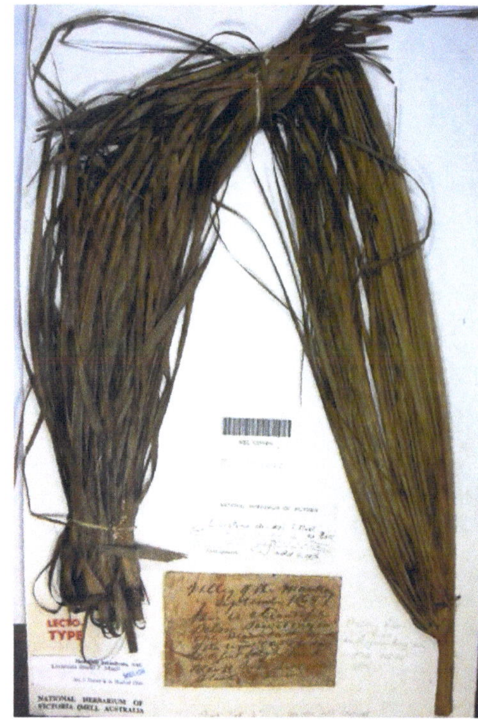

Figure 8.22 *Livistona drudei*. **Left:** Distribution based on herbarium specimen data. **Right:** Type specimen. Australia. Queensland. Rockingham Bay, Sept. 1865, *Dallachy s.n.*; holotype MEL [1 of 2 sheets].

Typification: The holotype for *L. drudei* is an unnumbered Dallachy [MEL] specimen dated Sept. 1865, labelled 'Valley of the Mackay, September 1865'. The location 'Valley of the Mackay' does not refer to the present city of Mackay but to the Tully River, which was for a short time referred to as the Mackay River (Fig. 8.22).

Etymology: '*Huic palmae nunc pro speciei vel varietatis distinctione nomen Doctoris Oscaris Drude impono*' (Mueller 1878, p. 55) [In his honor this distinctive species is named after Doctor Oscar Drude]: Named for [Carl George] Oscar Drude (b.1852, d.1933), botanist, Director of Dresden Botanic Gardens 1879–1920, co-author and illustrator of *Palmae Australasicae* (Wendland and Drude 1875) and *Über die australischen Livistona-Arten* (Drude 1893).

Notes: *Livistona drudei* most closely resembles *L. concinna* and *L. benthamii*, with which it shares a moderate-sized inflorescence and smallish fruit. It does not display the sexually dimorphic inflorescences of the former nor the retained petiole stubs of the latter. In preliminary molecular investigations *L. drudei* is sister to *L. concinna*.

7 Livistona eastonii

Livistona eastonii C.A.Gardner, *Forest Dept. Bull. W. Australia* 32: 36 (1923) [as *L. Eastoni*]. Type: Australia. Western Australia. Napier, Broome Bay, Lower King Edward River, 22 Aug. 1921, *C.A.Gardner 1544*; lectotype PERTH [2 sheets]; isolectotypes B, BH, CANB, K, MEL.

Functionally dioecious palm. **Stem** to 12 (21) m tall, 10–15 cm dbh; leafscars raised; internodes narrow, grey; leafbases often persistent in the lower 2–4 m. **Leaves** 9–20 in a globose crown; petiole 50–200 cm long, 1.4–2 cm wide, arcuate, margins with single or double curved brown spines throughout its length but largest and closer in the proximal portion; leafbase fibres moderately prominent, coarse, persistent; lamina regularly segmented, subcircular, 60–90 cm long, coriaceous, pale greyish green on the adaxial surface, lighter green on the abaxial surface, yellowish with age, pruinose and waxy. **Segments** 40–50, rigid, free for 50–90% of their length, depth of apical cleft 49–63% of the length of the free portion, apical lobes acute; parallel veinlets 5–9 each side of the midrib; transverse veinlets thinner than parallel veinlets. **Inflorescences** 100–200 cm long; partial inflorescences 5–6, branched to 4 orders; prophyll to 25 cm long, to 6 cm wide, thickly papyraceous, with scattered pubescent scales; peduncular bract 1, tubular, with scattered pubescent scales; rachis bracts tubular with scattered appressed scales densest toward the

Figure 8.23 *Livistona eastonii* in low semi-open woodland on laterite, Doongan, Mitchell Plateau, Western Australia.

apex; rachillae 1–9 cm long, glabrous. **Flowers** solitary or in clusters of 2–4, 1.6–1.9 mm long, campanulate in bud; sepals broadly ovate, 0.2–1 mm long, fleshy, acute, cream to yellow; petals narrowly ovate, 0.5–1.9 mm long, thick, acute, cream to yellow; stamens c. 1.5 mm long; anthers c. 1 mm long, yellow. **Fruit** ellipsoid to obovoid, 12–16 mm long, 8–9 mm diam., glossy, purple-black; epicarp smooth; pedicel 0.5–2 mm long. **Seed** ellipsoid, 10–13 mm long. Figures 8.23, 8.24.

Kimberley fan palm, Mitchell Plateau fan palm, Tdungana (Wunambal language)

Distribution and ecology: Endemic to Western Australia, in the Kimberley Region confined to the Mitchell Plateau as far south as Doongan Station, occurring in large or small populations or isolated groups or individuals, usually on flat sites or depressions in open woodland in association with *Eucalyptus tetrodonta*, *E. miniata* and *Corymbia nesophila*, 100–300 m asl (Fig. 8.25). The climate in this area is strongly monsoonal and soils are lateritic in origin. Hnatiuk (1977) estimated the entire population to number in the 'tens of millions' (all size classes). Flowering Apr.– Dec.; fruiting Dec.–Jun. Conservation status – *No present threats*.

Typification: Gardner did not cite any specimens in the protologue, and Rodd (1998) designated the collection *Gardner 1544* [PERTH], collected in Aug. 1921 from Lower King Edward River, Western Australia, as the lectotype. This specimen consists of a whole leaf and a partial inflorescence (Fig. 8.25).

Etymology: 'This species is named out of compliment to William Robert Easton, who commanded the Kimberley Exploration Expedition in 1921, of which the author was a member' (Gardner 1923, p. 37).

Notes: *Livistona eastonii* most closely resembles *L. humilis* but it is distinguished by its larger size, usually to c. 12 m tall but occasionally to c. 21 m tall, leaves to c. 90 cm long and light green below, and fruit ellipsoid to obovoid 12–16 mm long by 8–9 mm in diam. The inflorescences are not sexually dimorphic, although plants are functionally dioecious. Preliminary molecular investigations place *L. eastonii* as sister to *L. muelleri* (Fig. 8.4).

8 Livistona fulva

Livistona fulva Rodd, *Telopea* 8: 103, figs 4a, 12a–d, 13a (1998). Type: Australia. Queensland. Blackdown Tableland, 23°43'S, 149°07'E, 450 m alt., 11 May 1976, *A.N.Rodd 3062 with S.Jacobs*; holotype NSW [4 sheets]; isotypes BH, BRI, K.

Functionally dioecious palm. **Stem** to 13 m tall, 20–25 cm dbh; leafscars raised; internodes narrow, grey or brown; petiole remnants deciduous. **Leaves** 25–35 in a globose crown; petiole 150–250 cm long, 12–15 mm wide, moderately ridged on the adaxial surface, basal margins armed with single, curved black spines; leafbase fibres moderately prominent, fine, persistent; lamina regularly segmented, circular, 90–100 cm long, coriaceous, greyish green to glaucous on the adaxial surface, coppery brown floccose tomentum on the abaxial surface. **Segments** 60–66, rigid, free for 50–55% of their length, depth of apical cleft 3–5% of the length of the free portion, apical lobes acute to acuminate; parallel veinlets 8–9 each side of the midrib; transverse veinlets thinner than parallel veinlets. **Inflorescences** 100–230 cm long; partial inflorescences 7–9, branched to 4 orders; prophyll to 32 cm long, to 12 cm wide, woody, with orange-brown scales, apex coarsely fibrous; peduncular bract 1, with orange-brown

Figure 8.24 *Livistona eastonii*. **Top left:** Juvenile palm. **Top right:** Flowers at anthesis. **Bottom left:** Infructescence with mature fruit. **Bottom right:** Fruit, whole and in longitudinal section.

scales; rachis bracts with orange-brown scales; rachillae 5–16 mm long, papillose. **Flowers** solitary or in clusters of 2–3, funnel-shaped, 1.6–2 mm long in bud; sepals narrowly triangular, 1.0–1.3 mm long, membranous, acute, yellow; petals broadly ovate, 1.6–2 mm long, thick, acute, yellow; stamens c. 1.6 mm long. **Fruit** globose, 12–16 mm diam., dull black, pruinose; epicarp smooth; suture line not obvious; mesocarp thin, fibrous; endocarp thin, crustaceous. **Seed** globose, 10–13 mm wide; eophyll 5-ribbed. Figures 8.26, 8.27.

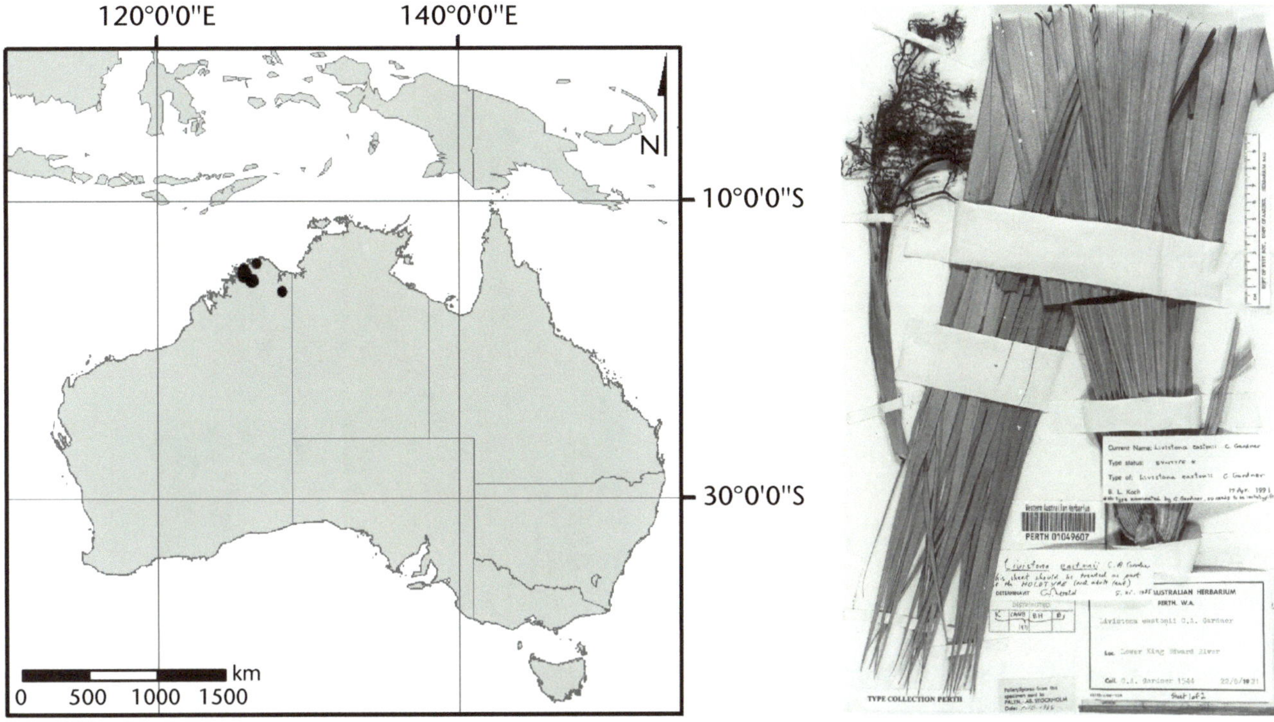

Figure 8.25 *Livistona eastonii*. **Left:** Distribution based on herbarium specimen data. **Right:** Type specimen. Australia. Western Australia, Napier, Broome Bay, Lower King Edward River, 22 Aug. 1921, *C.A.Gardner 1544*; lectotype PERTH [1 of 2 sheets].

Figure 8.26 *Livistona fulva* at the base of a sandstone escarpment, Blackdown Tableland, central Queensland.

Blackdown fan palm

Distribution and ecology: Endemic to Blackdown Tableland, central Queensland, occurring in large populations and small groups in moist sites in open forest and woodland, most common in gullies and gorges near streams and waterfalls, and at cliff bases at 400–800 m altitude, and in open forest on the tableland at 800–950 m altitude, in soils derived from eroded sandstone (Fig. 8.28). Flowering Sept.– Feb.; fruiting Dec.–May. Conservation status – *Rare* (Bostock and Holland 2007).

Typification: The holotype of *Livistona fulva* is *Rodd 3062* [NSW], collected in May 1976 from Blackdown Tableland in central Queensland. The collection consists of portions of leaf, petiole and inflorescence, and fruit (Fig. 8.28).

Etymology: 'The epithet *fulva*, Latin for 'tawny', draws attention to the striking golden-brown indumentum of the leaf undersides, a unique feature among Australian members of the genus' (Rodd 1998, p. 105).

Notes: The relationships of *L. fulva* are unclear, but it is morphologically similar to *L. muelleri* in having a flat rigid lamina, to *L. decora* in inflorescence size and morphology, and to *L. victoriae* in overall size and fruit morphology. It

Figure 8.27 *Livistona fulva*. **Top left:** Abaxial surface of leaf with characteristic coppery brown floccose tomentum. **Top right:** Crown with inflorescence. **Bottom left:** Flowers at anthesis. **Bottom right:** Mature fruit.

is readily distinguished by the unique golden-brown to coppery coloured indumentum on the underside of the leaf. In preliminary molecular investigations, *L. fulva* is sister to all other Australian *Livistona* species (Fig. 8.4).

9 Livistona humilis

Livistona humilis R.Br., *Prodr.*: 268 (1810); *Saribus humilis* (R.Br.) Kuntze, *Revis. Gen. Pl.* 2: 736 (1891). Type: Australia. Northern Territory. Gulf of Carpentaria, Blue Mud Bay, (Island S) Morgans Is., 20 Jan.1803, *R.Brown s.n.* (Bennett 5796); holotype BM [2 sheets]; isotypes FI, K.

Livistona leichhardtii F.Muell., *Fragm.* 8: 221 (1874); *Fragm.* 5: 49 (1865) *nomen*. Type: Australia. Northern Territory. McAdam Range, 1855, *F.Mueller s.n.*; lectotype MEL, *fide* Rodd (1998).

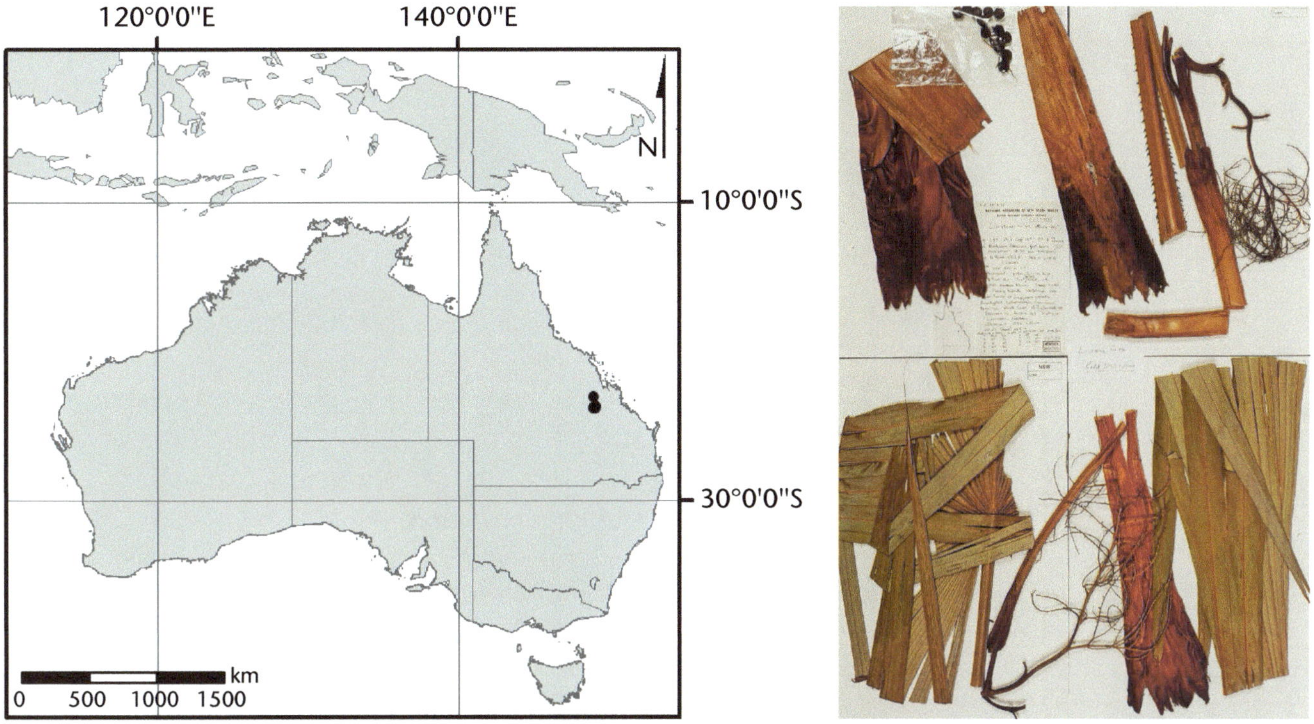

Figure 8.28 *Livistona fulva*. **Left:** Distribution based on herbarium specimen data. **Right:** Type specimen. Australia. Queensland. Blackdown Tableland, 23°43′S, 149°07′E, 450 m alt., 11 May 1976, *A.N.Rodd 3062 with S.Jacobs*; holotype NSW [4 of 16 sheets].

[*Livistona humilis* var. *minutiflora* Becc., *Ann. Roy. Bot. Gard. (Calcutta)* 13: pl. 7, Fig. 3a, (1931) *nom. inval.*].

Dioecious palm. **Stem** to 7 m tall, 5–8 cm dbh; leaf-scars raised; internodes narrow, grey; petiole remnants persistent or deciduous with age or fire. **Leaves** 8–15 in a ± globose crown; petiole 40–70 cm long, 6–14 mm wide, margins with small, single, curved dark red spines; leafbase fibres not prominent, coarse, persistent; lamina regularly segmented, circular, 30–50 cm long, chartaceous, dark green on the adaxial surface, lighter green on the abaxial surface, glossy. **Segments** 30–44, rigid, free for 60–87% of their length, depth of apical cleft 35–89% of the length of the free portion, apical lobes acuminate; parallel veinlets 6 each side of the midrib; transverse veinlets thinner than parallel veinlets. **Inflorescences** sexually dimorphic; those on fruit-bearing (female) plants erect, to 230 cm long, with a single distal partial inflorescence (infrequently with 1–2 lower partial inflorescences) branched to 3 orders; peduncular bracts 5–8, scurfy pubescent; inflorescences on non-fruit-bearing (male) plants arcuate, to 180 cm long, peduncular bract lacking; partial inflorescences 4–7, branched to 3 orders; rachis bracts scurfy pubescent; rachillae 3–12 mm long, pubescent. **Flowers** in clusters of 2–4, globose, 1.5–1.8 mm long in bud; sepals broadly ovate, 1.0–1.3 mm long, membranous, cuspidate, yellow; petals broadly ovate, 1.5–1.8 mm long, fleshy, acute, yellow; stamens c. 1.3 mm long. **Fruit** ellipsoid, pyriform or obovoid, 11–19 mm long, 8–10 mm diam.; epicarp c. 0.5 mm thick, shiny, purple-black with scattered lenticellular pores; suture line extends for full length of the fruit, marked with lip-like structures; mesocarp fleshy, c. 2 mm thick; endocarp thin, brittle, c. 0.5 mm thick; pedicel 0.5–1 mm long. **Seed** ellipsoid, 7–9 mm long, 4–6 mm wide; eophyll 3-ribbed. Figures 8.29, 8.30, 8.31.

Sand palm

Distribution and ecology: Endemic to the Northern Territory, Australia, from near Fitzmaurice River across the Top End to Cape Arnhem and islands in north-west Gulf of Carpentaria and as far inland as Katherine, mostly at low altitude, occurring in large populations or small groups in open forest and woodland, most common on deep sandy soils but also on various soil types and rocky areas, 0–240 m asl

Figure 8.29 *Livistona humilis* in open woodland habitat, Humpty Doo, Northern Territory.

(Fig. 8.32). Flowering May–Dec.; fruiting Oct.–May. Conservation status – *No present threats*.

Typification: The holotype of *L. humilis* is an unnumbered collection by Robert Brown [BM – Bennett 5796] collected in Jan. 1803 from Morgans Island in Blue Mud Bay, Gulf of Carpentaria. The collection consists of two sheets of two complete leaves, inflorescence portions and fruit (Fig. 8.32).

Etymology: *'caudice 4–6–pedali'* (Brown 1810, p. 268) [Stem 4–6 feet]: With reference to the small stature of the species, from the Latin *humilis*, low.

Notes: *Livistona humilis* was one of two species described by Brown (1810) when establishing *Livistona*. Beccari (1931) provided a thorough appraisal of the species, describing two varieties, *L. humilis* var. *sclerophylla* from north-east Queensland and *L. humilis* var. *novoguineensis* from Merauke, Indonesia, both of which are attributable to *L. muelleri* (see Notes under that species).

Mueller (1874b) described *L. leichhardtii* based on the collection *Mueller s.n.* [MEL], from Arnhem Land, Northern Territory. Beccari (1931) provided an illustration labelled as *L. humilis* var. *minutiflora*, a *nomen invalidum*, but did not provide a description or other taxonomic reference.

The closest relationship of *L. humilis* appears to be with *L. inermis*. *Livistona humilis* is distinguished by its smaller stature, to about 7 m tall, smaller leaves to 50 cm long and inflorescences that are strongly sexually dimorphic, with the female plants having a single partial inflorescence at the apex and the male having several partial inflorescences along the length of the primary axis. Preliminary molecular investigations place *L. humilis* as sister to *L. inermis*.

10 Livistona inermis

Livistona inermis R.Br., *Prodr.*: 268 (1810); *Saribus inermis* (R.Br.) Kuntze, *Revis. Gen. Pl.* 2: 736 (1891). Type: Australia. Northern Territory. Gulf of Carpentaria, Sir Edward Pellew Group, [Island h] North Is., 16? Dec. 1802, *R.Brown s.n.* (Bennett 5795); holotype BM [2 sheets]; isotypes FI, K.

Functionally dioecious palm. **Stem** to 10 m tall, 6–10 cm dbh; leafscars prominently raised; internodes narrow, grey; petiole remnants persistent in the basal portion. **Leaves** 10–30 in a ± globose crown; petiole arching, 60–90 cm long, 6–10 mm wide, flat or shallowly concave on the adaxial surface, margins with small single curved reddish to black spines to 5 mm long, usually smaller, restricted to the extreme proximal portion, distally smooth or with widely spaced small reddish to black prickles to c. 1 mm high; leafbase fibres moderately prominent, coarse, persistent; lamina regularly segmented, circular to subcircular, 30–70 cm long, coriaceous, light green to green-grey on the adaxial surface, lighter green-grey glossy to lightly pruinose on the abaxial surface. **Segments** 24–48, rigid, segments free for 80–97% of their length, depth of apical cleft 70–84% of the length of the free portion; apices pendulous, lobes acuminate to filiform; parallel veinlets 8–10 each side of the midrib; transverse veinlets thinner than parallel veinlets. **Inflorescences** not sexually dimorphic, 40–90 cm long; partial inflorescences about 3, branched to 3 orders; the most basal partial inflorescence about as long as the remainder of the inflorescence; prophyll 12–28 cm long, 2–3 cm wide, papyraceous, glabrous; peduncular bract lacking; rachis bracts loosely tubular, papyraceous, glabrous, slightly pubescent toward the apex; rudimentary bracts subtend the distal partial inflorescences; rachillae 1–9 cm long, glabrous. **Flowers** solitary or in clusters of 2–3, funnel-shaped, 1.5–2.3 mm

Figure 8.30 *Livistona humilis*. **Top left:** Petiole with characteristic spines. **Top right:** Leafbase fibres. **Bottom left:** Inflorescence of staminate plant with mature flowers. **Bottom right:** Staminate flowers at anthesis.

long in bud; sepals narrowly triangular, 0.7–1.4 mm long, membranous, acute, white to cream or yellow; petals triangular to broadly ovate, 1.5–1.9 mm long, thick, acute to apiculate, white to cream or yellow; stamens 1.5–1.6 mm long; anthers bright yellow. **Fruit** obovoid to pyriform, 10–13 mm long, 6–7 mm diam.; epicarp smooth, glossy black; suture line extends the length of the fruit, marked with lip-like structures; mesocarp fleshy; endocarp thin. **Seed** ellipsoid, 8–9 mm long; eophyll 3-ribbed. Figures 8.33, 8.34.

Figure 8.31 *Livistona humilis*, pistillate plant. **Top left:** Inflorescence with mature flowers. **Top right:** Flowers at anthesis. **Bottom left:** Infructescence with mature fruit. **Bottom right:** Mature fruit.

Small fan palm

Distribution and ecology: Endemic to northern Australia, in the northern part of the Northern Territory from the headwaters of the Daly River across the Top End inland to Katherine, and in north-west Queensland, including islands in the Gulf of Carpentaria, occurring in small populations and isolated groups or scattered individuals in open forest and woodland and in sandstone gorges and outcrops, 20–285 m asl (Fig. 8.35). Flowering Aug.–Mar.; fruiting Jun.–Oct. Conservation status – *No present threats*.

Typification: The holotype of *L. inermis* is an unnumbered Robert Brown collection [BM – Bennett 5795] collected in Dec. 1802 from North Island in the southern Gulf

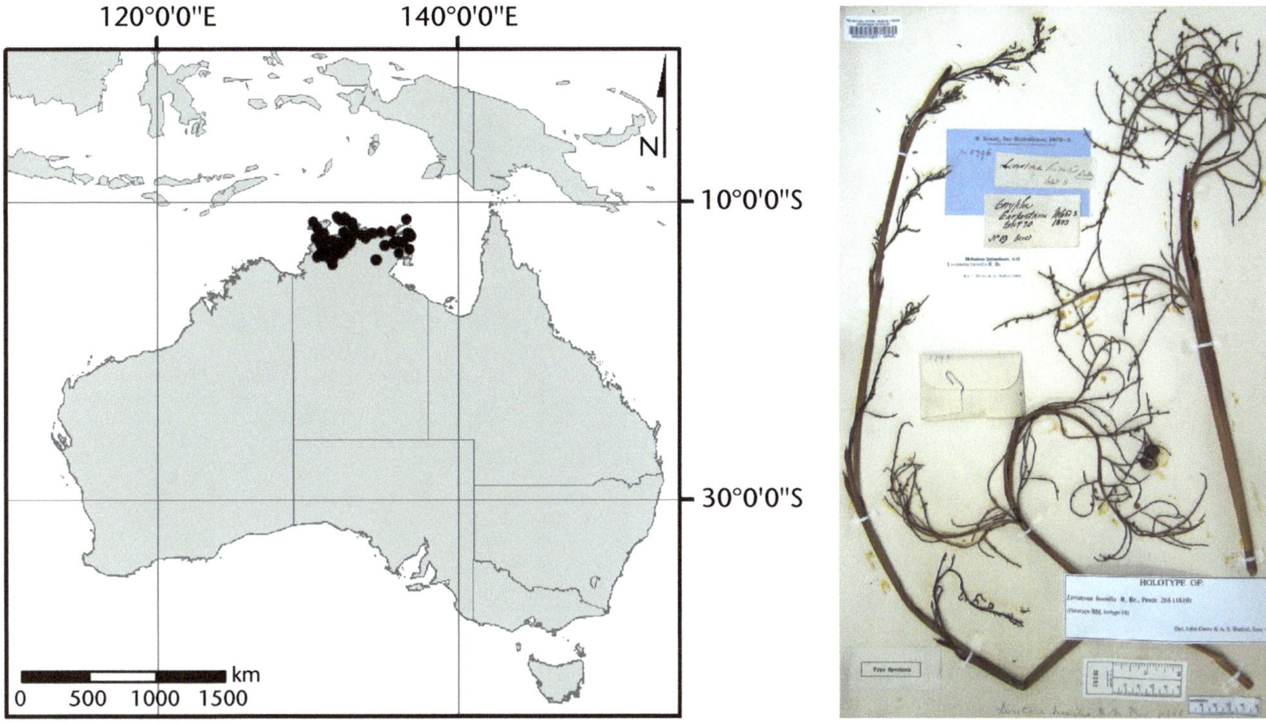

Figure 8.32 *Livistona humilis*. **Left:** Distribution based on herbarium specimen data. **Right:** Type specimen. Australia. Northern Territory. Gulf of Carpentaria, Blue Mud Bay, Morgans Island, 20 Jan. 1803, *R.Brown s.n.* (Bennett 5796); holotype BM [1 of 2 sheets].

Figure 8.33 *Livistona inermis* in open habitat on escarpment summit, Katherine Gorge, Nitmiluk National Park, Northern Territory.

of Carpentaria. The collection consists of two sheets with a complete leaf and inflorescence with flowers (Fig. 8.35).

Etymology: '*stipitibus inermibus*' (Brown 1810, p. 268) [leaf stalk unarmed]: With reference to the unarmed petioles, from the Latin *inermis*, unarmed.

Notes: *Livistona inermis* was one of two species described by Brown (1810) in establishing *Livistona*, and was named for the unarmed petioles. The other species was *L. humilis*. Both species were described with only eight words, and their identity caused some confusion for early taxonomists. For example, Wendland and Drude (1875) established their expanded description of *L. inermis* on specimens of what is now known to be *L. decora* while Bentham (1878) suggested 'it [*L. inermis*] may prove to be a variety only of *L. humilis*'. Drude (1893) placed it as a possible synonym of *L. leichhardtii* (= *L. humilis*). Beccari (1931) applied considerable discussion to the identity of *L. inermis* and provided the first account succinctly characterising the species. Preliminary molecular investigations place *L. inermis* as sister to *L. humilis*.

11 Livistona lanuginosa

Livistona lanuginosa Rodd, *Telopea* 8: 82, figs 3d, 7a–d, 9a (1998). Type: Australia. Queensland. 75 km S of

Figure 8.34 *Livistona inermis*. **Top left:** Petioles with characteristic armature. **Top right:** Crown with emerging inflorescence. **Bottom left:** Flowers at anthesis. **Bottom right:** Full-sized but immature fruit.

Ravenswood, Glenroy Creek, 20°35′S, 147°10′E, 25 Aug. 1978, *A.K.Irvine 1912*; holotype NSW [12 sheets]; isotypes BH, BRI, CNS, MEL.

[*Livistona mariae auct. non* F.Muell., F.M.Bailey, *Queensl. fl.* 5: 1684 (1902)].

Functionally dioecious palm. **Stem** to 18 m tall, 25–35 cm dbh; leafscars raised, compressed; internodes 2–10 cm wide, grey-brown; base with persistent petiole remnants. **Leaves** 35–45 in a globose crown; petiole 150–200 cm long, 30–35 mm wide, flat on the adaxial surface, margins with small single curved black spines confined to the proximal portion; leafbase fibres moderately prominent, coarse, persistent; lamina regularly segmented, circular, 130–190 cm long, coriaceous, pale

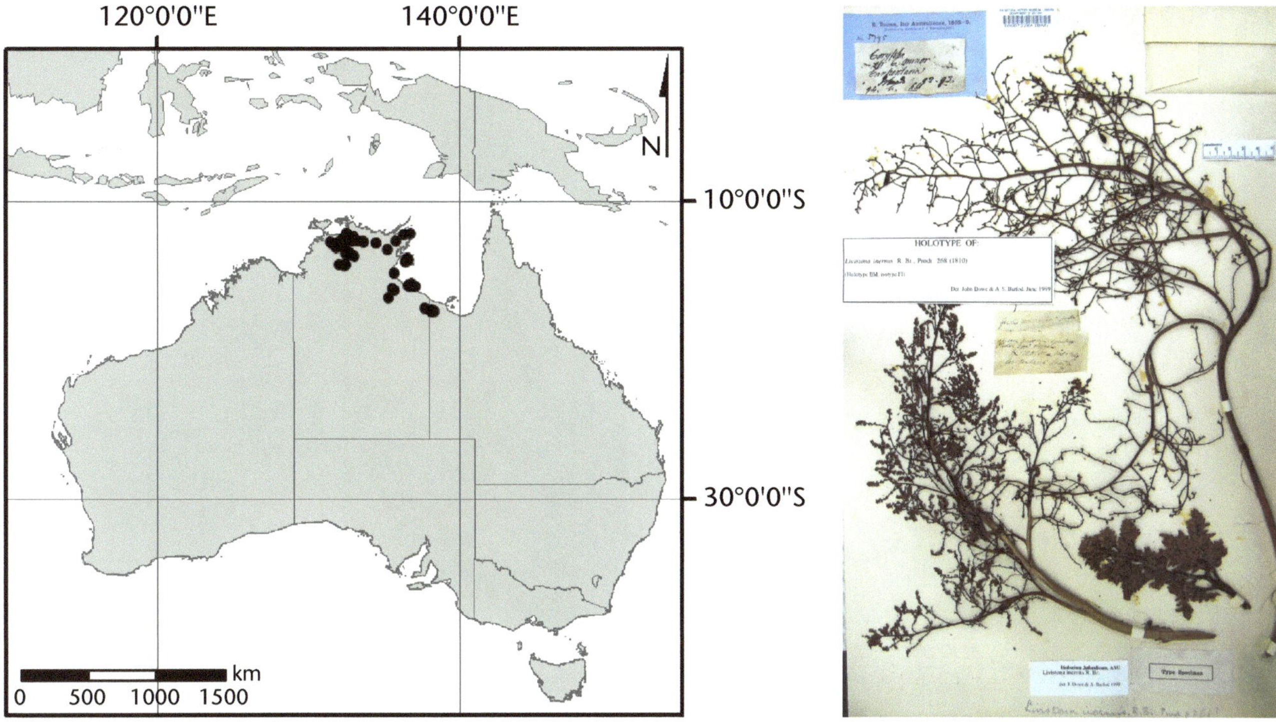

Figure 8.35 *Livistona inermis*. **Left:** Distribution based on herbarium specimen data. **Right:** Type specimen. Australia. Northern Territory. Gulf of Carpentaria, Sir Edward Pellew Group, North Island, Dec. 1802, *R.Brown s.n.* (Bennett 5795); holotype BM [1 of 2 sheets].

grey-green on the adaxial surface, lighter grey-green, waxy and pruinose on the abaxial surface. **Segments** 70–92, free for about 34% of their length, depth of the apical cleft about 24% of the length of the free portion; apical lobes semi-pendulous, acuminate-attenuate; parallel veinlets 8 each side of the midrib; transverse veinlets thinner than parallel veinlets. **Inflorescences** 140–220 cm long; partial inflorescences 8–12, branched to 4 orders; prophyll densely tomentose; peduncular bracts 1–2, densely tomentose-lanuginose; rachis bracts loosely sheathing, not disintegrating, densely lanuginose; rachillae 3–12 cm long, glabrous. **Flowers** solitary or in pairs, funnel-shaped, 2.8–3 mm long in bud; sepals narrowly triangular, c. 1.5 mm long, fleshy, acuminate to aristate, cream to yellow; petals triangular, 2.8–3 mm long, thick, acute, cream to yellow; stamens c. 2 mm long. **Fruit** globose, 25–36 mm diam.; epicarp purple-brown to black with scattered white flecks with large lenticellular pores; suture line extends for about half the length of the fruit, marked with lip-like structures; mesocarp fibrous, dry; endocarp 2–3 mm thick; pedicel to 1 mm long. **Seed** globose, 18–22 mm wide; eophyll 5-ribbed. Figures 8.36, 8.37.

Figure 8.36 *Livistona lanuginosa* in tall woodland habitat, Deep Creek, Dandenong Park, Burdekin River catchment, north-east Queensland.

Figure 8.37 *Livistona lanuginosa*. **Top left:** Prophyll and peduncular bracts with characteristic dense lanuginose tomentum. **Top right:** Flowers at anthesis. **Bottom left:** Infructescence with mature fruit. **Bottom right:** Mature fruit.

Cape River fan palm, Burdekin Livistona, Burdekin fan palm, woolly palm

Distribution and ecology: Endemic to north-east Queensland in the catchment of the Burdekin River, occurring in open to semi-closed riparian forest on seasonally flowing streams and less frequently on adjacent floodplains, in large populations, small groups or isolated individuals, as a canopy emergent on sandy alluvial soils, 150–250 m asl (Fig. 8.38). Pettit and Dowe (2003) provided a population study and estimated the total population as comprising fewer than 1000 mature individuals. Accordingly, the species has been designated a Vulnerable status. Flowering Mar.–Nov.; fruiting Sept.–Jan. Conservation status – *Vulnerable* (Bostock and Holland 2007).

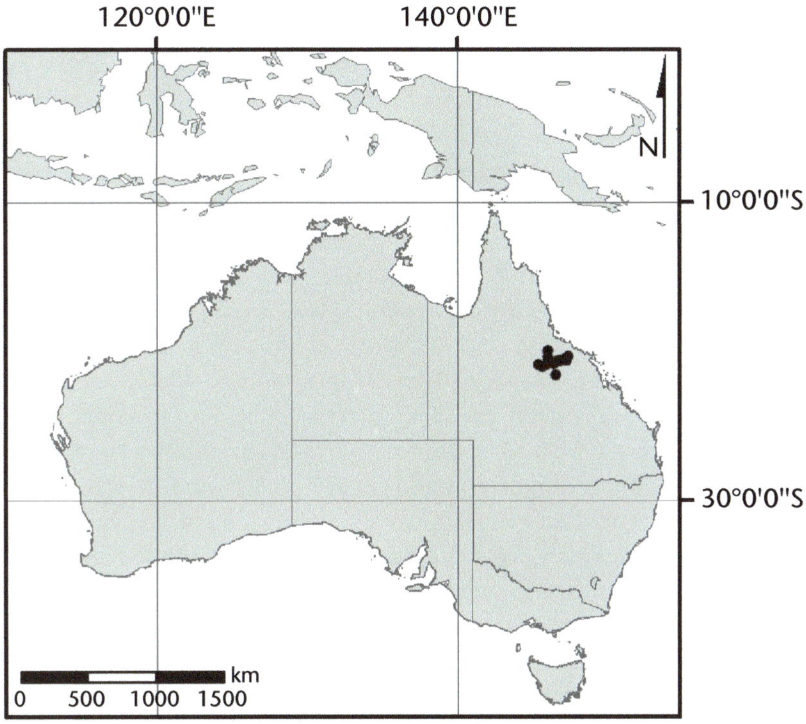

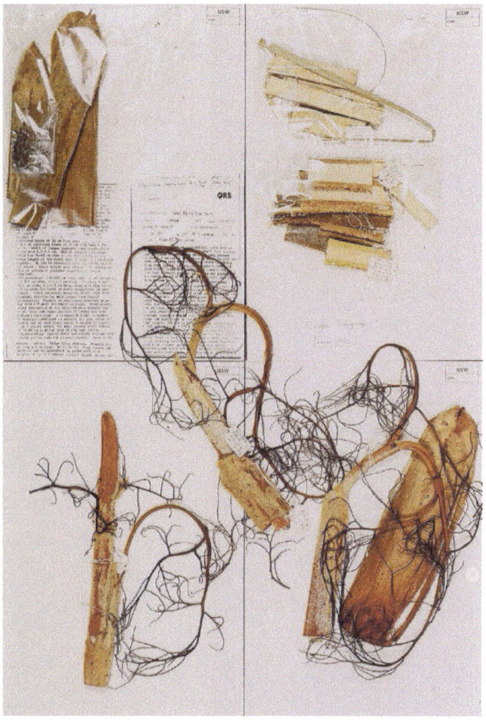

Figure 8.38 *Livistona lanuginosa*. **Left:** Distribution based on herbarium specimen data. **Right:** Type specimen. Australia. Queensland. 75 km south of Ravenswood, Glenroy Creek, 20°35'S, 147°10'E, 25 Aug. 1978, *A.K.Irvine 1912*; holotype NSW [4 of 12 sheets].

Typification: The holotype of *Livistona lanuginosa* is *Rodd 1912* [NSW] collected in Aug. 1978 from Glenroy Creek, a tributary of the Burdekin River, Queensland. The collection consists of two leaves cut into sections, a petiole base and entire petiole cut into small portions, a number of partial inflorescences and a prophyll and peduncular bracts (Fig. 8.38).

Etymology: '*L. lanuginosa* is a very distinct species, instantly recognisable by the abundant woolly scales on the petioles and rachis bracts' (Rodd 1998, p. 85): From the Latin *lanugo*, down or woolliness.

Notes: *Livistona lanuginosa* is most similar to other inland species, and in preliminary molecular investigations is sister to *L. rigida* and *L. mariae* (Fig. 8.4). It differs from these species by the densely tomentose inflorescence bracts and the large fruit, to 36 mm diam.

12 Livistona lorophylla

Livistona lorophylla Becc., *Webbia* 5: 18 (1921) [as *L. loriphylla*]. Type: Australia. Western Australia. Cambridge Gulf, 1885, *Johnston s.n.*; holotype FI; isotypes NSW, MEL [3 sheets].

Livistona kimberleyana Rodd, *Telopea* 8: 121, figs 4g, 17d (1998); *Livistona* sp. 'A', Wilson, *Fl. Kimberley Region*: 1250 (1992). Type: Australia. Western Australia. Kimberley Region, Durack Ra., Mt King, 24 Oct. 1974, *A.N.Rodd 2866*; holotype NSW; isotypes CANB, K, PERTH.

[*Livistona alfredii auct. non* F.Muell., W.V.Fitzgerald, *J. & Proc. Roy. Soc. W. Australia* 3: 24 (1918); C.A.Gardner, *Forest Dept. Bull. W. Australia* 32: 36 (1923)].

Functionally dioecious palm. **Stem** to 15 m tall, 8–20 cm dbh; leafscars raised; internodes narrow, grey; base with persistent petiole remnants. **Leaves** 25–40 in a ± globose crown; petiole arching, 70–200 cm long, 10–17 mm wide, flat on the adaxial surface, margins with small to moderate scattered single curved black spines; leafbase fibres moderately prominent, coarse, persistent; lamina regularly segmented, circular, 60–100 cm long, coriaceous, glossy mid green to pruinose grey on both surfaces. **Segments** 34–50, rigid, segments free for 85–98% of their length, depth of apical cleft 55–78% of the length of the free portion; apical lobes semi-pendulous, acuminate to filiform; parallel veinlets 8–10 each side of the midrib; transverse veinlets thinner than parallel veinlets.

Inflorescences 20–160 cm long; partial inflorescences 4–8, branched to 3 orders; prophyll 20–22 cm long, 3–4 cm wide, glabrous, apices papery lacerate; peduncular bract 1 or lacking; rachis bracts sparsely tomentose; terminal partial inflorescence bifurcating at the base, ebracteate; rachillae 10–35 mm long, glabrous. **Flowers** solitary or in clusters of 2–4, broadly funnel-shaped, 1.2–3 mm long, to c. 1 mm wide in bud; sepals triangular, 0.7–1.8 mm long, membranous, acute, white-cream; petals triangular to broadly ovate, 1.2–3 mm long, acute to cuspidate, apex curved inward, white-cream; stamens c. 1.6 mm long; anthers c. 0.5 mm long, subglobose to ovoid, cream. **Fruit** obovoid-pyriform, 8–14 mm long, 6–9 mm diam.; epicarp dull black pruinose with scattered lenticellular pores; mesocarp thin, fibrous; endocarp thin, crustaceous. **Seed** ellipsoid, 7–9 mm long, 4–7 mm wide; eophyll long and narrow, 3-ribbed. Figures 8.39, 8.40.

Figure 8.39 *Livistona lorophylla* in open woodland habitat, Explosion Gorge, El Questro, East Kimberley, Western Australia.

Fine-leafed Kimberley fan palm

Distribution and ecology: Western Australia and Northern Territory. In the Kimberley Region, from Sir Graham Moore Island to Sale River and Cambridge Gulf, and inland to the King Leopold and Durack Ranges, and in Northern Territory in the Victoria River District, just to the east of the WA/NT border, occurring as an emergent in open forest and woodlands, and on sandstone outcrops and gorges, in large populations, small groups or scattered individuals, 10–290 m asl (Fig. 8.41). Flowering Aug.–Dec.; fruiting Dec.–Jul. Conservation status – *No present threats*.

Typification: The holotype of *L. lorophylla* is an unnumbered collection by Johnson (Johnston on label, Johnson in protologue citation) [FI] collected in 1885 from near Cambridge Gulf, Western Australia. The isotype collection [MEL] consists of two complete leaves and an inflorescence with flowers (Fig. 8.41).

Etymology: '*Livistona loriphylla*': There was no explanation of the derivation of the name in the protologue, but with probable reference to the leaf segments resembling straps, as in the protologue they are described as thin, from the Latin *lorum*, strap, and *phylla*, leaf.

Notes: *Livistona kimberleyana* was described by Rodd (1998) based on *Rodd 2866* [NSW] collected from Mt King, Western Australia. This taxon was distinguished from *L. lorophylla* primarily by grey pruinose rather than glossy green leaves. In some populations, individuals exhibit intermediate leaf colour. Other than leaf colour, there are no characters to distinguish the taxa and *L. kimberleyana* is therefore placed as a synonym of *L. lorophylla*.

Livistona lorophylla morphologically resembles *L. inermis* but can be distinguished by its overall larger size, including greater height, stem diameter and leaf dimensions. The inflorescence of *L. lorophylla* has up to eight partial inflorescences but there are only three in *L. inermis*; it possesses a single peduncular bract, which is lacking in *L. inermis*. The fruit are semi-glossy in *L. lorophylla*, whereas they are glossy in *L. inermis*. Preliminary molecular investigations place *L. lorophylla* as sister to *L. victoriae* and *L. alfredii* (Fig. 8.4).

13 Livistona mariae

Livistona mariae F.Muell., *Fragm.* 11: 54 (1878); Mueller, *Fragm.* 8: 283 (1874) *nom. provis.*; Giles, *Geo. trav. central Austral.*: 222 (1875) *nom. ined.*; *Saribus mariae* (F.Muell.) Kuntze, *Revis. Gen. Pl.* 2: 736 (1891); *Livistona mariae* F.Muell. subsp. *mariae* Rodd, *Telopea* 8: 80, figs 3a, 5a (1998). Type: Australia. Northern Territory. Macdonnell Ranges, 'Glen of Palms', *E.Giles s.n.*; holotype MEL [3 sheets].

Figure 8.40 *Livistona lorophylla*. **Top left:** Leaf, showing depth of segment divisions. **Top right:** Petiole bases and leafbase fibres. **Bottom left:** Flowers at anthesis. **Bottom right:** Immature fruit.

Functionally dioecious palm. **Stem** to 30 m tall, 30–40 cm dbh; leafscars stepped, raised; internodes broad, grey; petiole remnants persistent in the basal 1 m or so. **Leaves** 30–50 in a globose crown; petiole erect to arching, 150–250 cm long, 20–45 mm wide, ridged on the adaxial surface, margins with small single curved black spines in the proximal portion, smooth distally; leafbase fibres moderately prominent, coarse, persistent; lamina regularly segmented, ± circular, folded into a broad 'v', 100–220 cm diam., rigidly coriaceous, grey-green and glossy on the adaxial surface, lighter grey-green and waxy pruinose on the abaxial surface. **Segments** 50–86, distally pendulous, proximal margins of outer segments with small spines; segments

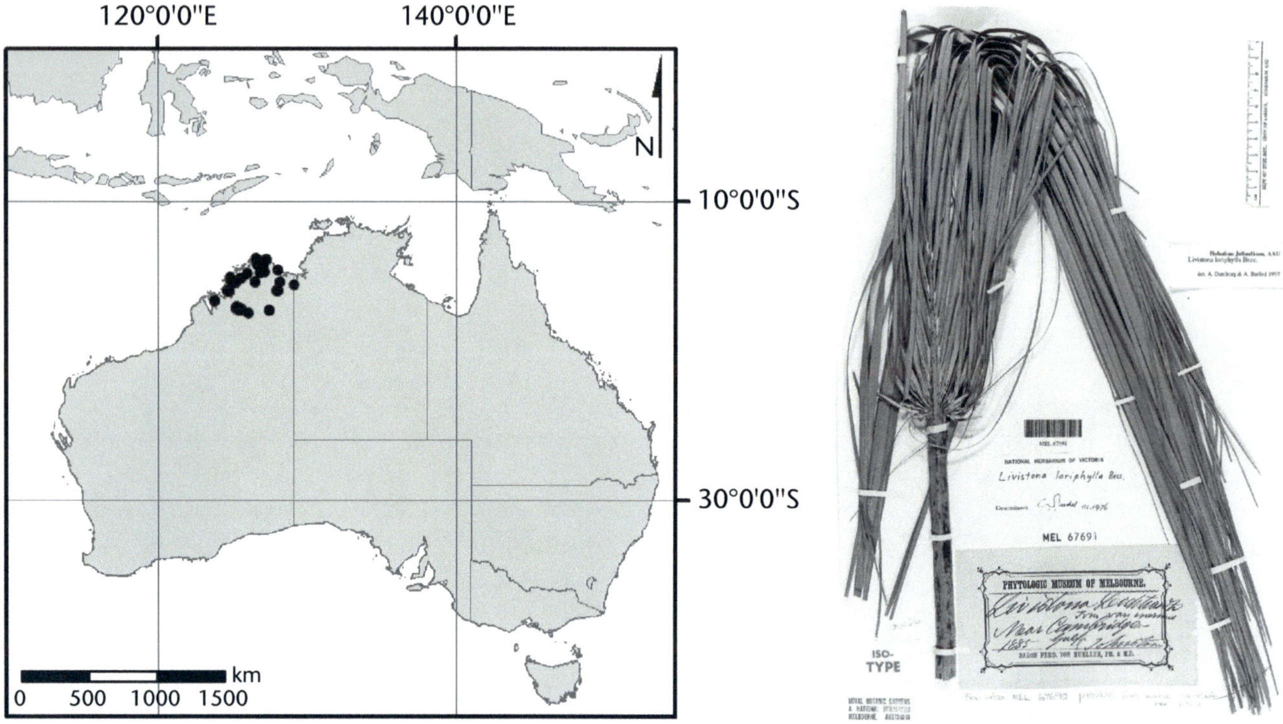

Figure 8.41 *Livistona lorophylla*. **Left:** Distribution based on herbarium specimen data. **Right:** Type specimen. Australia. Western Australia, Cambridge Gulf, 1885, *Johnston s.n.*; isotype MEL [1 of 3 sheets].

free for 45–55% of their length, depth of apical cleft 45–65% of the length of the free portion, apical lobes attenuate; distal segments usually overlap; parallel veinlets 7–9 each side of the midrib; transverse veinlets thinner than parallel veinlets. **Inflorescences** 125–250 cm long; partial inflorescences 10–14, branched to 4 orders; peduncular bracts 1, loosely sheathing with dense white-grey scales as are the rachis bracts; rachillae 3–8 cm long, glabrous. **Flowers** in clusters of 3–6, campanulate, 1.0–1.8 mm long in bud; sepals ovate, 0.8–1 mm long, membranous, bluntly acute, greenish cream to yellow; petals broadly oblong, 1.2–1.5 mm long, acute, greenish cream to yellow; stamens c. 1.2 mm long. **Fruit** globose, 12–18 mm diam.; epicarp semi-glossy black, with scattered lenticellular pores; suture line extends for about 75% the length of the fruit, marked with lip-like structures; mesocarp fibrous; endocarp thin, crustaceous. **Seed** globose, 8–12 mm wide; eophyll 3-ribbed. Figures 8.42, 8.43.

Central Australian fan palm, Central Australian cabbage palm, Palm Valley Livistona

Distribution and ecology: Endemic to the Finke River catchment in the Macdonnell Ranges, Northern Territory, where it grows along creek lines and watercourses

Figure 8.42 *Livistona mariae* in riparian open forest, Palm Valley, Finke River, Northern Territory. Photo by David Tanswell.

Figure 8.43 *Livistona mariae*. **Top:** Palm Valley, Finke River, Northern Territory. Photo by David Tanswell. **Bottom left:** Flowers at anthesis. **Bottom right:** Mature and maturing fruit.

with a permanent shallow water supply, and in sandstone gorges, 450–500 m asl (Fig. 8.44). The total population includes about 750 mature individuals, primarily along Finke River, Palm Creek, Kunara Creek and Little Palm Creek (Latz 1975). Wischusen *et al.* (2004) concluded that Palm Valley was a Pleistocene flora refuge, based on the slow but consistent rates of groundwater discharge into the valley. Flowering Jul.–Dec.; fruiting Nov.–Feb. Conservation status – *Vulnerable* (NRETAS 2009).

Typification: The holotype of *Livistona mariae* is an unnumbered and undated collection by E. Giles [MEL] from the Glen of Palms, Macdonnell Ranges, Northern Territory. The collection consists of two incomplete leaves and portions of cortex. Rodd (1998) previously designated this collection as the lectotype (Fig. 8.44).

Etymology: *'Palmam inter principes plantarum in geographia ordinis memorabilem summa observantia ductus tribui praecelsae Grandi-principi Mariae, Ducissae Edinensi, cultus hortorum nobilioris tam in Russia quam nunc in Britannia patronae imperiali'* [Palms are among the princes of the plant-world and there is no better memorial to honour the Grand Duchess Marie, Duchess

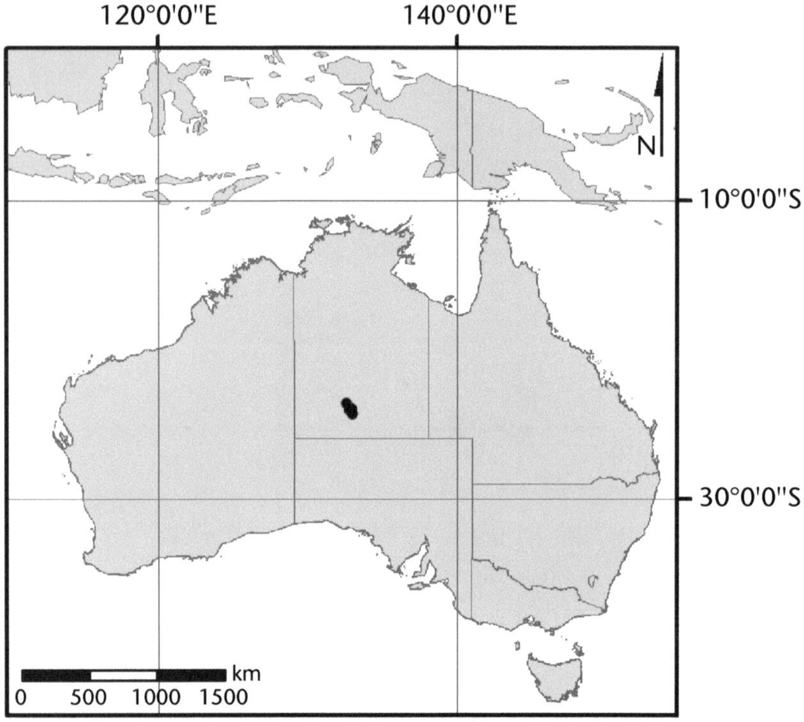

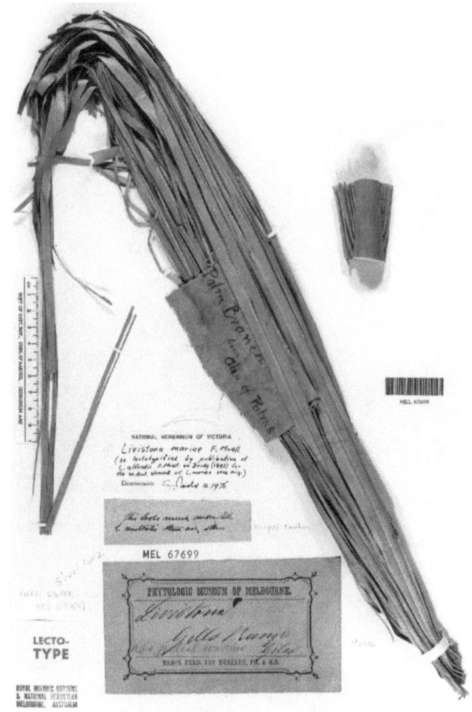

Figure 8.44 *Livistona mariae*. **Left:** Distribution based on herbarium specimen data. **Right:** Type specimen. Australia. Northern Territory. MacDonnell Ranges, 'Glen of Palms', *E.Giles s.n.*; holotype MEL [1 of 3 sheets].

of Edinburgh, formerly of the exalted Russian court now of the British royal throne]: Named for Grand Duchess Marie Alexandrovna (b.1853, d.1920), daughter of Tsar Alexander II of Russia, married Prince Albert, son of Queen Victoria, in 1874, to become Duchess of Edinburgh.

Notes: The name *Livistona mariae* was first used by Mueller (1874b) in an addenda to *L. leichhardtii*, with which he considered it may be conspecific. Previously, in the same volume, Mueller (1874a) noted it as the palm from the 'Glen of Palms' collected by Giles and provided a brief description of the leaf, but with insufficient detail to distinguish it. Chapman (1991) considered the former was the place of valid publication of the name *L. mariae*, but this must be rejected on the grounds that no significant distinguishing description was provided. The eventual complete description was provided by Mueller (1878), thus validating the name.

The close relationship of *L. mariae* to *L. rigida* has been recognised by some authors. Johnson (1981) wrote of the relationship of *L. rigida*: 'very closely related to the Fan-Leaved palm (*L. mariae*) of central Australia and may even be conspecific'. Rodd (1998) resolved this by placing *L. rigida* as a subspecies of *L. mariae*. Dowe (2009a) recognised it as a distinct species.

The leaves of *L. rigida* are comparatively smaller those of than *L. mariae*, the folding into a 'v' along the axis of the costa is considerably more pronounced and the segment apices are most often rigid rather than semi-pendulous. The fruit are comparatively smaller. Preliminary molecular investigations place *L. mariae* as sister to *L. rigida*.

14 Livistona muelleri

Livistona muelleri F.M.Bailey, *Queensl. fl.* 5: 1683 (1902); '*Livistona humilis* R.Br. var.', F.M.Bailey, *Queensland Agric. J.* 2: 130 (1898), *nom. illeg*. Type: Australia. Queensland. Cairns, Oct. 1900, *E.Cowley s.n.*; holotype BRI [2 parcels]; isotype FI.

Livistona humilis var. *sclerophylla* Becc., *Webbia* 5: 20 (1921) [also spelt as *sclirophylla*]. Type: Australia. Queensland. Bloomfield R., *E.Bauer 3*; lectotype FI.

Livistona humilis var. *novoguineensis* Becc., *Webbia* 5: 20 (1921). Type: Indonesia. Papua. Merauke Division, Merauke River, Apr. 1901, *Jaheri s.n.* (*HHBB no. 197*); holotype FI; isotype BO.

Livistona brassii Burret, *Notizbl. Bot. Gart. Berlin-Dahlem* 12: 309 (1935). Type: Papua New Guinea. Western

Province. Dagwa, Oriomo R., Feb. 1934, *L.J.Brass 5950*; holotype A; isotypes BM, BO, BRI, NY.

Livistona crustacea Burret, *J. Arnold Arbor.* 20: 189 (1939). Type: Papua New Guinea. Western Province. Middle Fly R., Lake Daviumbo, Sept. 1936, *L.J.Brass 7668*; holotype B, destroyed; isotypes A, BO, BM, BRI, L; photos BRI.

Functionally dioecious palm. **Stem** to 12 m tall, 15–25 cm dbh; leafscars narrow, raised; internodes narrow, grey; petiole remnants persistent, or deciduous with extreme age or fire. **Leaves** 25–35 in a globose crown, held erect; petiole 70–100 cm long, 14–20 mm wide, concave on the adaxial surface, margins with single curved reddish to black spines 2–12 mm long throughout, largest and closer-spaced in the proximal portion; both the adaxial and abaxial surfaces with rows of persistent corky scales, at first red-brown ageing to grey; leafbase fibres not prominent, fine, disintegrating; lamina regularly segmented, circular, 60–90 cm long, rigid, flat, chartaceous, olive-green to grey-green on the adaxial surface, dull bluish-green and glabrous except for a few scales on the ribs on the abaxial surface. **Segments** 48–60, rigid, free for 50–65% of their length, depth of apical cleft 5–14% of the length of the free portion; apical lobes acute, rigid; parallel veinlets 8 each side of the midrib; transverse veinlets thinner than parallel veinlets. **Inflorescences** not sexually dimorphic, 80–160 cm long; partial inflorescences 5–10, branched to 4 orders; peduncular bract lacking; rachis bracts loosely tubular, with silver scales, splitting and disintegrating with age, but margins not lacerate; rachillae 2–13 cm long, papillose. **Flowers** solitary or in clusters of 2–3, 1.3–1.6 mm long in bud; sepals broadly triangular, 0.8–1 mm long, maroon, fleshy, cuspidate; petals ovate, 1.3–1.6 mm long, subacute, yellow; stamens c. 1.4 mm long, yellow; carpels pink to maroon. **Fruit** ellipsoid, 10–12 mm long, 8.5–10 mm diam.; epicarp smooth, pruinose, powdery blue, reddish-black or bluish-black; suture line extends for full length of the fruit, marked with lip-like structures; mesocarp thin, dry; endocarp thin, brittle; pedicel 0.5–1 mm long. **Seed** globose, 8–9 mm wide; eophyll 3-ribbed. Figures 8.45, 8.46.

Cairns fan palm

Distribution and ecology: Co-occurring in Australia and New Guinea; in Queensland from the Torres Strait islands to near Innisfail and in southern New Guinea, growing in grassy open forest, woodland, moist sclerophyll forest, and less commonly on the margins of vine thickets, 0–300 m alt (Fig. 8.47). Flowering Sept.–Mar.; fruiting Nov.–May. Conservation status – *No present threats*.

Figure 8.45 *Livistona muelleri* in semi-open woodland habitat, Jardine River, Cape York Peninsula, north Queensland.

Typification: The holotype of *Livistona muelleri* is an unnumbered collection by E. Cowley [BRI], dated Oct. 1900 from Cairns, north-east Queensland. The collection consists of a complete leaf, a portion of petiole, a complete inflorescence and a packet of fruit (Fig. 8.47).

Etymology: 'after Baron Mueller' (Bailey 1902): Named for Baron Sir Ferdinand Jakob Heinrich von Mueller (b.1825, d.1896), Victorian Government Botanist 1853–96.

Notes: Beccari (1921) described two varieties of *L. humilis* that are attributable to *L. muelleri*: *L. humilis* var. *sclerophylla*, based on *Bauer 3* [FI] from the Bloomfield River, Queensland, and *L. humilis* var. *novoguineensis*, based on *Jaheri s.n.* [FI], from Merauke, Papua, Indonesia. Burret (1935, 1939) described *L. brassii* and *L. crustacea* from Western Province, Papua New Guinea, based on the collections *Brass 5950* [A] and *Brass 7668* [A] respectively. Both taxa were synonymised under *L. muelleri* by Rodd (1998).

Livistona muelleri bears some resemblance to *L. fulva* of central Queensland, with similarity in both leaf and inflorescence morphology. Preliminary molecular

Figure 8.46 *Livistona muelleri*. **Top left:** Crown with inflorescences. **Top right:** Flowers at anthesis. **Bottom left:** Mature fruit. **Bottom right:** Fruit, whole and in longitudinal section with seed coat intrusion and embryo.

investigations place *L. muelleri* as sister to *L. eastonii* (Fig. 8.4).

15 Livistona nasmophila

Livistona nasmophila Dowe & D.L.Jones, *Austrobaileya* 6: 980 (2004); *Livistona mariae* subsp. *occidentalis* Rodd, *Telopea* 8: 81, figs 3c, 5d (1998). Type: Australia. Western Australia. Durack Range, SE base of Mt King, Dec. 1974, *A.N.Rodd 2868*; holotype NSW [8 sheets]; isotypes BH, CNS, K, PERTH.

Functionally dioecious palm. **Stem** to 30 m tall, 30–60 cm dbh; leafscars raised; internodes broad, grey; petiole remnants persistent in basal 1 m or so. **Leaves** 35–55 in a ± globose crown; petiole 165–230 cm long,

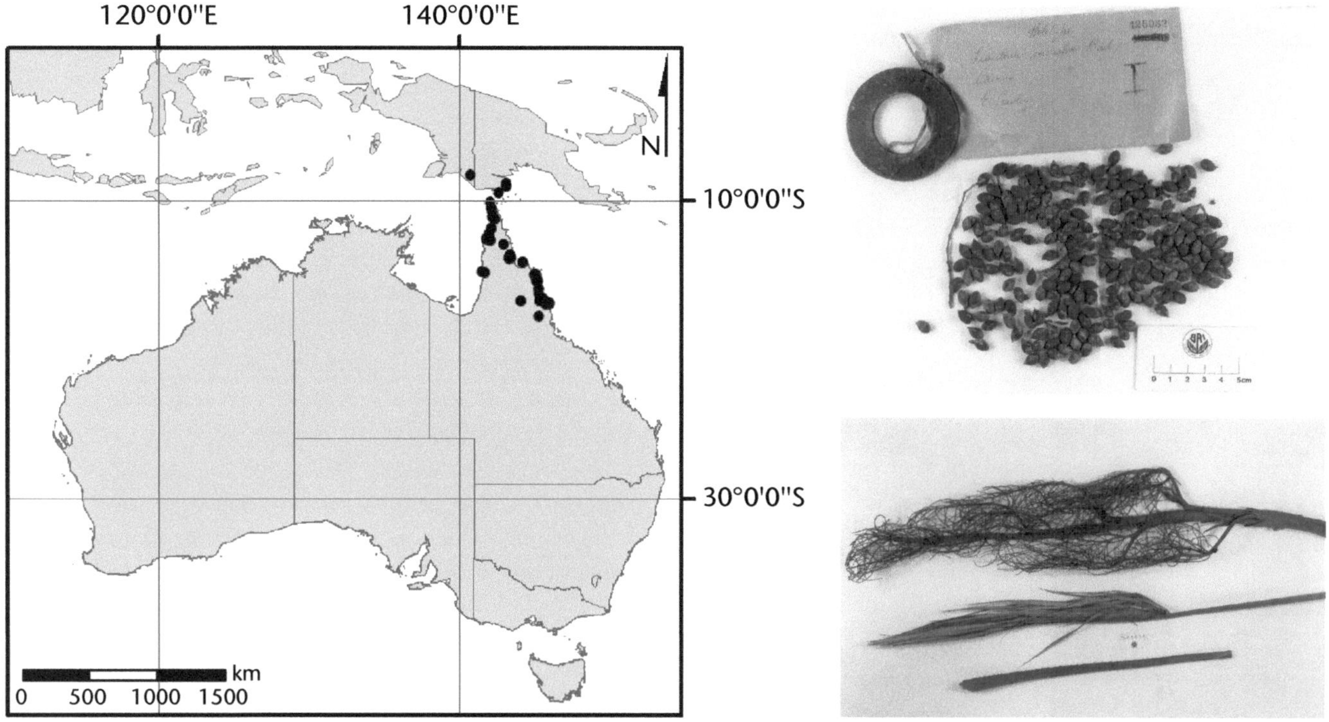

Figure 8.47 *Livistona muelleri*. **Left:** Distribution based on herbarium specimen data. **Right:** Type specimen. Australia. Queensland. Cairns, Oct. 1900, *E.Cowley s.n.*; holotype BRI.

4–10 cm wide, flat on the adaxial surface, margins with single or double retrorse reddish spines confined to the proximal portion, margin distally smooth; leafbase fibres moderately prominent, coarse, persistent; lamina regularly segmented, ± oval in outline, 130–175 cm diam., rigidly coriaceous, glossy grey-green on the adaxial surface, dull grey to glaucous on the abaxial surface. **Segments** 52–58; apices rigid, or becoming semi-pendulous with age or as the result of damage; segments free for about 48% of their length, depth of apical cleft about 48% of the length of the free portion; distal segments usually not overlapping; parallel veinlets 7–9 each side of the midrib; transverse veinlets thinner than parallel veinlets. **Inflorescences** not sexually dimorphic, 260–300 cm long; partial inflorescences 9–11, branched to 5 orders; prophyll 40–55 cm long, 10–15 cm wide, woody, with scattered long scales, apices fibrous-lacerate; peduncular bract 1 or lacking; rachis bracts loosely tubular, with scattered long scales, becoming fully glabrous with age, not disintegrating or becoming marginally lacerate; rachillae 5–9 cm long, glabrous. **Flowers** in clusters of 4–6, globose, angular in bud, 1.4–1.5 mm long; sepals broadly ovate, c. 1.2 mm long, thin, acute, cream to yellow; petals triangular, 1.4–1.5 mm long, obtuse, cream to yellow; stamens c. 0.9 mm long. **Fruit** globose, 11–14 mm diam.; epicarp purple-black with scattered lenticellular pores, semi-glossy; mesocarp fibrous, dryish; endocarp thin, crustaceous; pedicel to 1 mm long. **Seed** globose, 7–11 mm wide; eophyll 3-ribbed. Figures 8.48, 8.49.

Durack Range fan palm

Distribution and ecology: Endemic to Western Australia in the Durack Range and Cambridge Gulf area. Most common in large populations associated with permanent springs, also on intermittent or permanent watercourses, in open forest, 50–150 m asl (Fig. 8.50). Flowering Jul.–Oct.; fruiting Oct.–Jan. Conservation status – *No present threats*.

Typification: The holotype of *Livistona nasmophila* is *Rodd 2868* [NSW], collected Dec. 1974 from the base of Mt King in the Durack Ranges, Western Australia. The collection consists of a complete leaf, a portion of inflorescence and fruit (Fig. 8.50).

Etymology: 'The new name, '*nasmophila*', was chosen to illustrate the palm's habit of occupying permanent watercourses fed by springs through much of its range'

Figure 8.48 *Livistona nasmophila* in spring habitat, Zebedee Springs, El Questro, East Kimberley, Western Australia.

(Dowe and Jones 2004): With reference to the occurrence of this species near springs, from the Greek *nasmo*, spring, and *phila*, loving.

Notes: Rodd (1998) established this taxon as *L. mariae* subsp. *occidentalis*. Following field-work in the Kimberley Region, this taxon was considered distinct enough from *L. mariae* and *L. rigida* to be raised to species rank.

Livistona nasmophila bears close morphological resemblance to both *L. mariae* and *L. rigida*. Petiole armature in *L. nasmophila* is considerably less than in *L. mariae*. In *L. nasmophila*, the inflorescence bracts are glabrous or nearly so rather than densely tomentose, and branching of the partial inflorescences is to five rather than four orders. In *L. nasmophila*, fruit is purple-black rather than black as in *L. mariae*. Preliminary molecular investigation places *L. nasmophila* as sister to *L. eastonii* and *L. muelleri* (Fig. 8.4).

16 Livistona nitida

Livistona nitida Rodd, *Telopea* 8: 96, figs 2a, b, 3h, 8d, 9b, c (1998). Type: Australia. Queensland. Cracow, Dawson River, Delusion Creek, 25°11'S, 150°11'E, 9 May 1976, *A.N.Rodd 3055 with S.Jacobs*; holotype NSW [4 sheets]; isotypes BH, BRI, K.

Functionally dioecious palm. **Stem** to 35 m tall, 25–40 cm dbh; leafscars raised; internodes narrow, grey; petiole remnants persistent in the basal 1 m or so. **Leaves** 35–50 in a globose crown; petiole 170–200 cm long, 20–26 mm wide, ridged on the adaxial surface, margins distally smooth, proximally with single curved dark red spines; leafbase fibres moderately prominent, coarse, disintegrating; lamina regularly segmented, ± circular, 160–190 cm long, coriaceous, dark green and glossy on the adaxial surface, lighter green on the abaxial surface. **Segments** 68–80, distally pendulous; segments free for 63–70% of their length, depth of apical cleft 60–73% of the length of the free portion; parallel veinlets about 7 each side of the midrib; transverse veinlets thinner than parallel veinlets. **Inflorescences** 150–200 cm long; partial inflorescences 8–12, branched to 4 orders; peduncular bract lacking; rachis bracts densely scaly; rachillae 5–20 cm long, glabrous to papillose. **Flowers** solitary or in clusters of 2–5, funnel-shaped, 2.0–3.2 mm long in bud; sepals triangular, c. 1.5 mm long, fleshy, subacute, cream-yellow; petals broadly ovate, 2.0–2.2 mm long, thick, fleshy, acute, cream-yellow; stamens c. 16 mm long. **Fruit** globose, 13–20 mm diam.; epicarp glossy jet-black with scattered lenticellular pores, suture line extends for about half the length of the fruit, marked with lip-like structures; mesocarp fibrous, dry; endocarp thin, brittle; pedicel to 0.5 mm long. **Seed** globose, 10–12 mm wide; eophyll 5-ribbed. Figures 8.51, 8.52.

Carnarvon Gorge cabbage palm, Dawson River fan palm

Distribution and ecology: Endemic to the catchments of the Dawson, Comet and Burnett Rivers and Carnarvon Gorge in central Queensland, occurring along the margins of streams and floodplains, also associated with sandstone cliffs and gorges, 150–650 m asl (Fig. 8.53). Flowering Aug.– Dec.; fruiting Nov.–Mar. Conservation status – *Rare* (Bostock and Holland 2007).

Typification: The holotype of *Livistona nitida* is *Rodd 3055* [NSW], collected May 1976 from Delusion Creek, a tributary of the Dawson River in central Queensland. The collection consists of a portion of leaf, leafbase fibre, a prophyll and a portion of inflorescence (Fig. 8.53).

Etymology: 'meaning "shiny", refers to its most distinctive feature, *viz.* the highly glossy surface of ripe fruits' (Rodd 1998): From the Latin *nitidus*, bright or shining.

Figure 8.49 *Livistona nasmophila*. **Top left:** Crown with inflorescences. **Top right:** Leaf, abaxial surface. **Bottom left:** Flowers at anthesis. **Bottom right:** Immature fruit.

Notes: *Livistona nitida* is distinguished by large leaves with pendulous segments, and jet-black shiny fruits. Preliminary molecular investigations place *L. nitida* as sister to *L. decora* (Fig. 8.4).

17 Livistona rigida

Livistona rigida Becc., *Webbia* 5: 19 (1921); *Livistona mariae* subsp. *rigida* (Becc.) Rodd, *Telopea* 8: 80, figs 3b, 5b, c, 6a–e (1998). Type: Australia. Queensland. Gulf of Carpentaria, Albert River, Aug. 1855, *F.Mueller s.n.*; holotype K [1 sheet]; isotype FI.

Functionally dioecious palm. **Stem** to 28 m tall, 30–40 cm dbh; leafscars raised; internodes narrow, grey; petiole remnants persistent in the basal 1 m or so. **Leaves** 30–50 in a globose crown; petiole 150–250 cm long, 20–35 mm wide, concave on the adaxial surface, margins

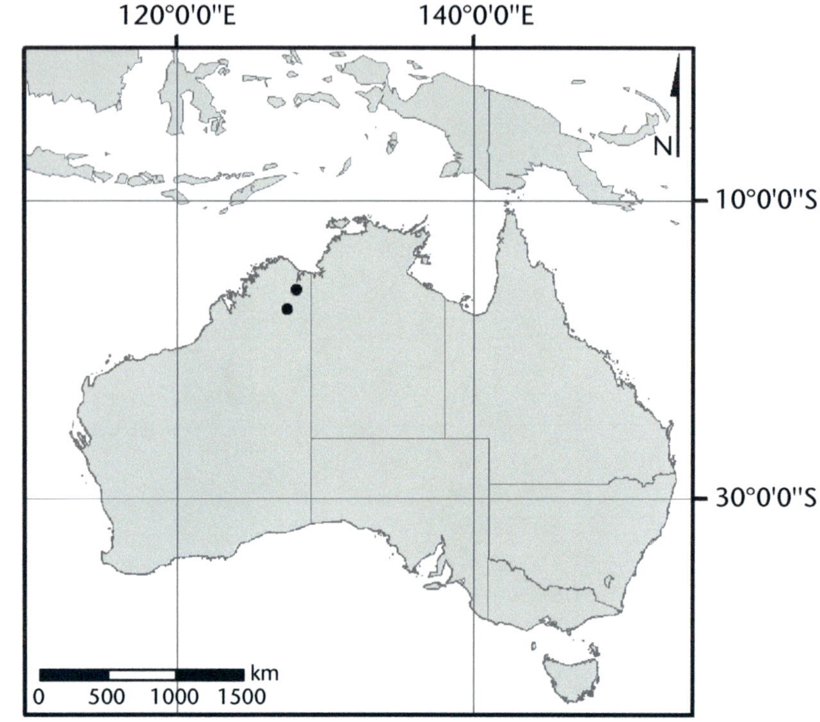

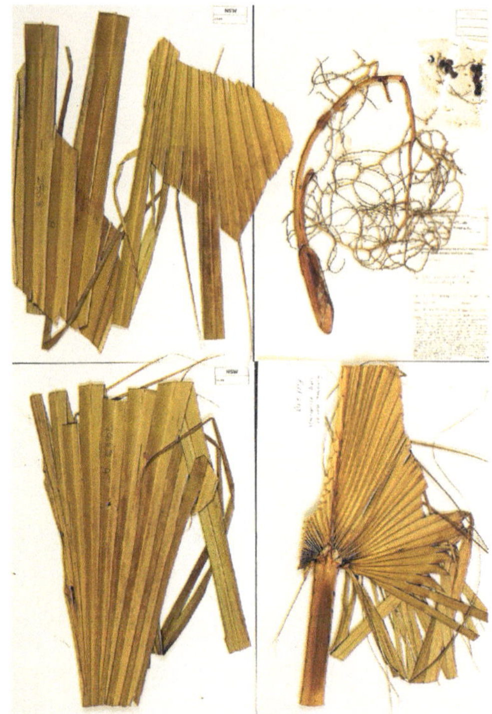

Figure 8.50 *Livistona nasmophila*. **Left:** Distribution based on herbarium specimen data. **Right:** Type specimen. Australia. Western Australia. Durack Range, south-east base of Mt King, Dec. 1974, *A.N.Rodd 2868*; holotype NSW [4 of 8 sheets].

Figure 8.51 *Livistona nitida* in tall semi-open forest, Palm Tree Creek, Dawson River catchment, central Queensland.

with single curved reddish-black spines to 6 mm long in the proximal portion, smooth in the distal portion; leafbase fibres moderately prominent, coarse, persistent; lamina regularly segmented, circular, folded into a narrow 'v', 125–170 cm long, 100–150 cm diam., rigidly coriaceous, grey to glaucous, mid-green and semi-glossy on the adaxial surface, lighter green on the abaxial surface. **Segments** 50–78, distally rigid or semi-pendulous, segments free for 30–35% of their length, depth of apical cleft 30–63% of the length of the free portion, apical lobes acuminate; parallel veinlets 5–7 each side of the midrib; transverse veinlets thinner than parallel veinlets. **Inflorescences** 100–250 cm long; partial inflorescences 9–14, branched to 4 orders; prophyll 25–38 cm long, 8–9 cm wide; peduncular bract 1, densely to moderately scaly; rachis bracts loosely tubular with sparse white scales; rachillae 3–8 cm long, glabrous. **Flowers** in clusters of 3–8, globose, c. 1.8 mm long; sepals broadly ovate, c. 2 mm long, thin, acute, cream to yellow; petals triangular, 1.0–1.4 mm long, obtuse, cream to yellow; stamens c. 1.2 mm long. **Fruit** globose, 12–14 mm diam.; epicarp black, semi-glossy with scattered lenticellular pores; suture line extends for about half the length of the

Figure 8.52 *Livistona nitida*. **Top left:** Leaf. **Top right:** Crown with inflorescences. **Bottom left:** Flowers at anthesis. **Bottom right:** Full-size but immature fruit.

fruit, marked with lip-like structures; mesocarp fibrous; endocarp thin; pedicel c. 2 mm long. **Seed** globose, 9–11 mm wide; eophyll 5-ribbed. Figures 8.54, 8.55.

Mataranka fan palm, Lawn Hill fan palm

Distribution and ecology: Occurs in north-west Queensland on the Gregory, Leichhardt, Nicolson and Albert Rivers, and in the Northern Territory on the Roper and South Alligator Rivers, as a riparian element along stream margins, on seasonally inundated banks, in creek lines and watercourses, sometimes with intermittent flow, but with a permanent shallow water supply, 10–250 m asl (Fig. 8.56). Flowering Jun.– Dec.; fruiting Nov.–May. Conservation status – *No present threats*.

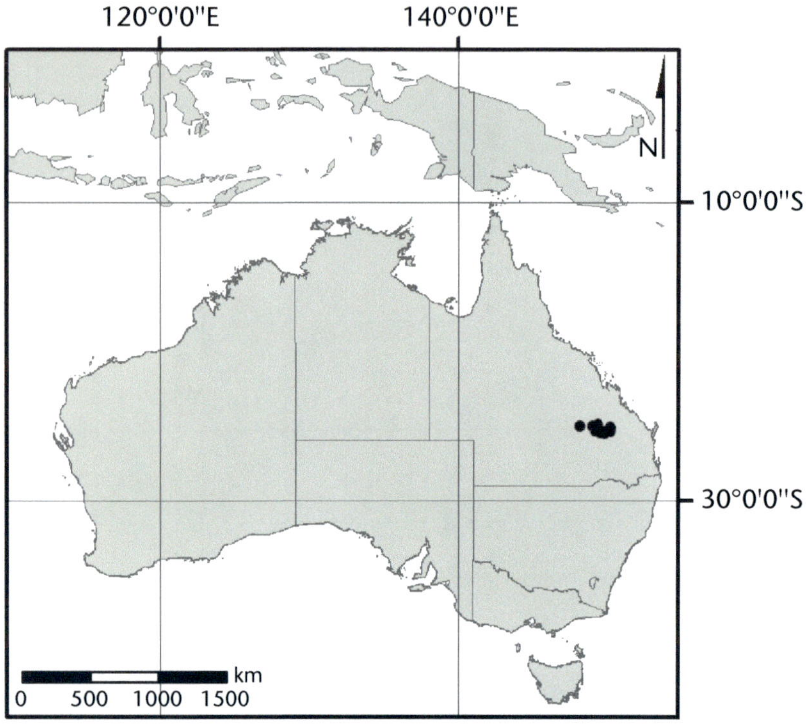

Figure 8.53 *Livistona nitida*. **Left:** Distribution based on herbarium specimen data. **Right:** Type specimen. Australia. Queensland. Cracow, Dawson River, Delusion Creek, 25°11′S, 150°11′E, 9 May 1976, *A.N.Rodd 3055 with S.Jacobs*; holotype NSW [1 of 4 sheets].

Figure 8.54 *Livistona rigida* in riparian habitat, Lawn Hill Gorge, north-west Queensland.

Typification: The holotype of *Livistona rigida* is an unnumbered and undated collection by F. Mueller [K] from Albert River, in north-west Queensland. The collection consists of a portion of leaf and a portion of inflorescence (Fig. 8.56).

Etymology: '*Spadix rigidus; spathis tenuiter coriaceis, glabris, cylindraceis; ramulis floriferis rigidis*' (Beccari 1921). 'Inflorescence rigid; spathes thin, leathery, glabrous, cylindrical; flower rachillae rigid': With probable reference to the rigid inflorescence and rachillae, from the Latin *rigidus*, stiff.

Notes: *Livistona rigida* is morphologically similar to *L. mariae* but is distinguished by smaller leaves, rigid or semi-pendulous segment apices and smaller fruits. See Notes under *L. mariae* for further discussion. Preliminary molecular investigations place *L. rigida* as sister to *L. mariae* (Fig. 8.4).

18 Livistona victoriae

Livistona victoriae Rodd, *Telopea* 8: 123, figs 16, 4f, 17c (1998); *Livistona* sp. 'B', Wilson, *Fl. Kimberley Region*: 1250 (1992). Type: Australia. Northern Territory. Katherine-Kununurra Rd, Victoria River Crossing, 3 Nov. 1974,

Figure 8.55 *Livistona rigida*. **Top left:** Riparian habitat, Lawn Hill Gorge, north-west Queensland. **Top right:** Crown with inflorescences. **Bottom left:** Flowers at anthesis. **Bottom right:** Fruit, whole and in longitudinal section to reveal seed coat intrusion.

A.N.Rodd 2934; holotype NSW [4 sheets]; isotype DNA, K, PERTH.

Functionally dioecious palm. **Stem** to 17 m tall, 15–30 cm dbh; leafscars narrow; internodes narrow, grey; petiole remnants not persistent; outer 'cortex' deciduous. **Leaves** 25–40 in a ± globose crown; petiole arching, 80–200 cm long, pruinose, flat on the adaxial surface, margins with single curved blunt black thorns, confined to the proximal portion; leafbase fibres moderately prominent, fine, persistent; lamina regularly segmented, ± subcircular in outline, 80–110 cm long, coriaceous, grey-green to bluish-green and strongly pruinose to dull on the adaxial surface, similar on the abaxial surface. **Segments** 40–56, rigid, free for 55–65% of

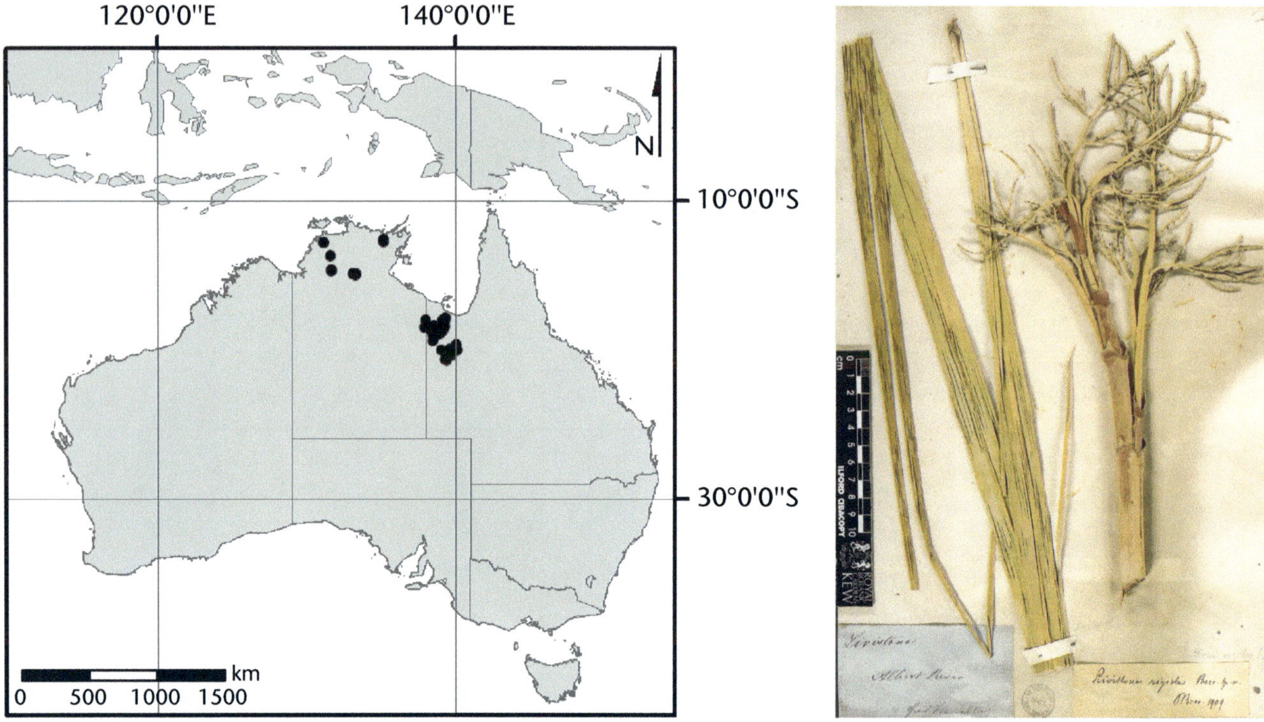

Figure 8.56 *Livistona rigida*. **Left:** Distribution based on herbarium specimen data. **Right:** Type specimen. Australia. Queensland. Gulf of Carpentaria, Albert River, Aug. 1855, *F.Mueller s.n.*; holotype K.

their length, depth of apical cleft 55–70% of the length of the free portion, apical lobes attenuate; parallel veinlets 9–11 each side of the midrib; transverse veinlets thinner than parallel veinlets. **Inflorescences** 50–150 cm long; partial inflorescences 5–9, branched to 4 orders; prophyll 20–25 cm long, 4–5 cm wide, moderately white to cream tomentose, apices fibrous-lacerate; peduncular bract 1 or lacking; rachis bracts subtend all branches to the apical partial inflorescence, loosely sheathing, fibrous with age, densely tomentose; rachillae 1–7 cm long, glabrous, light pink at floral maturity. **Flowers** solitary or paired, funnel-shaped, 1.5–4 mm long in bud; sepals triangular, 0.7–0.8 mm long, membranous, acute, cream to dark yellow, basally flushed pink; petals narrowly ovate, 1.2–4 mm long, acute, cream to pale yellow with scattered pink spots; stamens c. 0.9 mm long; anthers dark yellow. **Fruit** globose to subglobose and ellipsoid, 8–15 mm long, c. 10 mm diam.; epicarp c. 0.5 mm thick, smooth, dark reddish-brown to black, glossy; suture line extends for most of the length of the fruit; mesocarp c. 1 mm thick, fibrous; endocarp c. 0.25 mm thick, brittle; pedicel 0.5–2 mm long. **Seed** globose, 8–9 mm diam.; seed coat extends to c. halfway through the endosperm; embryo basal; eophyll 3-ribbed. Figures 8.57, 8.58.

Figure 8.57 *Livistona victoriae* in open woodland habitat, Echidna Gorge, Bungle Bungle Range, Purnululu National Park, Western Australia.

Figure 8.58 *Livistona victoriae*. **Top left:** Open woodland habitat, Echidna Gorge, Bungle Bungle Range, Western Australia. **Top right:** Crown with infructescences. **Bottom left:** Mature fruit. **Bottom right:** Fruit, whole, with mesocarp removed to show endocarp, and in longitudinal section to reveal seed coat intrusion and embryo.

Victoria River fan palm, Bungle Bungle Range fan palm

Distribution and ecology: Occurring in the Northern Territory and Western Australia, between the Victoria River basin and the Bungle Bungle Range, in open forest in sandstone gorges, in escarpment gullies and cliff bases and cliff tops, and along streams and intermittent water-courses, 50–240 m asl (Fig. 8.59). Flowering Mar.–Oct.; fruiting Oct.–Mar. Conservation status – *No present threats*.

Typification: The holotype of *Livistona victoriae* is *Rodd 2934* [NSW], collected in Nov. 1974 from the Victoria River Crossing area in north-west Northern Territory.

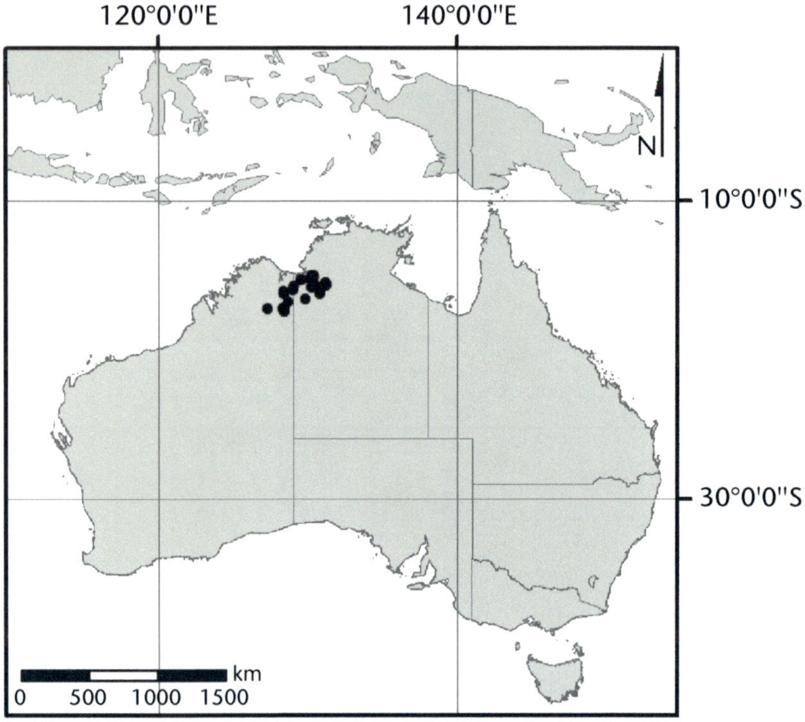

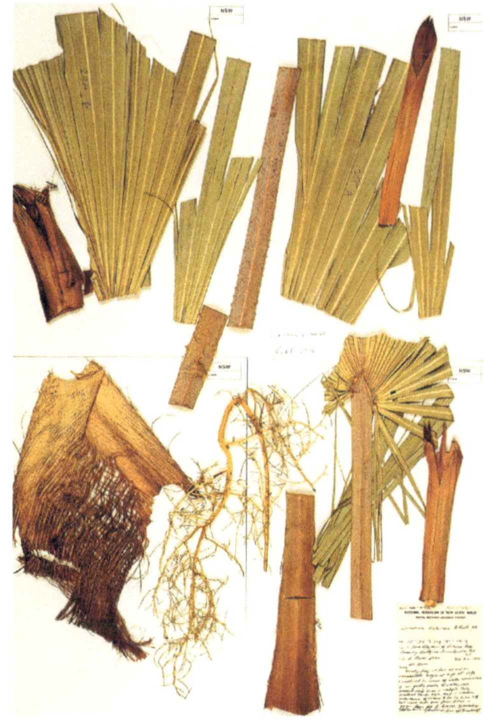

Figure 8.59 *Livistona victoriae*. **Left:** Distribution based on herbarium specimen data. **Right:** Type specimen. Australia. Northern Territory. Katherine–Kununurra Rd, Victoria River Crossing, 3 Nov. 1974, *A.N.Rodd 2934*; holotype NSW [4 of 16 sheets].

The collection consists of leaf sections, petiole portions, leafbase fibre, a prophyll and a portion of inflorescence (Fig. 8.59).

Etymology: 'The epithet *victoriae* alludes to its occurrence in the vicinity of the Victoria River, but the genetive form is used to complete a trio with *L. alfredii* and *L. mariae*, named respectively for Queen Victoria's son and daughter-in-law and occurring more or less equidistant from the present species in similarly semi-arid habitats' (Rodd 1998): Named for Queen Victoria (b.1819, d.1901), Queen of Great Britain 1837–1901, and for the occurrence of the species in the Victoria River District of north-west Australia.

Notes: *Livistona victoriae* most closely resembles *L. alfredii* but is distinguished by its smaller stature, open rather than dense crown, shorter inflorescence, partial inflorescences branched to four rather than three orders, and smaller fruit. Preliminary molecular investigations place *L. victoriae* as sister to *L. alfredii* (Fig. 8.4).

Licuala

Licuala Wurmb, *Verh. Batav. Genootsch. Kunsten* 2: 469 (1780). *Licuala* sect. *Eulicuala* Becc., *Webbia* 5: 22 [24] (1921). Type: *Licuala spinosa* Wurmb.

Pericycla Blume, *Rumphia* 2: 47 ('1836', published 1838). Type: *Pericycla penduliflora* Blume = *Licuala penduliflora* (Blume) Zipp. ex Blume.

NB: Generic synonyms which do not have a taxonomic association with Australian taxa are not included in this treatment. A full list of generic synonyms is provided in Dransfield *et al.* (2008).

Solitary or clustering, small to tall, acaulescent or erect, pleonanthic hermaphroditic or dioecious palms. **Stems** short and subterranean, creeping or erect, slender to moderate, cylindrical, broader at the base, ringed with leafscars; internodes narrow; leafsheath remnants sometimes persistent. **Leaves** palmate, when senescent sometimes forming a persistent skirt; leafsheaths forming a fibrous network, usually ligulate; petioles long, concave, flat or ridged on the adaxial surface, rounded on the abaxial surface, margins smooth or spiny; hastula present on adaxial surface, lacking on the abaxial surface; lamina entire or segmented. **Segments** reduplicate, single or united into regular or irregular multiples; individual segments wedge-shaped, widening toward the apex, split along the folds on the lower side of the blade, apices shallowly cleft, rigid or semi-pendulous; central segment infrequently on a stalk-like extension; midrib prominent, few to many prominent parallel veinlets each side of the midrib; transverse veinlets equal thickness or thinner than parallel veinlets. **Inflorescences** interfoliar, short to long, spicate to paniculate and branched to 3 orders, with partial inflorescences positioned along the length; peduncle short to long; prophyll tubular, 2-keeled; peduncular bracts 0–5, loosely or tightly tubular, glabrous or tomentose; rachis bracts tubular, subtending each partial inflorescence, secondary rachis bracts lacking or present; rachillae few or crowded, straight or flexuous, glabrous or tomentose, papillose, pubescent or pruinose. **Flowers** hermaphroditic, solitary or in sympodial pairs or clusters of 3–4, spirally arranged, sessile or on stalks, distantly spaced or crowded, subtended by a single minute bract, variously coloured; sepals triangular, fused at the base with apices regularly or irregularly lobed, glabrous or hairy; petals basally connate or tubular, apically free, longer than the sepals, triangular, thick, apically valvate, glabrous or hairy; stamens 6, epipetalous, shorter than the petals; filaments distinct or united; anthers erect or pendulous, didymous, roundish to oblong, medifixed, versatile, latrorse; gynoecium tricarpellate, glabrous or hairy; carpels wedge-shaped, basally distinct; styles united, slender, columnar, shorter than the stamens; stigma apex rounded. **Pollen** ellipsoidal; aperture monosulcate; exine tectate, smooth, rugulate or pitted, reticulate or foveate. **Fruit** globose, ovoid, narrowly ellipsoid, spindle-shaped or curved, variously coloured, 1-seeded; stigmatic remains apical; epicarp thin or thick, dull or shiny, smooth or infrequently corky-warted, variously coloured; mesocarp fleshy, thin to thick, with or without embedded fibres; endocarp thin, crustaceous; perianth persistent or not on fruit. **Seed** basally attached, globose, ovoid or spindle-shaped; endosperm homogeneous, intruded by the testa from one side with spongy-crystalline tissue to displace part of the endosperm; embryo lateral; germination remote-tubular; eophyll simple, with longitudinal ribs or plicate.

Licuala is a genus of c. 135 species, distributed from Bhutan throughout south-east Asia east to Indochina, the Philippines, Malesia east to Vanuatu, and Australia where one endemic species and one subspecies occur.

Etymology: *'Namnet Licuala ar gifvit detta trad af Macasfarene, under hvilket namn de tock ar I Indien allmannaft bekant och kandt bade af infodde och dar boende Europeer'* ['The name *Licuala* has been given to this plant by the Macassans, under which name they made it familiar in India and known by both the indigenous people and the Europeans']: Thunberg (1782) noted the name *Licuala* had been previously used by Rumphius (1741), derived from the Macassar native name *leko wala*, for *Licuala spinosa* Thunb., in eastern Indonesia.

Licuala ramsayi

Licuala ramsayi (F.Muell.) Domin, *Biblioth. Bot.* 85: 500 (1915); *Livistona ramsayi* F.Muell. *Fragm.* 8: 221 (1874). Type: Australia. Queensland. Rockingham Bay, [post-dated 1878 in Mueller's handwriting] 1874, *E.P.Ramsay s.n.*; holotype MEL [3 sheets]; isotype K.

Licuala muelleri H.Wendl. & Drude, *Linnaea* 39: 223 (1875); *Pericycla muelleri* (H.Wendl. & Drude) Drude ex Salomon, *Die Palmen*: 138 (1887). Type: Australia. Queensland. Dalrymple Gap, Jan. 1866, *J.Dallachy s.n.*; holotype MEL [2 sheets]; isotype BO.

Stems solitary or rarely clustered, to 10 m tall, 7–20 cm dbh; leafscars rough, to 2 cm wide; internodes to 10 cm long, grey; petiole remnants persistent in juveniles and subadults. **Leaves** 9–12 in an irregular crown, erect to semi-drooping; leafsheath up to 100 cm long disintegrating into a light to dark brown fibrous mesh; ligule to 50 cm long, light brown; petiole to 2 m long, to 1.8 cm diam., glabrous or scaly, greenish-beige basally to green distally, slightly ridged or concave on the adaxial surface, to flat below the blade, lower quarter to third armed or unarmed throughout; when armed, spines to 5 mm long, straight to recurved; lamina broadly elliptic to almost circular, with scattered scales near central portion, otherwise glabrous, dark green on the adaxial surface, light green on the abaxial surface, to 1.8 m diam.; transverse veinlets obscure; hastula slightly eccentric, pointed. **Segments** 17–23, rigid; distal segments 1–3, 60–80 cm long, 20–40 cm wide, inserted on a thick and conspicuous extension of the costa; midleaf segments single, or united with 7–23 segments; lateral segments 50–75 cm long, 8–18 cm wide united with 2–5 segments; basal segments 50–65 cm long, 5–10 cm wide, united with 2–5 segments; divisions between united segments 1–5 cm deep; apical clefts of segments to 5 mm deep. **Inflorescences** to 3.5 m long, arching; partial inflorescences 8–10, branched to 2 orders; prophyll 20–30 cm long, green at the base to brown and chartaceous distally; peduncular bract 50–70 cm long; bracts subtending partial inflorescences to 30 cm long, tubular, decreasing in size towards the apex, glabrous or with scattered scales, apices moderately fibrous-lacerate; partial inflorescences 25–55 cm long, decreasing in length towards the apex; rachillae 15–30 cm long, glabrous to sparsely scaly. **Flowers** solitary, in pairs or clusters of 3–4, subsessile to shortly pedicellate, subtending bract deltoid to strap-shaped, up to 1 mm long, sepals 2.5–3.5 mm long, urn-shaped to cyathiform, divided half to third of its length in 3 rounded to obtuse lobes, glabrous, cream with brown edges, receptacle and sepals fused for 0.3–0.5 mm; petals 3.3–3.5 mm long, cream, glabrous; androecium 2–2.3 mm long, fused to corolla for 1.5–1.7 mm, staminal ring cream at anthesis, truncate, filaments subulate, about 0.2 mm long, anthers 0.3–0.6 mm long, rounded to elliptic; ovary c. 1 mm long, glabrous, turbinate, rounded to truncate apically, style 1.2–1.3 mm long, filiform. **Fruit** ovoid to globose, 8–15 mm long, 8–13 mm diam., epicarp c. 0.2 mm thick, smooth, orange to red at maturity; mesocarp 1–1.5 mm thick; pedicel 1.5–3 mm long; perianth persistent on fruit. **Seed** ovoid to globose, c. 5 mm long, c. 4 mm diam.; endosperm homogeneous, intruded by seed coat.

Notes: *Licuala ramsayi* is an outlier to a genus with bicentric diversity in west Malesia and New Guinea. Similarly tall species occur in New Guinea, although many of the west Malesian taxa tend to be small understorey palmlets. The differences that distinguish the varieties are slight; leaf arrangement and armature are the most obvious characters.

Key to varieties of Licuala ramsayi

1 Leaf segments united into irregular multiples; leaf-sheath disintegrates into a fibrous mesh that eventually detaches from the leaf; petiole conspicuously armed with spines to 5 mm long; flowers in clusters of 3–4 to solitary at the tip of the rachillae .... **1. *L. ramsayi* var. *ramsayi***

1: Leaf segments united into ± regular multiples; leaf-sheath remains attached to the leaf; petiole unarmed or with a few spines at the base of the petiole; flowers mostly solitary .... **2. *L. ramsayi* var. *tuckeri***

1 Licuala ramsayi var. ramsayi

Solitary or rarely clustered palm. **Stem** with persistent fibrous mesh from disintegrated leafsheath eventually detached from the leaf and often sliding down the stem; petiole conspicuously armed with spines to 5 mm long mostly in the basal half. **Segments** united into irregular multiples. **Flowers** in clusters of 3–4 to solitary at the apex of the rachillae; flowers rounded to globose in bud; sepals with apex rounded to obtuse; anthers 0.5–0.6 mm long. Figures 8.60, 8.61.

Mission Beach fan palm

Distribution and ecology: Occurs in north-east Queensland, from near Cooktown south to near Ingham, in rainforest, swamp forest, littoral forest, riparian forest, in dense almost monospecific populations, in large or small groups and scattered individuals, as a subcanopy to canopy emergent, on various soils, 0–1100 m asl (Fig. 8.62). Flowering Feb.–Jul.; fruiting Apr.–Jan. Conservation status – *No present threats*.

Typification: The holotype of *Licuala ramsayi* var. *ramsayi* (basionym *Livistona ramsayi*) is an unnumbered collection by Edward Ramsay [MEL], collected in 1874 from Rockingham Bay, north-east Queensland (Fig. 8.62).

Etymology: '*In silvis montium pone Rockingham's Bay: Ed. P. Ramsay*' ['In mountain forest behind Rockingham Bay: Ed. P. Ramsay'] (Mueller 1874a): Named for the collector of the type specimen, Edward Pearson Ramsay (b.1842, d.1916), zoologist and ornithologist, Curator of

Figure 8.60 *Licuala ramsayi* var. *ramsayi* in lowland rainforest, Licuala State Forest, Mission Beach, north-east Queensland.

Figure 8.61 *Licuala ramsayi* var. *ramsayi*. **Top left:** Leaf, adaxial surface. **Top right:** Inflorescence. **Bottom left:** Flowers at anthesis. **Bottom right:** Mature fruit.

the Australian Museum 1874–94, founder of the Linnean Society of New South Wales in 1874.

2 Licuala ramsayi var. tuckeri

Licuala ramsayi var. ***tuckeri*** Barfod & Dowe, *Palms* 49: 21, figs 5, 6 (2005). Type: Australia. Queensland. Lockerbie Scrub, south of Pajinka Wilderness Lodge, 10°45′S, 142°33′E, 29 Sept. 1999, *J.L.Dowe 602 with A.S.Barfod*; holotype BRI AQ681978 [3 sheets]; isotype AAU.

Stem solitary; leafsheath remains attached to the leaf; leaf petiole unarmed or with a few small spines at the base. **Segments** united into ± regular multiples. **Flowers**

Figure 8.62 *Licuala ramsayi* var. *ramsayi*. **Left:** Distribution based on herbarium specimen data. **Right:** Type specimen. Australia. Queensland. Rockingham Bay, 1874, *E.P.Ramsay s.n.*; holotype MEL [1 of 3 sheets].

mostly solitary throughout the rachiallae; flowers bullet-shaped in bud; sepals with apex acuminate; anthers 0.3–0.4 mm long. Figures 8.63, 8.64.

Tucker's Licuala, Cape York Licuala

Distribution and ecology: Occurs in north Queensland, from Moa Island, Torres Strait south to near Cooktown, in rainforest, swamp forest, littoral forest, riverine forest, as a canopy element in large to small populations and scattered individuals on various soils, 0–300 m asl (Fig. 8.65). Flowering Jul.–Sept.; fruiting Dec.–Mar. Conservation status – *No present threats.*

Typification: The holotype of *Licuala ramsayi* var. *tuckeri* is *Dowe 602 & Barfod* [BRI], collected in 1999 from Lockerbie Scrub, Cape York Peninsula, Queensland. The collection consists of a leaf portion, inflorescence and infructescence with immature fruit (Fig. 8.65).

Etymology: 'Named for horticulturist, botanical illustrator and taxonomist (Pandanaceae) Robert James Thomas Tucker (1955–92), landscape designer for Townsville Botanic Gardens, 1982–92, and designer of The Palmetum, Townsville, a botanic garden devoted to palms' (Barfod and Dowe 2005): Tucker collected palms and pandanus throughout Cape York Peninsula [deposited in

Figure 8.63 *Licuala ramsayi* var. *tuckeri* in lowland rainforest, inland of Captain Billy Landing, Cape York Peninsula, north Queensland.

Figure 8.64 *Licuala ramsayi* var. *tuckeri*. **Top left:** Petiole bases and leafbase fibres. **Top right:** Leaf, abaxial view. **Bottom left:** Flowers at anthesis. **Bottom right:** Mature fruit.

BRI and CNS]. He was the author of *The Palms of Subequatorial Queensland* (Tucker 1988) and taxonomic works on the Pandanaceae (Tucker 1994; Dowe 1992).

■ CARYOTEAE

Caryota

Caryota L., *Sp. pl.* 2: 1189 (1753). Type: *Caryota urens* L.

NB: Generic synonyms which do not have a taxonomic association with Australian taxa are not included in this treatment. A full list of generic synonyms is provided in Dransfield *et al.* (2008).

Moderate to large, solitary or clustering, hapaxanthic, monoecious palms. **Stems** slender to robust, leaning or erect; leafscars prominent; internodes short to long. **Leaves** bipinnate, persistent, rarely cleanly deciduous;

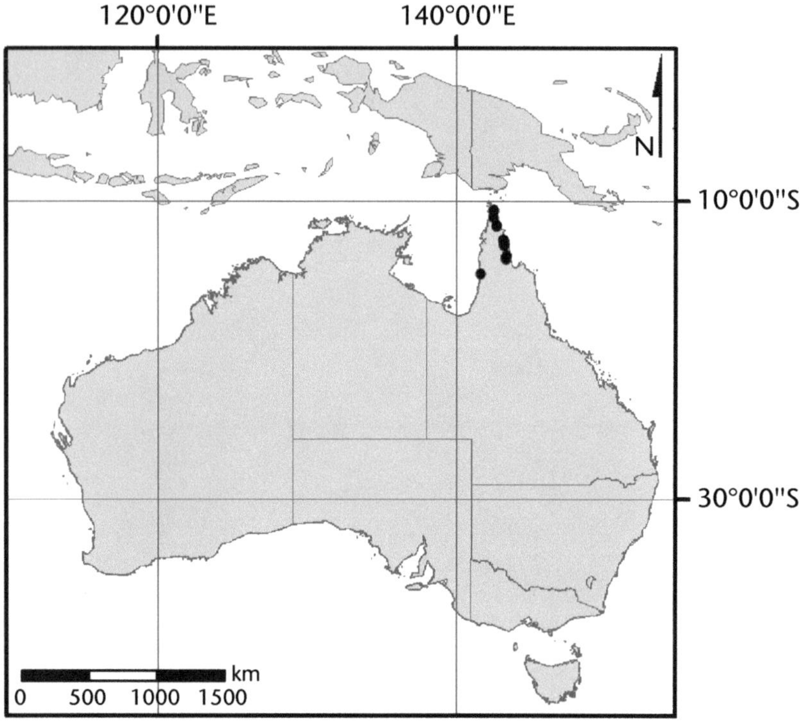

Figure 8.65 *Licuala ramsayi* var. *tuckeri*. **Left:** Distribution based on herbarium specimen data. **Right:** Type specimen. Australia. Queensland. Lockerbie Scrub, 10°45′S, 142°33′E, 29 Sept. 1999, *J.L.Dowe 602 with A.S.Barfod*; holotype BRI [1 of 3 sheets].

leafbases partially enclosing the stem, split opposite the petiole with disintegrating margins and producing a fibrous mass, often with long spine-like protuberances, frequently ligulate, surface with dense felty scales; petiole lacking to long, channelled on the adaxial surface, rounded on the abaxial surface, all surfaces with dense felty scales; primary and secondary rachises similar, arranged regularly, sometimes basally congested. **Pinnae** in a single plane on the secondary rachises, regularly arranged, obliquely wedge-shaped, induplicate, midrib lacking, several thick veinlets radiating from the base, truncate, apically praemorse, concolourous. **Inflorescence** interfoliar or infrequently infrafoliar, rarely spicate, usually paniculate and branched to 1 or 2 orders, produced basipetally, pendulous; prophyll tubular, 2-keeled, densely scaly; peduncular bracts to 8, large, spirally arranged, enclosing the inflorescence, tubular then splitting as inflorescence emerges, densely scaly; peduncle short to long, terete, densely scaly; rachis shorter or longer than the peduncle; rachillae spirally arranged, congested, scaly, with low subtending bracts. **Flowers** in spirally arranged triads in proximal portion with one pistillate flower flanked by two staminate flowers or in pairs or single staminate flowers in distal portion. **Staminate flowers** elongate and symmetrical in bud; sepals 3, coriaceous, rounded, imbricate; petals 3, long, connate at the base, coriaceous, valvate; stamens 6–100; filaments short, basally connate; anthers linear, latrorse, connective elongate, pointed; pistillode lacking. **Pistillate flowers** globular in bud; sepals 3, coriaceous, rounded, basally connate, imbricate; petals 3, coriaceous, basally connate forming a brief tube, apically valvate; staminodes 0–6; ovary trilocular with 1–3 fertile locules; stigma trilobed at the apex; ovule hemianatropous. **Pollen** ellipsoidal to circular; aperture monosulcate or trichotomosulcate; exine intectate, with spines or protuberances. **Fruit** globose to subglobose, 1–3-seeded, stigmatic remains apical or subapical; epicarp smooth, dull or brightly coloured; mesocarp fleshy to succulent with irritant oxalate crystals; endocarp not differentiated; perianth persistent. **Seed** basally attached, tear-shaped, hemispherical or globose, smooth or grooved, black; endosperm ruminate or homogeneous; embryo lateral to subbasal; germination remote-tubular; eophyll bifid.

A genus of c. 13 species distributed from India throughout south-east Asia to the Philippines, Malesia

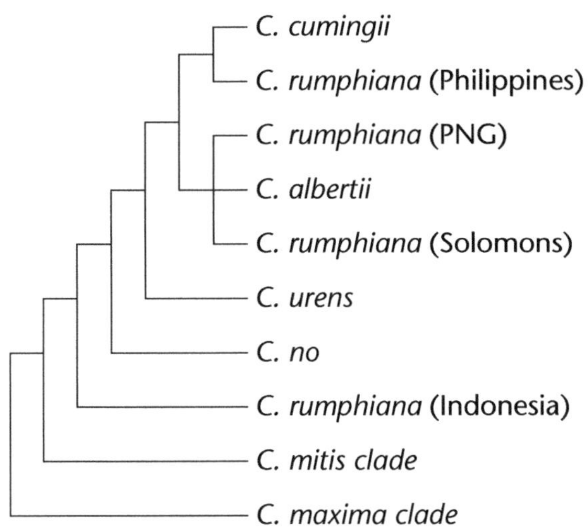

Figure 8.66 Phylogenetic tree summarising the estimated relationships of species of *Caryota*. Adapted from Hahn and Sytsma (1999).

and east to Vanuatu. One endemic species in Australia restricted to northern Cape York Peninsula. Hahn and Sytsma (1999) assessed the phylogeny and biogeography of *Caryota* and placed *C. albertii* in a trichotomy with populations of *C. rumphiana* from New Guinea and the Solomon Islands (Fig. 8.66).

Etymology: *'Caryota'*: Derived from the Greek for nut, *karyon*.

Key to genera of Caryoteae

1 Leaves bipinnate .... 1. *Caryota*
1: Leaves pinnate or flabellate 2
2 Sepals of staminate flowers free, imbricate
.... 2. ***Arenga***
2: Sepals of staminate flowers connate in a tube
.... 3. ***Wallichia*** (not in Australia)

Caryota albertii

Caryota albertii F.Muell. ex H.Wendl., *in* H.Wendland & O.Drude, (as *C. Alberti*) *Linnaea* 39: 221 (1875); *Caryota rumphiana* var. *australiensis* Becc., *Malesia* 1: 74 (1877); *Caryota rumphiana* var. *albertii* (F.Muell. ex H.Wendl.) F.M.Bailey, *Queensland Agric. J.* 1: 233 (1897). Type: Australia. Queensland. Cape York, Somerset, *E.Daemel*; holotype MEL [1 sheet]; isotypes BO, GOET.

[*Caryota urens* L., *in* F.Mueller, *Essay Pl. Collect. Fitzalan*: 18 (1860); *Caryota rumphiana* Mart., *in* G.Bentham, *Fl. Austral.* 7: 144 (1878)].

Stem solitary, to 20 m tall, to 40 cm dbh, slightly expanded at the base or not; leafscars to 3 cm wide, irregular; internodes to 30 cm long, at first with thick white-grey tomentum, eventually patchily or fully deciduous. **Leaves** 5–15, to 7 m long, to 5 m wide, held horizontal; secondary rachises 12–15 on each side of primary rachis; c. 25 pinnules each side of secondary rachis, 8–30 cm long, regularly arranged; leafbases with lacerate margins; petiole to 150 cm long, with dense whitish tomentum; rachis to 550 cm long, densely white tomentose. **Inflorescence** 2–3 m long, branched to 2 orders, axes green; prophyll to c. 13 cm long, grey, densely tomentose; peduncular bracts 5–7, to 25 cm long, grey, densely tomentose, caducous; peduncle to c. 50 cm long; rachis to c. 75 cm long; rachillae 80–120 (250) cm long, straight, pendulous; floral pits sunken, margins rounded. **Flower** triads spirally arranged throughout rachillae, protandrous. **Staminate flowers** 12–20 mm long, 4–8 mm wide in bud, slightly curved, opening to c. 22 mm wide at anthesis; sepals imbricate, margins fimbriate, c. 3 mm high, c. 6 mm wide, light green-cream; petals thick, 13–18 mm long, 4–8 mm wide, one slightly shorter than other two, apically rounded, symmetrical, cream suffused with pink; stamens 22–36, 8–9 mm long, spirally arranged on a column 4–5 mm high, widely and irregularly spreading at anthesis; connective broad at the base, narrowing toward the anther, 3–4 mm long; anthers basifixed, not versatile, 7–8 mm long, yellow. **Pistillate flowers** 3–9 mm long, 4–6 mm wide in bud; sepals 3–4 mm long, 4–6 mm wide, light green; petals 5–8 mm long, 5–7 mm wide, yellow-green; stigmas c. 0.8 mm long, extending to c. 1.5 mm wide at receptivity; ovary globose, 5–6 mm diam. **Fruit** subglobose, to 22 mm long, to 28 mm wide; stigmatic remains apical; epicarp smooth, 0.25 mm thick, dull crimson to black at maturity; mesocarp to 5 mm thick, with few thick fibres. **Seeds** 1–3, hemispherical to 25 mm long, to 25 mm wide; endosperm deeply ruminate; embryo subbasal. Figures 8.67, 8.68.

Native fish-tail palm, Australian fish-tail palm

Distribution and ecology: Endemic to Cape York Peninsula, Queensland, from Cape York (Lockerbie Scrub) south to MacIlwraith Range. Occurs as scattered individuals and small groups, but never as dense populations, most common in lowland rainforest but less common in vine forest, swamp forest, mangroves and riparian forests, on various soils, 0–300 m asl (Fig. 8.69). Flowering Sept.–Dec.; fruiting Sept.–Jan. Conservation status – *No present threats*.

Typification: The holotype of *Caryota albertii* is an unnumbered and undated collection by 'Dalmel' [correctly Daemel] [MEL] from Cape York, Queensland.

Figure 8.67 *Caryota albertii* in lowland rainforest, West Claudie River, Iron Range, north Queensland.

The collection consists of a portion of leaf, although the protologue described inflorescence and flowers in detail (Fig. 8.69).

Etymology: There was no explanation of the derivation of the epithet in the protologue, but named for Prince Albert (b.1819, d.1861), Prince of Saxe-Coburg Gotha, married to Queen Victoria in 1840.

Notes: The first documentation of *Caryota albertii* (as *Caryota urens*) was provided by MacGillivray during the *Rattlesnake* voyage in October 1848:

> *One of the most interesting features in the botany of Cape York, is the occurrence of a palm, not hitherto mentioned as Australian. It is the* Caryota urens *(found also in India and the Indian archipelago), one of the noblest of the family, combining the foliage of the tree-fern with a trunk a foot in diameter, and sixty in height. It is found in the dense brushes along with three other palms,* Seaforthia, Corypha, *and* Calamus.

Mueller (1866) used the name *C. urens* again in an account of Australian palms in *Fragmenta Phytographiae Australiae*. In a treatment of New Guinea palms, Beccari (1877) included *C. albertii* as a synonym of *C. rumphiana* var. *australiensis*. Bentham (1878) listed it as *C. rumphiana*, citing *C. obtusa* as another synonym. Mueller (1889) subsequently followed Bentham's taxonomy, while Bailey (1897, 1909) listed it as *C. rumphiana* var. *alberti*, as a nomenclatural correction. Dransfield (1974) considered the large solitary *Caryota* from the Moluccas to be *C. rumphiana* and that some of the named subspecies should be considered distinct.

Caryota albertii is readily distinguished in habitat in Australia by the bipinnate leaves. At the specific level, it is closely related to *C. rumphiana*, a species distributed in the Moluccas and New Guinea, but is distinct with a more horizontally oriented leaf, more symmetrical arrangement of secondary rachises and smaller staminate flowers with fewer stamens.

Arenga

Arenga Labill. ex DC, *Bull. Sci. Soc. Philom. Paris* 2: 162 (1800), *nom. cons.* [Greuter *et al.* 1994); *Arenga* Labill. ex DC section *Arenga*, *Bull. Soc. Philom. Paris* 2: 162 (1800). Type: *Arenga saccharifera* Labill. ex DC. = *Arenga pinnata* (Wurmb) Merr.

Saguerus Steck, *Sagu*: 15 (1757), *nom. rej.* [Rickett and Stafleu (1959), ICBN Proposal to reject No. 575]. Type: *Saguerus pinnatus* Wurmb = *Arenga pinnata* (Wurmb) Merr.

Didymosperma H.Wendl. & Drude ex Hook.f., *in* G.Bentham and J.D.Hooker, *Gen. pl.* 3: 917 (1883); *Arenga* section *Didymosperma* (H.Wendl. & Drude ex. Hook.f.) H.E.Moore, *Principes* 4: 114 (1960). Type: *Didymosperma porphyrocarpum* (Blume.) H.Wendl. & Drude ex Hook.f. = *Arenga porphyrocarpa* (Blume) H.E.Moore.

NB: Generic synonyms which do not have a taxonomic association with Australian taxa are not included in this treatment. A full list of generic synonyms is provided in Dransfield *et al.* (2008).

Small to large, solitary or clustering, pleonanthic or hapaxanthic, monoecious palms. **Stems** slender to robust, leaning or erect; leafscars prominent; internodes congested to elongate. **Leaves** flabellate, paripinnate or imparipinnate, persistent, rarely cleanly deciduous; leafbases partially enclosing the stem, split opposite the petiole, disintegrating into a fibrous mass, often with long spine-like protuberances, frequently ligulate; ligule tubular or

Figure 8.68 *Caryota albertii*. **Top left:** Leaf, abaxial view. **Top right:** Staminate flowers at anthesis. **Bottom left:** Pistillate flowers at anthesis. **Bottom right:** Mature fruit.

open; petiole lacking to long, channelled or ridged on the adaxial surface, rounded on the abaxial surface, frequently densely scaly; rachis rounded or angled on the adaxial surface, flat to rounded on the abaxial surface. **Pinnae** in 1 or 2 planes along the rachis, terminal pinnae flabellately united, pinnae opposite or subopposite or grouped, margins wavy or lobed, or entire, induplicate, apically praemorse, sometimes basally stalked, or sometimes with a basal auricle facing the rachis; the adaxial surface glabrescent, usually densely covered in pale green, grey to silvery indumentum; midribs prominent; ramenta lacking. **Inflorescence** interfoliar or infrafoliar, infrequently spicate, usually paniculate and branched to 1 or 2 orders, erect to pendulous, produced acropetally in pleonanthic species,

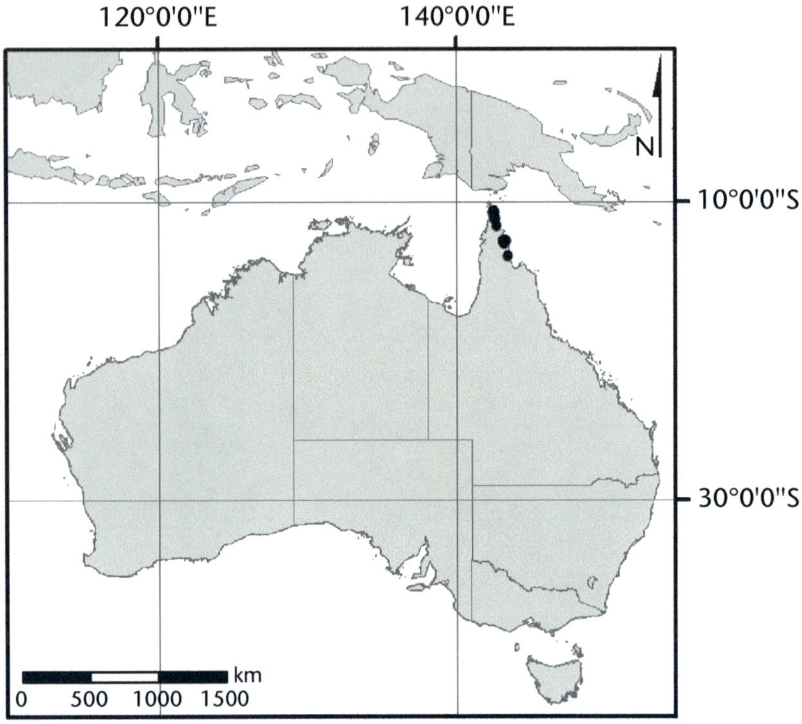

Figure 8.69 *Caryota albertii.* **Left:** Distribution based on herbarium specimen data. **Right:** Type specimen. Australia. Queensland. Cape York, Somerset, *E.Daemel s.n.*; holotype MEL.

basipetally in hapaxanthic species, distal inflorescences subtended by reduced leaves; prophyll inconspicuous, 2-keeled; peduncular bracts several, conspicuous, covered with dense indumentum; rachis bracts inconspicuous; peduncle usually short to moderate, narrow to broad; rachis usually longer than peduncle, infrequently shorter; rachillae erect or pendulous, well-spaced or congested, thin to massive, straight to sinuous, frequently tomentose. **Flowers** unisexual, variously coloured, in triads spirally arranged mostly in the proximal portion, infrequently throughout, subtended by low bracts, or sometimes with staminate and pistillate flowers on separate inflorescences. **Staminate flowers** with 3 rounded, imbricate coriaceous sepals; 3 petals, briefly united at the base, valvate, ovate to oblong, triangular, coriaceous; stamens 6–50+; filaments short; anthers erect, basifixed, elongate, connective apically pointed, latrorse; pistillode lacking. **Pistillate flower** globose, smaller or similar in size to staminate flower, infrequently much larger; 3 sepals, rounded, coriaceous, imbricate; 3 petals, connate basally, valvate, triangular; staminodes 0–3; ovary globose, trilocular, triovulate; stigmas 2–3; ovules hemianatropous. **Pollen** ellipsoidal or circular; aperture monosulcate; exine tectate with scattered spines. **Fruit** globose to ellipsoidal, or angled, 1–3-seeded; stigmatic remains apical; epicarp thick, smooth, yellow, red or black at maturity; mesocarp fleshy with irritant crystals; endocarp not differentiated. **Seed** basally attached, smooth, globose, tear-shaped to hemispherical, or angled or ridged; endosperm homogeneous; embryo lateral; germination remote-tubular; eophyll ovate to elliptical, or bifid.

Arenga is a genus of c. 20 species, distributed from India, throughout south-east Asia and southern China, and Malesia to as far east as the Solomon Islands, north to the Ryukyu Islands and the Philippines, and south to Australia where there are three species, two of which are endemic. There is considerable morphological diversity in the genus, with moderate clustering species to large solitary-stemmed palms which are among the world's largest palms, occurring mostly in high-rainfall areas, in littoral forest, rainforest and swamp forest.

Etymology: '*Il constitue un genre nouveau que le C. Labillardière nomme Arenga (du nom Areng qu'on lui donne dans les Molluques)*' (Labillardière 1800) [It constitutes a new genus that C. Labillardière names *Arenga* (after the name *Areng* given to it in the Moluccas)].

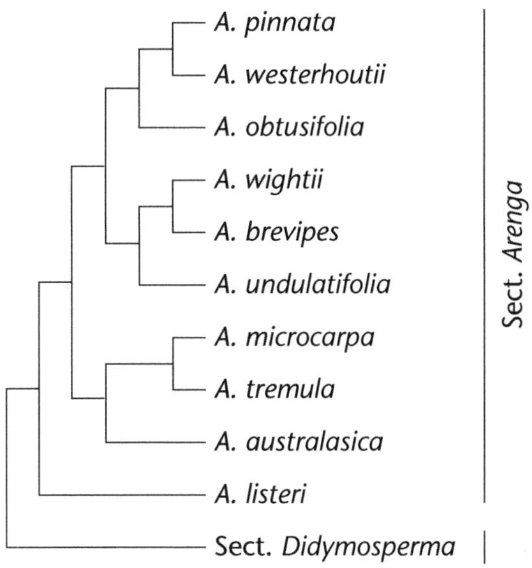

Figure 8.70 Phenogram summarising the estimated relationships of species in *Arenga*. Adapted from Mogea (1999).

Classification and relationships

Mogea (1999) provided an estimate of relationships of species (Fig. 8.70). In this, the Australia species are placed in Section *Arenga*, which is distinguished by moderate to large palms, lacking trullate leaves, stamens fewer than 50, and more than 25 pinnae each side of the rachis. The Australian species, as well as *A. tremula* (Philippine Islands), possess a tubular ligule. Section *Arenga* has globose to ellipsoid fruit to c. 1 cm diam. Species in Section *Didymosperma*, which do not occur in Australia, are characterised as understorey palms with trullate leaves, and fewer than 25 pinnae each side of the rachis.

Key to species of Arenga in Australia

1 Stem solitary .... **2. *A. listeri***
1: Stems basally clustered 2
2 Large palm to 20 m tall; pinnae with dense silver-grey indumentum on abaxial surface; fruit 8–28 mm wide .... **1. *A. australasica***
2: Moderate palm to 7 m tall; pinnae with moderately dense grey-green indumentum on the abaxial surface; fruit 15–18 mm wide .... **3. *A. microcarpa***

1 Arenga australasica

Arenga australasica (H.Wendl. & Drude) S.T.Blake ex H.E.Moore, *Gentes Herb.* 9: 268 (1963); *Saguerus australasicus* H.Wendl. & Drude, *Linnaea* 39: 219, tab. 3, Fig. 1 (1875). Type: Australia. Queensland. Rockingham Bay, Garden Island, Aug. 1865, *J.Dallachy*; holotype FI [1 sheet]; isotype BO [1 sheet].

[*Arenga saccharifera* Labill. ex DC, *in* O.Beccari, *Malesia* 1: 78 (1877); *Arenga microcarpa* Becc., *in* U.Martelli, *Nuovo Gior. Bot. Ital.* 42: 29 (1935), *pro parte*].

Moderate to large, hapaxanthic palms. **Stems** to 20 m tall, tightly clustering, 2–6 dominant, 15–30 cm dbh; leafscars prominent, 1–3 cm wide; internodes 10–40 cm long, with persistent dense white-grey pubescent scales. **Leaves** 10–18, 2.5–4 m long, with pinnae in 2 planes, 50–85 pinnae per side; leafbase margins broadly fibrous, papery; sheathing fibres soft, black, eventually deciduous with age; petiole to 60 cm long; rachis to 350 cm long; ligule tubular. **Pinnae** paired or single, regularly arranged or grouped, 40–80 cm long, to 6.5 cm wide, dark green on the adaxial surface, with dense silver-grey indumentum on the abaxial surface, margins with vein endings protruding, producing small sharp extensions; apex rounded, toothed or notched; midrib prominent on the abaxial surface; an auricle at the base on the proximal side faces back to the rachis on most pinnae, sometimes absent. **Inflorescences** to 2 m long, paniculate, produced basipetally on elongated aerial stems with flowers in triads, much reduced pistillate inflorescences frequently produced on basal leafy shoots, branched to 1 order, distally interfoliar, proximally infrafoliar, axes green to grey; prophyll to 40 cm long, to 8 cm wide, grey; peduncular bracts to 50 cm long, green with white tomentum; peduncle to 80 cm long, by 10 cm diam.; rachis to 80 cm long, basally to 10 cm diam.; rachillae to 80 cm long, pendulous in fruit. **Flowers** in triads on aerial inflorescences, solitary pistillate flowers on leafy basal branches. **Staminate flowers** 14–16 mm long, 7–10 mm wide in bud, opening 12–15 mm wide at anthesis; sepals 3–4 mm long, dark green, glabrous; petals 10–12 mm long, mid-green, darker green toward the apex, glabrous; stamens 18–24, 6–7 mm long; filaments c. 1 mm long; anthers 5–6 mm long, yellow. **Pistillate flowers** subglobose, sessile, 4–6 mm high, 6–7 mm wide in bud; sepals 1–1.5 mm high, 3–5 mm wide, imbricate, margins smooth, light green; petals basally connate, 4–6 mm long, c. 5 mm wide, lobes apically rounded, thick, margins smooth, cream-yellow; ovary subglobose, 4–6 mm high, 5–6 mm wide, light green, glabrous; stigmas discreet, not protruding, flush with surface. **Fruit** globose to subglobose, 8–26 mm long, 8–28 mm wide; epicarp smooth to slightly roughened, c. 0.5 mm thick, red at maturity; mesocarp 0.8–1 mm thick; perianth not persistent on the fruit.

Figure 8.71 *Arenga australasica*, Centenary Lakes, Cairns, north-east Queensland.

Seeds 2–3, hemispherical when 2-seeded, tear-shaped or angled when 3-seeded, 10–15 mm long, to 10 mm diam. Figures 8.71, 8.72.

Native sugar palm

Distribution and ecology: Endemic to Queensland, occurring from islands in Torres Strait south to Hinchinbrook Island, in littoral forest or near-coastal rainforest. It is occasionally gregarious on near-shore continental islands off the coast of north-east Queensland, and most commonly occurs in moderate-sized populations or scattered groups, as an understorey or subcanopy species, 0–400 m asl (Fig. 8.73). Some areas of former distribution have been cleared for urban development (Mission Beach, Kurrimine Beach) but the species is adequately conserved in national parks and conservation reserves. Flowering Aug.–Dec.; fruiting Oct. – Feb. Conservation status – *Vulnerable* (Bostock and Holland 2007).

Typification: The protologue of *Arenga australasica* (basionym *Saguerus australasicus*) includes *'Palmam leg. Dallachy Aug. mens 1865, neque flores neque fructus observans'* (Wendland and Drude 1875). [Leaves collected by Dallachy Aug. 1865, neither flowers nor fruits seen]. The holotype at FI consists of a section of leaf and portion of a single pinnae, and is labelled *'Garden Island, North Queensland'* (Fig. 8.73).

Etymology: *'Australasiae ora orientalis tropica'* (Wendland and Drude 1875). [On the shores of eastern tropical Australia]: From the Latin *australis*, southern.

Notes: The nomenclature of *A. australasica* has been problematic, with placement in the genera *Saguerus* and *Normanbya* (Bentham 1878; Baillon 1895). Moore (1963) resolved the anomalies by adopting a proposal by Blake, with *Arenga australasica* as the correct name.

The species is recorded from the Northern Territory and further field-work is required to verify this. The available specimens collected from the Northern Territory are closest to *A. microcarpa* (see below), although there is morphological overlap between it and *A. australasica*.

Arenga australasica is distinguished by its large size, in which dominant stems may be up to 20 m tall, and red fruits to 28 mm wide.

2 Arenga listeri

Arenga listeri Becc. ex Oliver, *Hooker's Icon. Pl.* 20: tab. 1985 (1891). Type: Christmas Island (Indian Ocean). 3 Oct 1887, *J.J.Lister s.n.*; holotype K [3 sheets, as sheets ii–iv]; isotype FI.

[*Didymosperma* sp., *in* W.B.Hemsley, *Journ. Linn. Soc.* 25: 359 (1890)].

Large, hapaxanthic palms. **Stems** solitary, to 20 m tall, 23–30 cm dbh; leafscars prominent, 1–3 cm wide; internodes to 30 cm long, grey, with dense grey-white tomentum; leafbases initially persistent but deciduous with age. **Leaves** 7–14, 4–7 m long, with most pinnae in a single plane, with an occasional pinnae inserted at a different angle from most, 40–90 pinnae per side; leafbase margins smoothly fibrous, papery; remnant sheathing fibres soft, black; petiole to 180 cm long; rachis to 550 cm long; ligule open. **Pinnae** mostly regularly arranged, sometimes a few grouped in the proximal portion, to 120 cm long, to 8 cm wide; dark green on the adaxial surface, grey indumentum on the abaxial surface; margins with vein endings

Figure 8.72 *Arenga australasica*. **Top left:** Pinnae. **Top right:** Staminate flowers at anthesis. **Bottom left:** Pistillate flowers at anthesis. **Bottom right:** Mature fruit.

protruding, producing small sharp extensions; apex pointed, toothed; midrib prominent on the abaxial surface; an auricle at the base on the proximal side faces back to the rachis, or lacking. **Inflorescences** to 2 m long, branched to 2 orders; peduncular bracts 9–11. **Staminate flowers** to 8 mm long, 3.5 mm wide in bud; sepals c. 2 mm long, green to white; petals to c. 7 mm long, oblong, obtuse, green, buff within; stamens c. 20, anthers bright yellow. **Pistillate flowers** with triangular petals c. 5 mm long. **Fruit** ellipsoidal, 13–17 mm long; epicarp to 0.5 mm thick, smooth, ripening white to red; mesocarp to 2 mm thick, fleshy. **Seeds** 2–3, hemispherical

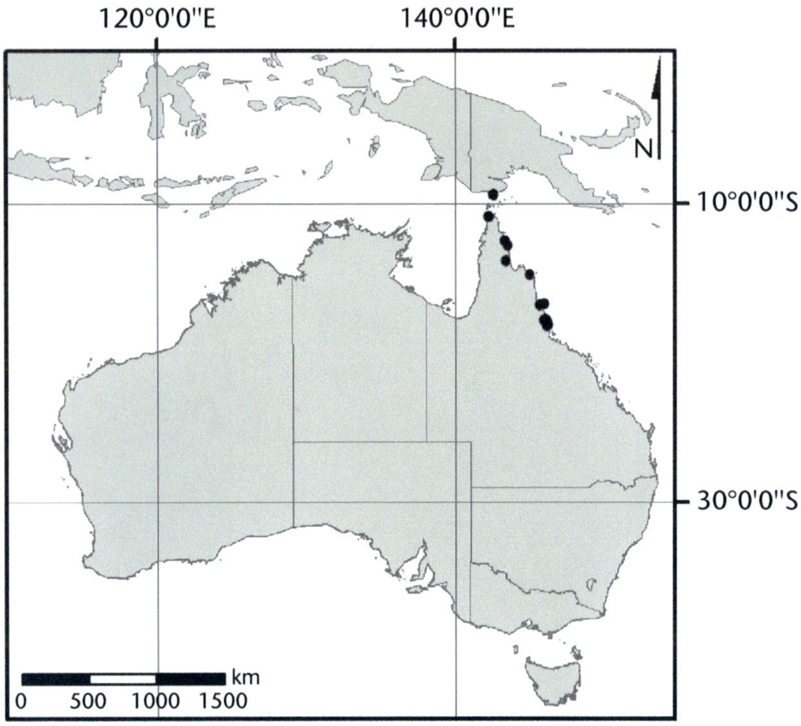

Figure 8.73 *Arenga australasica*. **Left:** Distribution based on herbarium specimen data. **Right:** Type specimen. Australia. Queensland. Rockingham Bay, Garden Island, Aug. 1865, *J.Dallachy*; holotype FI.

Figure 8.74 *Arenga listeri*, lowland rainforest, Christmas Island. Photo by David Tanswell.

when 2-seeded, tear-shaped when 3-seeded, 10–15 mm long, to 10 mm diam. Figures 8.74, 8.75.

Christmas Island palm, Lister's palm

Distribution and ecology: Endemic to Christmas Island, where it is abundant on moist basaltic soils sometimes forming almost pure stands as a canopy element, and a common understorey shrub on the plateau and terraces, 0–340 m asl (Fig. 8.76). Flowering all months; fruiting all months. Conservation status – *Vulnerable* (Johnson 1996).

Typification: The holotype of *Arenga listeri* is an unnumbered collection by *J.J. Lister* [K], collected in Oct. 1887 from Christmas Island. The collection consists of leaf portions with attached pinnae, and a portion of inflorescence with flowers (Fig. 8.76). An illustration, presumably drawn from the type specimen, accompanies the protologue.

Etymology: *'Hab. Christmas Island, J.J.Lister, 1887'* (Oliver 1891): Named for Joseph Jackson Lister (b.1857, d.1927), zoologist and naturalist, and collector of the type specimen. Lister provided an account of his visit to Christmas Island, where he spent 30 Sept.–9 Oct. 1887 as naturalist on board H.M. Surveying Vessel *Egeria* (Lister 1888).

Figure 8.75 *Arenga listeri*. **Top left:** Crown from below. Photo by David Tanswell. **Top right:** Leafbase fibres. Photo by David Tanswell. **Bottom left:** Newly emerging leaf. Photo by David Tanswell. **Bottom right upper:** Staminate flower at anthesis. **Bottom right lower:** Mature fruit with seeds removed. Photos by Jeff Clausen.

3 Arenga microcarpa

Arenga microcarpa Becc., *in* K.Schumann and M.Hollrung, *Fl. Kais. Wilh. Land*: 16 (1889); *Didymosperma microcarpum* (Becc.) Warb. ex K.Schum. & Lauterb., *in* K.Schumann & K.Lauterbach, *Fl. Schutzgeb. Südsee*: 204 (1901). Type: Papua New Guinea. Morobe Province. Finchaven, *U.M.Hollrung 302*; holotype B, destroyed; isotype FI [3 sheets].

Arenga gracilicaulis F.M.Bailey, *Queensland. Agric. J.* 2: 203 (1898). Type: Papua New Guinea. Oro Province, foot

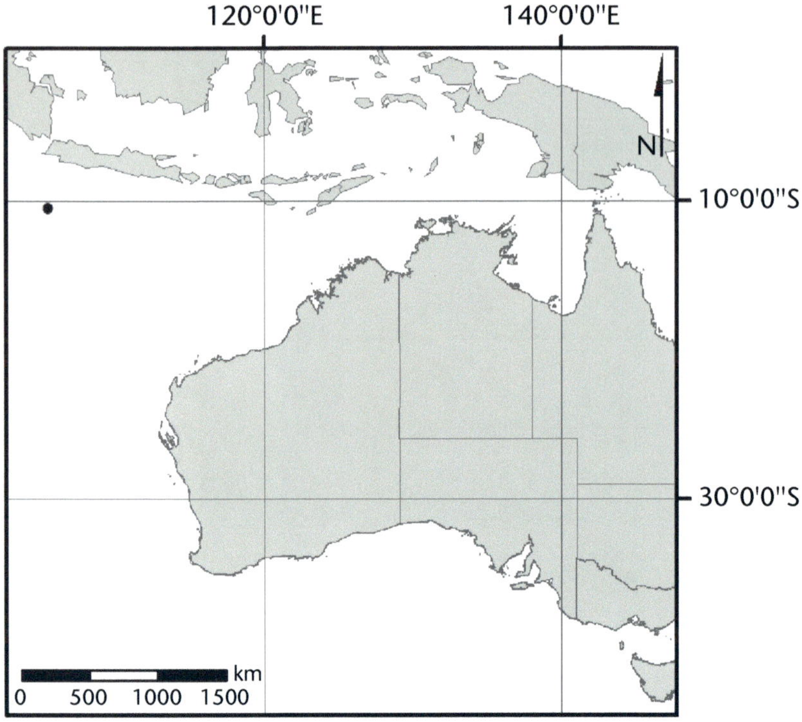

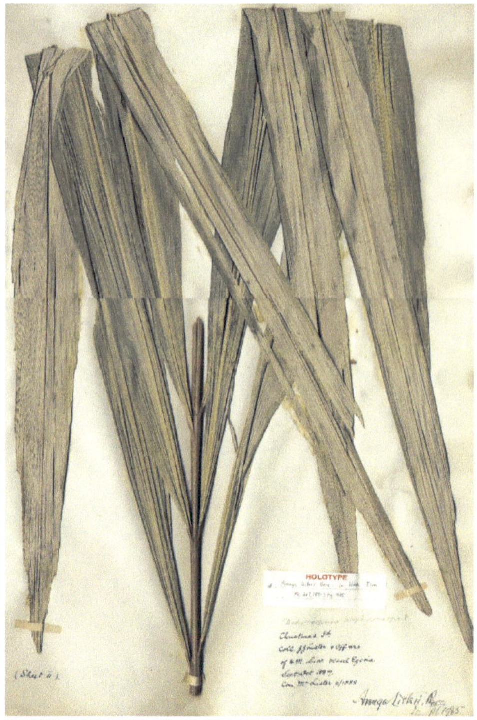

Figure 8.76 *Arenga listeri*. **Left:** Distribution based on herbarium specimen data. **Right:** Type specimen. Christmas Island (Indian Ocean). 3 Oct 1887, *J.J.Lister s.n.*; holotype K [1 of 3 sheets].

of Mt Trafalgar, 24 May 1898, *F.M.Bailey s.n.*; holotype BRI; isotype K.

Didymosperma novoguineense Warb. ex K.Schum. & Lauterb., *in* K.Schumann & K.Lauterbach, *Fl. Schutzgeb. Südsee*: 204 (1901). Type: Papua New Guinea. Madang Province, Hatzfeldhafen, *O.Warburg s.n.*; holotype B, destroyed?

Didymosperma humile Lauterb. & K.Schum., *in* K.Schumann & K.Lauterbach, *Fl. Schutzgeb. Südsee*: 204 (1901). Type: Papua New Guinea. Morobe Province, Finchaven, *K.A.G.Lauterbach 840*; holotype B, destroyed?

Arenga microcarpa var. *keyensis* Becc., *Nuovo Giorn. Bot. Ital.* n.s. 42: 29 (1935). Type: not designated.

[*Arenga australasica* Benth. ex Martelli, *Nuovo Giorn. Bot. Ital.* n.s. 42: 29 (1935)].

Moderate, hapaxanthic palms. **Stems** tightly clustering, 1–3 stems dominant, to 7 m tall, 10–15 cm dbh; leafscars to 2 cm wide; internodes to 45 cm long. **Leaves** 5–18, 2.5–4.0 m long, with pinnae in 2 planes; 30–70 pinnae each side; leafbase to 50 cm long, glabrous, green, with scattered dark scales, fibres on the margins enclosing the stem; petiole to 130 cm long; rachis to 270 cm long; ligule tubular. **Pinnae** 40–110 cm long, dark green on the adaxial surface, with moderately dense grey-green indumentum on the abaxial surface; margins with vein endings protruding, producing small sharp extensions; apex rounded or notched, toothed; midrib prominent on the abaxial surface; an auricle at the base on the proximal side faces back to the rachis on occasional pinnae, or lacking. **Inflorescences** 150–170 cm long, produced basipetally, branched to 1 order; prophyll to 30 cm long, to 10 cm wide; peduncular bracts 5–7, to 40 cm long, with dense indumentum, persistent; peduncle to 40 cm long, by 3 cm diam.; rachis to 50 cm long; rachillae to 40 cm long, green. **Staminate flowers** 6–9 mm long, 4–8 mm wide in bud; sepals 1.6–2 mm long, to 2 mm wide, white-cream to dirty yellow; petals 4.5–8 mm long, 4–6 mm wide, orange-yellow; stamens 14–24, 4–4.5 mm long; filaments to 0.5 mm long; anthers 3.5–4 mm long, c. 1 mm wide, yellow. **Pistillate flowers** to 6 mm long, to 8 mm wide in bud; sepals to 2 mm long, to 4 mm wide, green; petals to 5 mm long, to 4 mm wide, lemon yellow; stigmas to c. 0.5 mm long, lemon yellow. **Fruit** globose, 15–18 mm wide, ripening white to red. **Seeds** 1–3, globose to angled. Figures 8.77, 8.78.

Figure 8.77 *Arenga microcarpa*, closed coastal forest, Camel Creek, western Gulf of Carpentaria, Northern Territory.

Arenga

Distribution and ecology: In Australia, known from a few small scattered and isolated populations in coastal Northern Territory (see Notes for *A. australasica* above), from Adelaide River east to Camel Creek, as a subcanopy or understorey element in monsoon forest or moist vine forest, 0–150 m asl. It is widely distributed and common throughout New Guinea and eastern Indonesia (Maluka), on various soils but usually in wet areas, 0–700 m asl (Fig. 8.79). Flowering Jul.–Nov; fruiting Sept.–Jan. Conservation status – *Vulnerable* (NRETAS 2009).

Typification: The holotype of *Arenga microcarpa* (basionym *Didymosperma microcarpum*) is *Hollrung 302* [B], undated, from Finchaven in Papua New Guinea, but it appears to have been lost or destroyed. An isotype at FI consists of two sheets of leaf portions with complete pinnae, and a packet of fruit (Fig. 8.79).

Etymology: *'Fructu exacte spherico, 15–18 mm diametro'* [fruit exactly spherical, 15–18 mm diameter]: With reference to the relatively smaller fruit of this species compared to others in the genus, from the Latin *micro*, small, and *carpus*, fruit.

Notes: *Arenga microcarpa* is one of the most common palms in lowland areas of New Guinea. In Australia, the species is confined to the Northern Territory where it is rare and occurs as isolated individuals and small groups. No significant populations are known in Australia.

■ CORYPHEAE

Corypha

Corypha L., *Sp. pl.* 1187 (1753). Type: *Corypha umbraculifera* L.

NB: Generic synonyms which do not have a taxonomic association with Australian taxa are not included in this treatment. A full list of generic synonyms is provided in Dransfield *et al*. (2008).

Solitary, large to massive, erect, hapaxanthic, hermaphroditic palms. **Stem** robust, cylindrical to slightly ventricose, ringed with leafscars; internodes narrow to broad. **Leaves** costapalmate, persistent and eventually deciduous; sheath clasping the stem, basally cleft, or entire with lateral appendages, and with margin fibrous or smooth; petiole massive, long, with caducous indumentum, channelled on the adaxial surface, with numerous sharp irregular teeth or recurved or upcurved spines on the margins; hastula prominant on the adaxial surface, poorly developed on the abaxial surface; lamina regularly divided into single-fold segments. **Segments** shallowly divided along the abaxial folds, linear to lanceolate, induplicate, tapered to apex, with veinlets prominent or not, caducous floccose indumentum along folds. **Inflorescences** borne in a lateral position on an extension of the stem above the crown; individual inflorescences subtended by a reduced leaf or bract, branched 3–4 orders; prophylls on first-order branches tubular, 2-keeled; peduncular bracts tubular; rachis bracts small, triangular; rachillae rigid, erect to spreading. **Flowers** bisexual, in spirally arranged cincinni of 2–10, shortly stalked; 3 basally connate triangular sepals; 3 basally imbricate, marginally inrolled spreading petals; stamens 6, 3 antesepalous and free, 3 antepetalous and adnate to a petal; filament tapered; anthers short, basally sagittate, medifixed, latrorse; gynoecium tricarpelate, triovulate, ovary globose; style long, 3-grooved, stigma indistinct, ovule hemianatropous. **Pollen** ellipsoidal; aperture monosulcate or trichotomosulcate; exine

Figure 8.78 *Arenga microcarpa*. **Top left:** Closed coastal forest, Camel Creek, Northern Territory. **Top right:** Staminate flowers at anthesis. **Bottom left:** Pistillate flowers at anthesis. **Bottom right:** Mature fruit. Fruit and flowers from cultivated plant, provenance unknown, Centenary Lakes, Cairns, north-east Queensland.

tectate, finely reticulate. **Fruit** globose, 1-seeded; stigmatic remains basal; epicarp thick, smooth or minutely dotted; mesocarp fleshy; endocarp thin, adherent to seed; perianth not persistent. **Seed** globose, hilum basal, shallowly grooved; endosperm homogeneous, sometimes with a central cavity; embryo small, apical; germination remote-tubular; eophyll simple, lanceolate.

Corypha is a genus of six species, distributed from India, through south-east Asia to the Philippines, Malesia and northern Australia, where one species occurs.

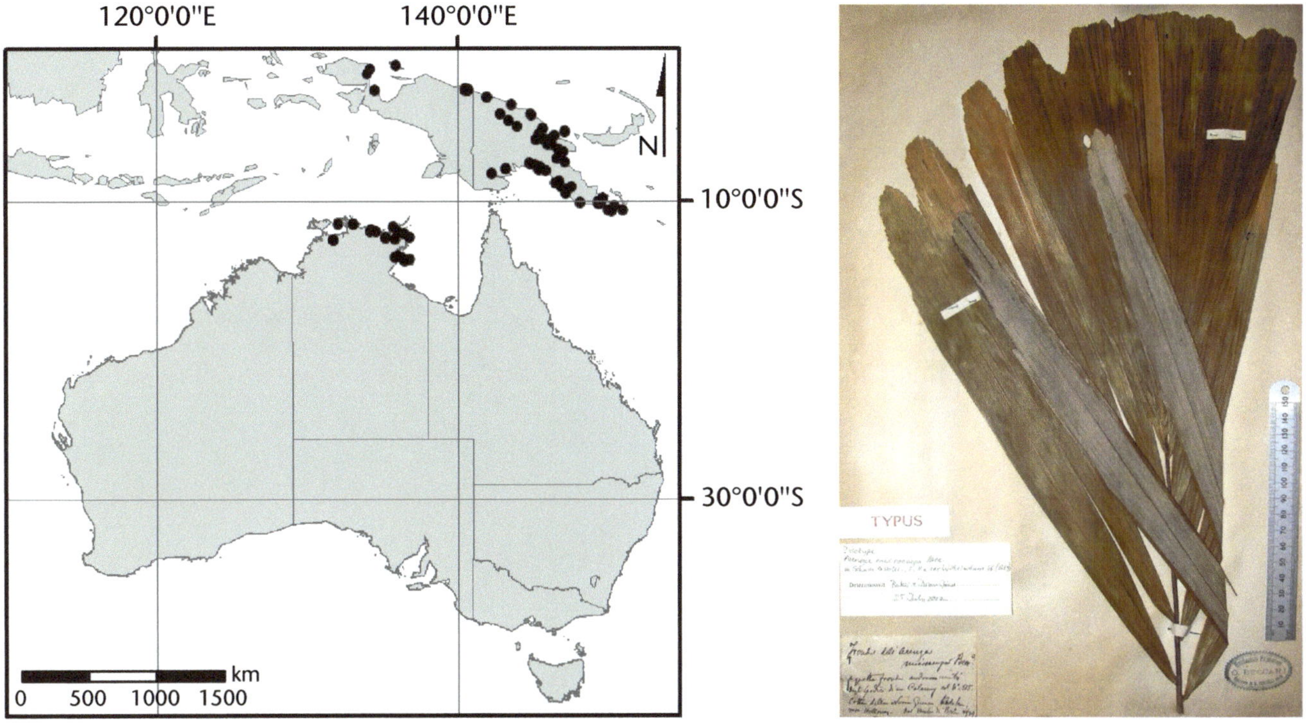

Figure 8.79 *Arenga microcarpa*. **Left:** Distribution based on herbarium specimen data. **Right:** Type specimen. Papua New Guinea. Morobe Province, Finchaven, *U.M.Hollrung 302*; isotype FI [1 of 3 sheets].

Etymology: '*Corypha*' (Linnaeus 1753): Derived from the Greek for summit or peak, *koryphe*, with reference to the emergence of the inflorescences from the apex of the stem.

Corypha utan

Corypha utan Lam., *Encycl.* 2: 131 (1786); *Borassus sylvestris* Giseke, *Prael. Ord. Nat. Pl.*: 86 (1792); *Taliera sylvestris* (Giseke) Blume, *in* J.A.Schultes & J.H. Schultes, *Syst.* 7: 1307 (1830); *Corypha sylvestris* (Giseke) Mart., *Hist. Nat. Palm. Edn 1*, 3: 233 (1838). Type: lectotype, illustration of *Lontarus silvestris* in G.E.Rumphius, *Herb. Amboin.*1: tab.11 (1741).

Corypha elata Roxb., *Fl. Ind.* 2: 176 (1824). Type: not designated.

[*Lontarus silvestris* Rumph., *Herb. Amboin.* 1: 53, tab.11 (1741)].

Stem to 25 m tall, moderately ventricose or with parallel sides, 0.5–1 m dbh; leafscars raised, rough, irregular, 1–5 cm wide; internodes to 10 cm long, grey; petiole remnants persistent in the basal portion. **Leaves** 15–30 in an elongate to rounded crown; petiole arching, 2.5–4 m long, 2–14 cm wide, yellowish, senescent leaves with petiole bent down near the middle to an angle of c. 90°; lamina ± suborbicular, 2–3 m diam., folded lengthwise, curved in profile, dull to grey-green on the adaxial surface, lighter green on the abaxial surface. **Segments** 60–80, 100–150 cm long, 5–8 cm wide, coriaceous; the midrib sunken on the adaxial surface, prominant on the abaxial surface; segments free for c. 30% or more of their length, depth of apical cleft 4–10% of the length of the free portion; apices rigid, lobes irregularly acuminate; parallel veinlets barely visible, >20 each side of the midrib; transverse veinlets not visible. **Inflorescence** structure 4–8 m high, pyramidal; individual inflorescences spirally arranged on an erect axis, branched to 5 orders, reducing in size toward the apex; basal inflorescence to 2.8 m long, distal inflorescence to 0.6 m long; first-order branches 30–50 on the basal inflorescence; prophyll on each inflorescence, to 17 cm long on the basal inflorescence, to 12 cm long on the distal inflorescence, bicarinate, papery, densely tomentose, apex extended to a point on one side;

Figure 8.80 *Corypha utan*, fruiting, Centenary Lakes, Cairns, north-east Queensland.

peduncular bract 1 on the basal inflorescence, absent on the distal inflorescence, to 15 cm long, tubular, papery, densely tomentose, apex extended to a point on one side; first-order branches with a much reduced bract at the base, scale-like, papery, glabrous; subsequent branches lacking bracts; peduncle in the basal inflorescence 20–30 cm long, terete in cross-section; subsequent branches subterete to angular, glabrous; rachillae more or less similar in length, c. 25 cm long, terete, narrowing to the apex, glabrous. **Flowers** 2–10 in elongate cincinni, spirally and regularly arranged along the rachillae; bracteoles subtend each flower, and each cincinni by a small ridge-like bract; sepals triangular, apex attenuate, 1–1.5 mm long, 1–1.5 mm wide, cream to white; petals boat-shaped, 2–2.6 mm long, 1.5–2.4 mm wide, cream to white; stamens 2.5–3.3 mm long; anthers c. 1 mm long, cream. **Fruit** 15–30 mm diam.; epicarp c. 0.2 mm thick, green to green-brown to brown with scattered lenticellular pores; mesocarp 2.5–5 mm thick; pedicel c. 5 mm long, 2–3 mm diam.; calyx not persistent on fruit, but persistent on rachillae. **Seed** 10–22 mm diam.; central cavity to 1 mm wide, or lacking; eophyll 5-ribbed. Figures 8.80, 8.81.

Corypha palm

Distribution and ecology: A widespread species, distributed from southern India throughout south-east Asia to Malesia, the Philippines and southern New Guinea, and in Australia in the Top End of the Northern Territory and Cape York Peninsula, Queensland. In the Northern Territory occurs on the floodplains of the Tomkinson and Liverpool Rivers and the Arafura Swamp, and in Queensland in Cape York Peninsula from the lower reaches of Mitchell River in the west, throughout the peninsula (except the far north) to Normanby River in the east. Occurs in

Figure 8.81 *Corypha utan*. **Top left:** Leafbases with characteristic basal cleft. **Top right:** Armature on petiole margin. **Bottom left:** Fallen flowers at anthesis. **Bottom right:** Mature fruit.

groves, small groups or scattered individuals associated with floodplain channels, anabranches, billabongs and seasonally wet depressions, and as a riparian element on the banks of seasonally flooded streams and rivers, as the dominant tree or as a canopy element where large trees are present, 10–80 m asl (Fig. 8.82). Flowering all months; fruiting all months. Conservation status – *No present threats*.

Typification: In the absence of a specimen, the illustration of *Lontarus silvestris* in plate 11 in volume 1 of

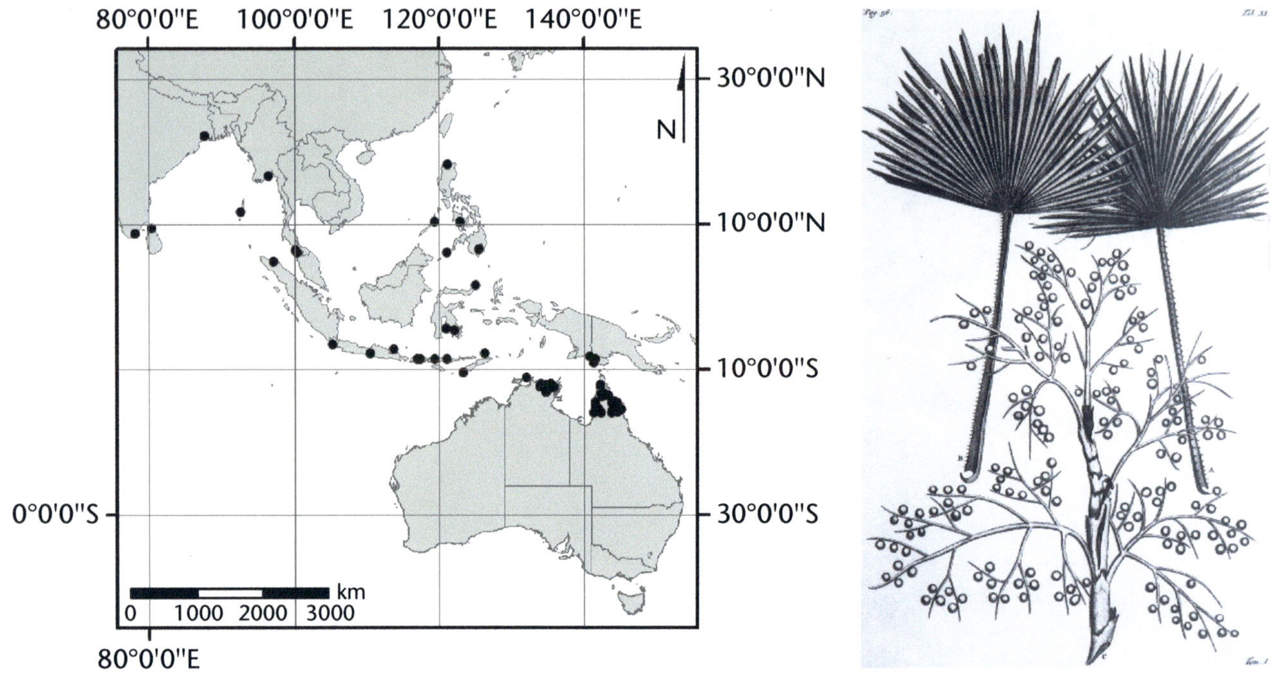

Figure 8.82 *Corypha utan*. **Left:** Distribution based on herbarium specimen data. **Right:** Type specimen. Lectotype, illustration of *Lontarus silvestris in* G.E.Rumphius, *Herbarium Amboinense* 1: plate 11 (1741).

Rumphius's *Herbarium Amboinensis* (1741) has been chosen as the lectotype. This illustration includes leaf and infructescence details (Fig. 8.82).

Etymology: *'Corypha (utan)'*: Lamarck (1786) did not provide an explanation of the epithet origin, but it is derived from the vernacular name cited by Rumphius in 1741 in *Herbarium Amboinensis*, *'Lontar Utan'*, which may be interpreted as the 'forest fan palm'.

Notes: *Corypha utan* was first recorded in Australia by Leichhardt (1847) in his account of the Overland Expedition from Darling Downs to Port Essington in 1844–45, in the lower Mitchell River area in the south-east Gulf of Carpentaria. The first specific recognition of this taxon as *Corypha elata* (= *C. utan*) was provided by Ewart *et al.* (1912) from the Lower Gilbert River, Queensland.

9. Subfamily 4: Ceroxyloideae

Ceroxyloideae Drude, *Bot. Zeit.* 35: 632 (1877), as *Ceroxylinae*.

Bisexual, monoecious or dioecious, unarmed pleonanthic palms. **Leaves** pinnate. **Pinnae** reduplicate. **Inflorescence** male and female similar or markedly dimorphic, paniculate with 1 to numerous peduncular bracts. **Flowers** dimorphic, solitary or in monopodial clusters; perianth segments free; stamens 6, or numerous; ovary smooth; gynoecium syncarpous and triovulate. **Fruit** 1- or 2–10-seeded; stigmatic remains apical to basal; epicarp smooth or cracking to form irregular fibrous pyramidal warts; mesocarp fleshy; endocarp thin or obsolescent. **Seed** not adhering to the endocarp, not sarcotestal; endosperm homogeneous; eophyll bifid or pinnate.

A subfamily of 8 genera and c. 42 species, with disjunct distribution in the Americas, Juan Fernandez Island and Madagascar, and in north-east Queensland with one endemic monotypic genus.

Classification and relationships

The Ceroxyloideae is monophyletic and resolves in most phylogenetic studies as sister to the Arecoideae (Asmussen *et al.* 2006; Trénel *et al.* 2007; Dransfield *et al.* 2008). Although one of the smallest subfamilies, with only eight genera and 42 species, the Ceroxyloideae is among the most vegetatively and reproductively diverse subfamilies in the Arecaceae. The centre of diversity of the Ceroxyloideae is the Americas, with isolated genera occurring in Madagascar and Australia. The subfamily has three clearly monophyletic tribes – the Cyclospatheae, Phytelepheae and Ceroxyleae – of which only the last has representatives in Australia (Fig. 9.1). The Ceroxyleae has four morphologically similar genera – *Ceroxylon* (11 species in South America), *Ravenea* (18 species in Madagascar), *Juania* (one species on Juan Fernandez Island) and the monotypic *Oraniopsis* in Australia (Table 9.1).

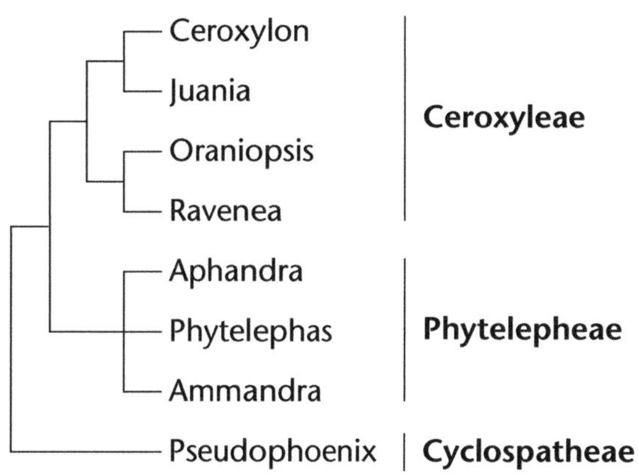

Figure 9.1 Phylogenetic tree summarising the estimated relationships of genera in the Ceroxyloideae. Adapted from Asmussen *et al.* (2000, 2006), Trénel *et al.* (2007) and Dransfield *et al.* (2008).

Table 9.1 Synopsis of the Ceroxyloideae tribes and genera in the Australian palm flora

Ceroxyleae Satake, *Hikobia* 3: 125 (1962)
Oraniopsis

Oraniopsis

Oraniopsis (Becc.) J.Dransf., A.K.Irvine & N.W.Uhl, *Principes* 29: 57 (1985); *Orania* Zipp. subg. *Oraniopsis* Becc., *in* O.Beccari & R.E.G.Pichi-Sermolli, *Webbia* 11: 172 (1955). Type: *Areca appendiculata* F.M.Bailey = *Oraniopsis appendiculata* (F.M.Bailey) J.Dransf., A.K.Irvine & N.W.Uhl.

Solitary, moderate to large, pleonanthic, dioecious palms. **Stem** partially procumbent to erect, leafscars not conspicuous; internodes compressed. **Leaves** paripinnate, erect to arching, marcescent, senescent leaves sometimes forming a skirt; leafbases clasping and not forming a tubular crownshaft, persistent, later caducous, deeply split opposite the petiole, margins fibrous; petiole short, channelled and glabrous on the adaxial surface, rounded and densely scaly on the abaxial surface, margins sharp; rachis rigid, sparsely scaly, flattened to channelled on the adaxial surface and distally angled, with a raised flange in the middle portion, approximately where the flattened and angled areas of the rachis meet, rounded on the abaxial surface. **Pinnae** in a single plane on the rachis, opposite to subopposite, linear, reduplicate, rigid, apex unevenly acute or acuminate, basal pinnae narrow and congested; adaxial surface glabrous with scattered scales along the midrib; midrib prominent; abaxial surface with a dense, silver-coppery felt-like indumentum; ramenta absent. **Inflorescences** interfoliar, paniculate, branched to 4 orders, erect to horizontal, branches divaricate; staminate and pistillate inflorescences similar; prophyll short, obscured by leafbases, incompletely tubular, 2-keeled, fibrous and distally disintegrating, sparsely tomentose; peduncular bracts 3–5, elongate, the most basal inserted near the prophyll, others regularly attached along the peduncle, the distal 2 or 3 enclosing the inflorescence in bud, beaked, leathery, tubular, splitting longitudinally, sparsely to densely tomentose, persistent, leaving prominent scars; rachis bracts short, triangular, acute or acuminate, membranous, subtending first-order branches; peduncle long, basally winged, distally elliptical in cross-section, tomentose; first-order branches with a basal bare portion; rachillae congested, twisted or flexuous. **Staminate and pistillate flowers** on different plants; flowers solitary, spirally arranged or subdistichous, stalked, each subtended by a membranous triangular bracteole. **Staminate flowers** symmetrical, or angled from congestion, congenitally open at bract abscission; 3 small triangular membranous sepals, basally connate forming a cup; 3 fleshy narrow triangular petals longer than the sepals; stamens 6, about as long as or longer than the petals; 3 antesepalous stamens inserted between the petals in the same whorl, the other 3 epipetalous; filaments fleshy with conical swollen bases, tapered; anthers oblong, versatile, basally sagittate, latrorse; pistillode shorter than the filaments, 3-angled, apically trifid. **Pistillate flowers** similar to the staminate with broader sepals and petals; staminodes as the stamens; anthers empty, large; gynoecium tricarpellate, triovulate, 3-lobed; 3 apical short recurved stigmas; ovules laterally attached, hemianatropous. **Pollen** ellipsoidal; aperture monosulcate; exine semitectate, the distal face perforate, the proximal face reticulate. **Fruit** globose to subglobose, 1-seeded; stigmatic remains basal; epicarp smooth, yellow-orange at maturity; mesocarp fleshy, with horizontal fibres and hardened cells; endocarp obsolescent; perianth not persistent on fruit. **Seed** globose with a short basal spur, and a few vascular strand impressions; endosperm homogeneous with a narrow central hollow; embryo subbasal; germination adjacent-ligular; eophyll bifid with entire tips.

Etymology: 'We have chosen to use Beccari's subgeneric name *Oraniopsis* for the new genus, although we must emphasise that there is no relationship between the new genus and Orania, except that caused by misunderstanding' (Dransfield *et al.* 1985, p. 57). *Orania* Zipp. is a genus distributed in Madagascar, the Philippines, Malay Peninsula, Indonesia and New Guinea, and is superficially similar but unrelated to *Oraniopsis*.

Oraniopsis appendiculata

Oraniopsis appendiculata (F.M.Bailey) J.Dransf., A.K.Irvine & N.W.Uhl, *Principes* 29: 61, figs 58, 60 (1985); *Areca appendiculata* F.M.Bailey, *Queensland Bot. Bull.* 4: 18 (1891); *Orania beccarii* F.M.Bailey, *Queensland Agric. J.* 23: 35 (1909) *nom. illeg.*; *Orania appendiculata* (F.M.Bailey) Domin, *Biblioth. Bot.* 85: 498 (1915). Type: Australia. Queensland. Bellenden Ker Range, 4000 ft [Palm Camp], 1889, *F.M.Bailey s.n.*; holotype BRI – AQ024162 [1 sheet].

Stem to 20 m tall, 20–45 cm dbh, grey, basally procumbent, enlarged in the procumbent portion; leafscars irregular, congested, to 10 mm wide; internodes compressed, indiscernible from leafscars. **Leaves** 8–15, to 6 m long; erectly arching to horizontal, 80–110 pinnae each side of rachis; leafbases c. 20 cm wide at the base, tapering to c. 15 cm wide, white-brownish, scurfy; petiole to 65 cm long, to 7 cm wide; marginal wings 1–2 cm long, sharp, brown tomentose on the abaxial surface; rachis to 5 m long, becoming ridged at ± midleaf on the adaxial surface; flange on the adaxial surface 1–5 mm high. **Pinnae** to 100 cm long, to 4.5 cm wide, rigid, the adaxial surface dark green, the abaxial surface with silver-coppery indumentum, tinged brown, with numerous small dark brown scales; apical pinnae basally united. **Staminate inflorescence** first-order branches c. 70, to 5 cm long, to 1.5 mm thick. **Pistillate inflorescence** first-order branches 30–40, to

10 cm long, to 2 mm thick. **Staminate and pistillate inflorescences** otherwise similar, to 120 cm long; peduncle to c. 75 cm long, to c. 4 cm in cross-section, strongly flattened and winged at the base; prophyll borne near the base of the peduncle, to 25 cm long, to 6 cm wide, open on the abaxial surface to the base, splitting irregularly; peduncular bracts 4–6, inserted at 12–20 cm intervals on the peduncle, to 135 cm long, to 12 cm wide; rachis bracts to 4 cm long, to 1 cm wide, triangular, membranous, disintegrating; distal rachis bracts small and inconspicuous; rachillae erect, rigid, to 35 cm long; floral pit margins scarcely raised. **Staminate flowers** to c. 7 mm long, to 5 mm wide in bud, opening to c. 12 mm wide at anthesis; sepals to c. 1 mm long, to 0.5 mm wide, white-cream; petals to c. 6 mm long, to c. 2 mm wide, white-cream; stamens to c. 6 mm long; filaments to 3.5 mm long, c. 1 mm diam. at the base; anthers to 5 mm long; pistillode c. 1 mm high; pedicels c. 1.2 long, to 1.2 mm thick. **Pistillate flowers** similar to staminate; sepals to c. 1 mm long, to 0.5 mm wide, white-cream; petals to 5 mm long, to 2.2 mm wide, white-cream; filaments of staminodes to c. 1.5 mm long; empty anthers to c. 3 mm long; ovary strongly 3-lobed, c. 2 mm high, to 3.5 mm wide; stigmas to c. 0.8 mm long; pedicels c. 2 mm long, to 2 mm thick. **Fruit** to 3.4 cm long, to 3 cm diam.; stigmatic remains basal, to 4 mm high; epicarp c. 0.8 mm thick, smooth, yellow-orange; mesocarp c. 3 mm thick; pedicel to 3 mm long, to 4 mm thick. **Seed** to c. 2.2 cm diam.; integument black, c. 0.5 mm thick; endosperm cavity to 2 mm wide, or lacking. Figures 9.2, 9.3.

Figure 9.2 *Oraniopsis appendiculata* in lowland rainforest habitat, Campbell Creek, Walter Hill Ranges, north-east Queensland, c. 100 m altitude.

Oraniopsis palm

Distribution and ecology: Endemic to north-east Queensland, from Mt Spurgeon south to Walter Hill Ranges, most common in high-rainfall locations at moderate to high altitude 400–1440 m, and infrequently as low as 80 m asl in the footslopes of Mt Tyson, occurring in small populations or isolated groups or individuals as an understorey or subcanopy element in rainforest, most common as a riparian species, also in gallery forest in moist sclerophyll forest (Fig. 9.4). Flowering Feb.–Oct.; fruiting Apr.– Feb. Conservation status – *No present threats*.

Typification: The holotype of *Oraniopsis appendiculata* (basionym *Areca appendiculata*) is an unnumbered collection made by F.M.Bailey [BRI] in 1889 from Bellenden Ker Range in north-east Queensland. The collection consists of a section of leaf with two pinnae, an outer peduncular bract, three portions of inflorescence and a packet of flowers (Fig. 9.4).

Etymology: 'referring to the appendages on petals' (Bailey 1891, p. 18): From the Latin *appendicula*, small addition, probably referring to the epipetalous stamens, as there are no 'abnormal' structures on the petals.

Notes: *Oraniopsis appendiculata* is distinguished by the pinnae with a cream tomentum on the abaxial surface, the inflorescence with multiple overlapping bracts and large round yellow fruits. The raised flange (Fig. 9.3) on the adaxial surface of the leaf rachis is a unique characteristic of the species. For a summary of historical biogeography see Chapters 3 and 4.

Figure 9.3 *Oraniopsis appendiculata*. **Top left:** Raised flange in the middle portion of the adaxial surface of the leaf rachis, where the flattened and angled areas of the rachis meet. **Top right:** Staminate flowers at anthesis. **Bottom left:** Crown with infructescences. **Bottom right:** Immature and mature fruit.

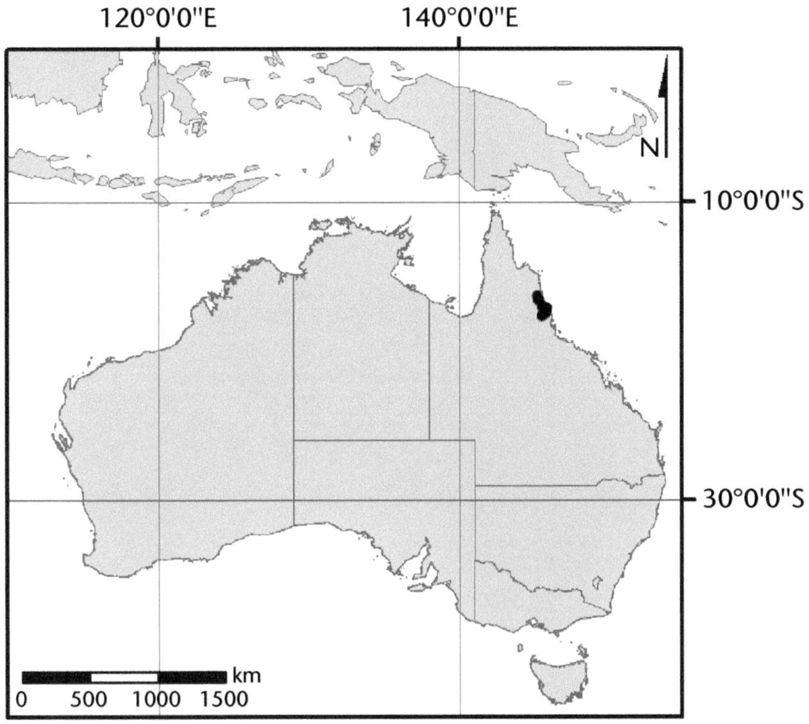

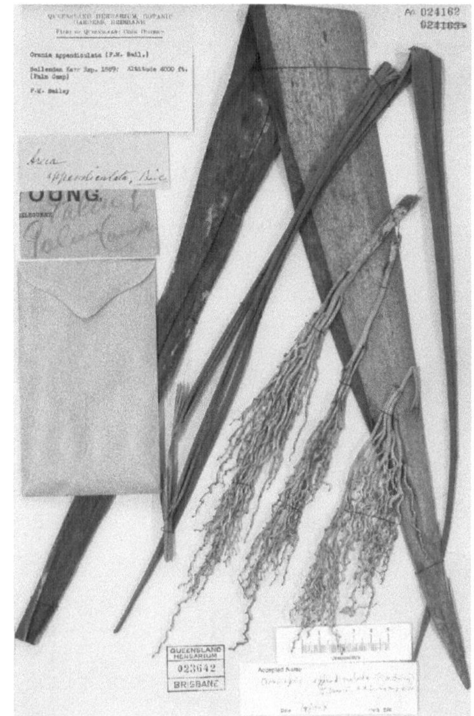

Figure 9.4 *Oraniopsis appendiculata*. **Left:** Distribution based on herbarium specimen data. **Right:** Type specimen. Australia. Queensland. Bellenden Ker Range, 4000 ft [Palm Camp], 1889, *F.M.Bailey s.n.*; holotype BRI.

10. Subfamily 5: Arecoideae

Arecoideae Burnett, *Outl. Bot.*: 401 (1835), as *Arecidae*.

Monoecious or dioecious, armed or unarmed, pleonanthic palms. **Leaves** pinnate, pinnae reduplicate. **Inflorescence** spicate or paniculate; peduncular bracts usually 1 or infrequently more. **Flowers** solitary, or borne in triads or units derived from triads; perianth segments free; stamens 3, 6 or more; ovary smooth; carpels pseudomonomerous, triovulate or trilocular. **Fruit** small to large, 1–8-seeded; epicarp smooth, or spiny, corky warty, or variously rugose; mesocarp fibrous or fleshy; endocarp thin or thick. **Seed** not sarcotestal; endosperm ruminate or homogeneous; eophyll simple, bifid or pinnate.

A subfamily with c. 107 genera and c. 1250 species in 13 tribes, with world-wide distribution; in Australia with two tribes, the Cocoseae with a single species (*Cocos nucifera*), and the Areceae with five subtribes with 12 genera (eight endemic) with 26 species (23 endemic) (Table 10.1).

Classification and relationships

The Arecoideae is monophyletic and resolves in most phylogenetic studies as sister to the Ceroxyloideae (Asmussen *et al.* 2006; Dransfield *et al.* 2008). It has the largest number of genera and species of all the subfamilies, and is characterised by pinnate and pinnately veined simple leaves, and flowers mostly in triads or derived from triads, although other arrangements occur. It has generally equal diversity in the Americas and in the south-east Asia to western Pacific region (Fig. 10.1). Aspects related to the fossil record of Arecoid-like species in Australia and New Zealand are discussed in Chapter 3. In summary, Arecoid-like palms were present in the region from at least the Palaeocene and were widespread in Australia and New Zealand. They were extirpated through the effects of climate and environmental changes by the Oligocene. Some lines, putatively related to the extant Areceae, may have persisted, although molecular evidence suggests a radiation from west Malesian stock may account for the extant distribution of this tribe.

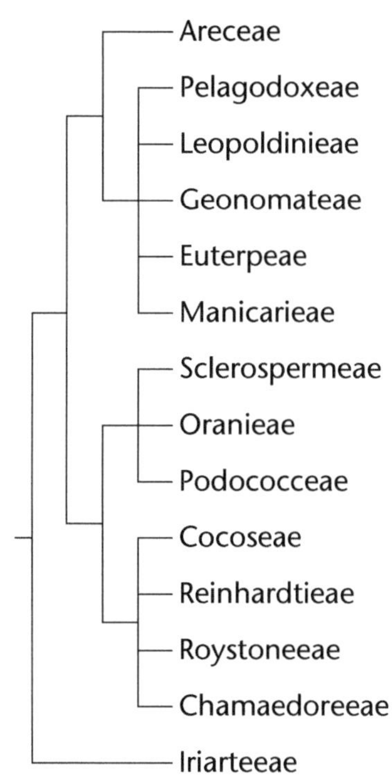

Figure 10.1 Phylogenetic tree summarising the estimated relationships of tribes in the Arecoideae. Adapted from Asmussen *et al.* (2000, 2006), Asmussen and Chase (2001), Lewis and Doyle (2002), Hahn (2002a), Norup *et al.* (2006) and Dransfield *et al.* (2008).

■ COCOSEAE

Relationships

The Cocoseae, with three subtribes and 18 genera, is predominantly American but is also in Africa (two genera, two species) and Madagascar (two genera, three species), and

Table 10.1 Synopsis of the Arecoideae tribes, subtribes and genera in the Australian palm flora

Cocoseae Mart. ex Dumort., *Anal. Fam. Pl.*: 56 (1829), as *Cocoinae*
Attaleinae Drude, *in* C.F.P. von Martius, *Fl. Brasil.* 3(2): 395 (1882)
Cocos
Areceae Mart. ex Dumort., *Anal. Fam. Pl.*: 55 (1829), as *Arecinae*
Archontophoenicinae J.Dransf. & N.W.Uhl, *Principes* 30: 8 (1986)
Archontophoenix
Basseliniinae J.Dransf., N.W.Uhl, C.Asmussen, W.J.Baker, M.M.Harley & C.Lewis, *Kew Bull.* 60: 563 (2005)
Lepidorrhachis
Linospadicinae Hook.f., *in* G.Bentham and J.D.Hooker, *Gen. pl.* 3: 872 [876] (1883), as *Linospadiceae*
Linospadix
Howea
Laccospadix
Ptychospermatinae Hook.f., *in* G.Bentham and J.D.Hooker, *Gen. pl.* 3: 872 [874] (1883), as *Ptychospermeae*
Ptychosperma
Carpentaria
Wodyetia
Normanbya
Rhopalostylidinae J.Dransf., N.W.Uhl, C.Asmussen, W.J.Baker, M.M.Harley & C.Lewis, *Kew Bull.* 60: 563 (2005)
Rhopalostylis
Hedyscepe
Unplaced Areceae
Hydriastele

has the monotypic genus, *Cocos*, from subtribe Attaleinae, in the Indian Ocean, Malesia, western Pacific and Australia (Fig. 10.2). The fossil evidence indicates that Cocosoid relatives were in Australia and New Zealand in the Miocene and Pliocene, but these appear to have been extirpated by the Late Pliocene (see Chapter 3). The Cocoseae is monophyletic, and is variously sister to the Reinhardtieae, the Roystoneeae or the Oranieae in the available molecular analyses (Dransfield *et al.* 2008).

Cocos

Cocos L. *Sp. pl.* 1188 (1753). Type: *Cocos nucifera* L.

NB: Generic synonyms that do not have a taxonomic association with Australian taxa are not included in this treatment. A full list of generic synonyms is provided in Dransfield *et al.* (2008).

[*Cocos* L., *in* G.Bentham, *Fl. austral.* 7: 142 (1878); *in* F.M.Bailey, *Queensl. Fl.* 5: 1687 (1902); *in* D.J. Du Puy and I.R.H. Telford, *Flora of Australia* 50: 433 (1993)].

Moderate to tall, solitary, leaning to erect, pleonanthic, monoecious palms. **Stems** slender to moderate, base expanded; leafscars prominent; internodes short to long. **Leaves** paripinnate, cleanly deciduous; leafbases clasping, not forming a crownshaft, splitting opposite the petiole, fibrous at the margins with a ligulate disintegrating extension; petiole short to long, channelled on the adaxial surface, rounded on the abaxial surface and tomentose; rachis elongate, straight or slightly curved, angled on the

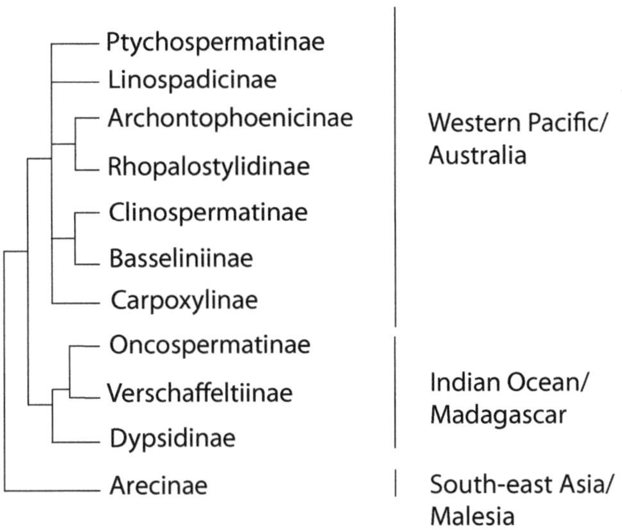

Figure 10.2 Phylogenetic tree summarising the estimated relationships of genera in the Attaleinae. Adapted from Hahn (2002a), Gun (2004), Asmussen *et al.* (2006) and Dransfield *et al.* (2008).

adaxial surface, rounded on the abaxial surface and tomentose. **Pinnae** in a single plane on the rachis, subopposite to opposite, linear-acuminate, reduplicate, erect to horizontal to slightly pendulous, glabrous on the adaxial surface, with dot-like scales on the abaxial surface, apex acute to tapered or bilobed; midrib prominent, transverse veinlets obvious, ramenta small, scale-like on abaxial midrib. **Inflorescence** interfoliar, paniculate, branched to 1 order, erect to pendulous, branches divaricate, protandrous; prophyll tubular, 2-keeled, opening apically, persistent, fibrous, within leafbases; peduncular bract large, tubular, woody, beaked, with longitudinal grooves, attached near prophyll; peduncle short to long, elliptic in cross-section, with scattered scales; rachis as long as peduncle; rachillae thick, arranged spirally, subtended by triangular bracts, erect, becoming pendulous in fruit. **Flowers** unisexual, sessile, in well-spaced triads of a single pistillate flower subtended by a pair of staminate flowers either side, borne spirally on the rachillae, triads in proximal portion, paired or solitary staminate flowers in distal portion. **Staminate flowers** asymmetrical, ovoid in bud; 3 imbricate sepals, unequal, triangular; petals much longer than sepals, thick, leathery, acute, valvate; stamens 6; filaments short; anthers medifixed, versatile, elongate, basally sagittate; pistillode thin, 3-lobed. **Pistillate flowers** large, globose in bud; 3 sepals, imbricate, rounded; 3 petals, rounded, longer than the sepals, thick, imbricate; staminodal ring not lobed; gynoecium basally trilocular, triovulate, ovoid, 3-angled; 3 short stigmas, ovule anatropous, small, laterally attached. **Pollen** ellipsoidal; aperture monosulcate or trichotomosulcate; exine tectate, reticulate. **Fruit** large, ellipsoidal to ovoid, 3-angled, 1-seeded; stigmatic remains apical; epicarp smooth; mesocarp thick, fibrous, dry; endocarp thick, woody, spherical to ovoid, with 3 basal operculate pores of which one covers the embryo; perianth enlarged, not persistent on fruit. **Seed** large; endosperm homogeneous, with central cavity filled partially with fluid; embryo basal; eophyll simple; germination adjacent-ligular.

Cocos is a monotypic genus native to the arc of islands between southern India and Malesia (Buckley and Harries 1984), but with relatives in South America, South Africa and Madagascar. The oil extracted from the endosperm of *C. nucifera* is a major world commodity used in cooking and the manufacture of soaps and cosmetics. It is a primary product of many indigenous cultures in Malesia, Melanesia and Polynesia.

Etymology: '*Cocos*': Linnaeus (1753) did not provide an explanation of the derivation of the name *Cocos*, although it had been used in numerous pre-Linnean publications. Derived from the Portugese word for coconut palm, *coco*, from the Latin for berry or kernal, *coccum*, which is from the Greek *kokkos*.

Cocos nucifera

Cocos nucifera L. *Sp. Pl.* 2: 1188 (1753). Type: lectotype, illustration of '*Tenga*', *in* H.A.Rheede, *Hort. Malab.* 1: 1–8, tab. 1–4 (1678), *fide* Moore and Dransfield (1979).

[*Cocos nucifera* L., *in* F.Mueller, *Essay Pl. Collect. Fitzalan*: 18 (1860); *in* F.Mueller, *Fragm.* 5: 49 (1865); *in* H.Wendland & O.Drude, *Linnaea* 39: 233 (1875); *in* Bentham, *Fl. austral.* 7: 143 (1878)].

Stem to 35 m tall, 30 cm dbh, enlarged at the base to 50 cm diam.; leafscars 1–7 cm wide, raised, irregular; internodes to 15 cm long, irregular, green to grey. **Leaves** 10–20, to 7 m long, held erect to horizontal, or below the horizontal, straight to gently arching, 70–110 pinnae each side of rachis; leafbase to 40 cm long; petiole 1–1.3 m long; rachis 4–5 m long. **Pinnae** to 130 cm long, to 5 cm wide, regularly arranged, mid green on the adaxial surface, light green on the abaxial surface; midrib raised prominently on adaxial surface, raised but less prominent on abaxial surface; ramenta scale-like, flat and coarse; c. 6 minor parallel veinlets each side of midrib, margins not thickened. **Inflorescence** 1–2 m long, to 60 cm wide, axes cream to green initially, becoming green to brown in fruit, branched to 1 order; prophyll to 70 cm long, to 15 cm wide, strongly winged, curved, laterally twisted, woody, apex acuminate, glabrous or with scattered minute pubescence; peduncular bract and inflorescence emergent from abaxial opening; peduncular bract to 135 cm long, to 12 cm wide, basally tubular tightly enclosing peduncle, hooded, woody, longitudinally grooved, internally smooth, apex elongately beaked; peduncle to 60 cm long, 3–5 cm wide, elliptical in cross-section, green with scattered dark scales; rachis to 40 cm long, basally c. 3 cm wide, green with scattered dark scales; rachillae c. 25, spirally arranged, 40–50 cm long, straight to moderately flexuous, angular, glabrous; staminate flowers 40–60 on each rachilla, spirally arranged in distal three-quarters, solitary or in pairs; pistillate flowers 1–3 on each rachilla, in proximal quarter; floral pits shallow, margins low, rounded. **Staminate flowers** 7–15 mm long, 3–9 mm wide in bud, asymmetric, opening 15–20 mm wide at anthesis; sepals to 3 mm long, to 5 mm wide, imbricate, margins entire, cream-yellow, glabrous; petals to 15 mm long, to 4 mm wide, asymmetric, thick, apex pointed, cream-yellow, glabrous; stamens 6; filaments to 8 mm long, straight, cream-yellow; anthers to 4 mm long, cream-yellow, medifixed, versatile; pistillode trifid, lobes to 6 mm long, recurved, cream-yellow. **Pistillate flowers** 20–40 mm high, 30–35 mm wide in bud; floral bracts sepal-like, to 15 mm long, to 25 mm wide, thick, green, glabrous; sepals 15–30 mm long, 25–32 mm wide, imbricate, margins irregularly indented, cream-yellow to green, glabrous; petals 25–35 mm long, 30–40 mm wide, imbricate, margins smooth, cream-yellow to green, glabrous; stigmas erect to recurved, to 4 mm long, opening 4–5 mm wide at receptivity, glistening white. **Fruit** 10–35 cm long, 8–20 cm diam., 3-angled; stigmatic remains apical; epicarp smooth, to 1.3 mm thick, green, brown, yellow or orange; mesocarp 1–5 cm thick, dryish, fibrous; endocarp to 8 mm thick, with 3 basal pores; perianth persistent on rachillae after fruit fall. **Seed** globose to subglobose, 6–25 cm diam.; endosperm 4–15 mm thick; cavity 5–22 cm wide; embryo to 7 mm long, apically truncate. Figures 10.3, 10.4.

Figure 10.3 *Cocos nucifera*, Centenary Lakes, Cairns, north-east Queensland.

Coconut palm

Distribution and ecology: In Australia, the coconut grows in tropical areas as a cultivated plant or as tide-dispersed strand-line progeny, from about Broome in Western Australia, along the northern coast east to about Gladstone in Queensland, on Christmas Island and the Cocos (Keeling) Islands (Du Puy and Telford 1993) and the Torres Strait islands (Fig. 10.5). Populations purportedly occurring prior to European exploration in 1770 have been recognised in a few localities, such as islands in central and northern Torres Strait, and Lizard Island and Frankland Islands off the north-east coast of Queensland (Dowe and Smith 2002). Occurs most often in groves or as scattered individuals or small groups in littoral forest and mangrove margins, 0–20 m asl, but cultivated up to 1200 m asl. Flowering all months; fruiting all months. Conservation status – *No present threats*.

Typification: In the absence of specimens, the illustrations of *'Tenga'* in H.A. Rheede's *Hortus Indicus Malabaricus* Vol. 1, plates 1–4 (1678), were chosen by Moore and Dransfield (1979) as the lectotype of *Cocos nucifera*. These illustrations show the habit, and details of leaf, inflorescence, flower and fruit (Fig. 10.5).

Figure 10.4 *Cocos nucifera*. **Top left:** Staminate flowers at anthesis. **Top right:** Pistillate flower at anthesis. **Bottom left:** Crown with infructescences. **Bottom right:** Fruit, whole, in longitudinal section and endocarp.

Etymology: Linnaeus did not provide an explanation of the epithet's derivation, but cited Bauhin's name *Palma indica nucifera* from the pre-Linnean book *Historia Plantarum Universalis* (Bauhin 1650–51), one of the earliest known uses of the name, as part of the taxonomic citation. With reference to the fruit being nut-like, from the Latin *nux*, a nut, and *ferre*, to bear.

Notes: Although the distinction between wild-type coconuts and cultivated coconuts is difficult to determine, natural populations consisting of wild-type forms have been reported from a few localities in eastern Australia, such as Lizard Island (Buckley and Harries 1984) and Frankland Islands (Dowe and Smith 2002). The first European explorers and navigators to explore the Australian coasts did not

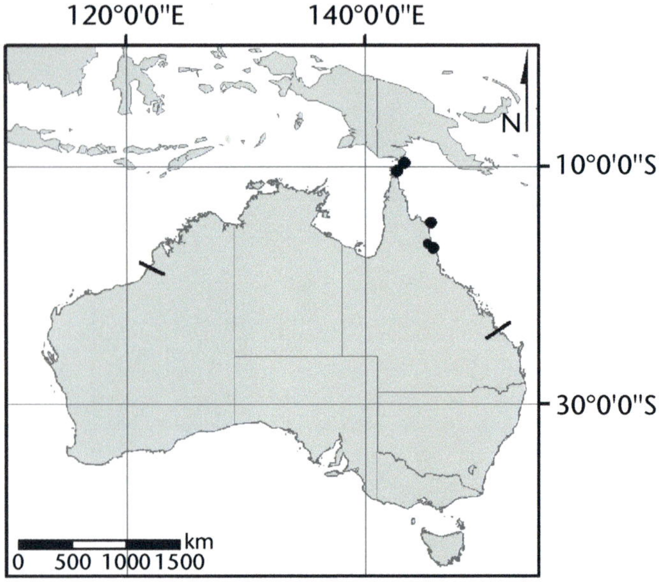

Figure 10.5 *Cocos nucifera*. **Left:** Distribution based on herbarium specimen data, and the southern limits of spontaneous regeneration on the west and east coasts, indicated by short lines. Extant distribution is more or less continuous on the northern coast of Australia and is more extensive than indicated by herbarium collections. **Right:** Type specimen. Lectotype, illustration of '*Tenga*', *in* H.A. van Rheede, *Hortus Malabaricus* 1: Tab. 1–4 (1678).

report seeing coconuts growing on the shores of mainland Australia, only floating fruits or open and decayed endocarps among the shoreline flotsam (Flinders 1814; Banks 1962). The first report of mature fruit-bearing coconuts was of a small population on the Frankland Islands off the coast of north-east Queensland, observed during the voyage of the *Rattlesnake* in 1848 (Brierley 1848; MacGillivray 1852; Dowe and Smith 2002). Fossil fruits dated to the late Pliocene have been found in the Chinchilla Sands of southern Queensland, the location interpreted as an ancient shoreline (Rigby 1995) (see Chapter 3).

■ ARECEAE

Relationships

The Areceae is monophyletic in most phylogenetic studies (Asmussen *et al.* 2006; Dransfield *et al.* 2008), but its relationships to other tribes is unresolved. A preliminary summary is presented here (Fig. 10.1). Of the 11 subtribes (and 10 unplaced genera) recognised in the Areceae, five (and one unplaced genus) have representatives in Australia (Fig. 10.6).

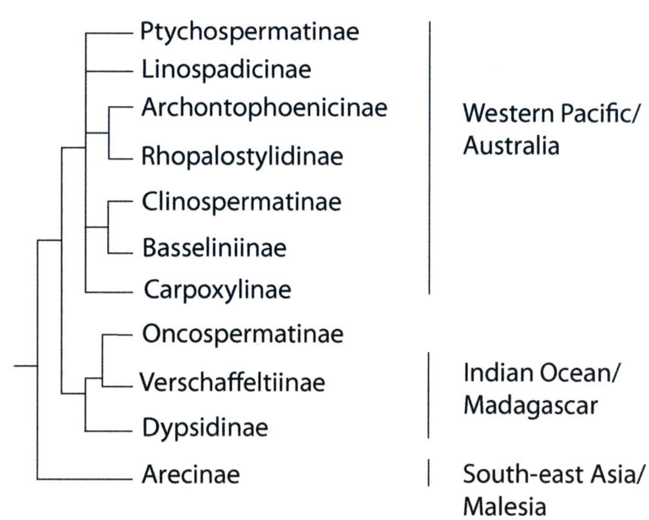

Figure 10.6 Phylogenetic tree summarising the estimated relationships of subtribes in the Areceae. Adapted from Hahn (2002a), Asmussen *et al.* (2006), Norup *et al.* (2006) and Dransfield *et al.* (2008).

■ ARCHONTOPHOENICINAE

Relationships

The Archontophoenicinae resolves as monophyletic in most analyses (Hahn 2002a; Norup *et al.* 2006; Dransfield *et al.* 2008) and sister to the Rhopalostylidinae, but relationships to other subtribes remain unclear. Morphologically and anatomically, *Archontophoenix* remains distinct but is closely related to *Actinokentia*, *Chambeyronia* and *Kentiopsis* (Pintaud 1999; Essig and Hernandez 2002). The relationships of *Actinorhytis* to other genera in the

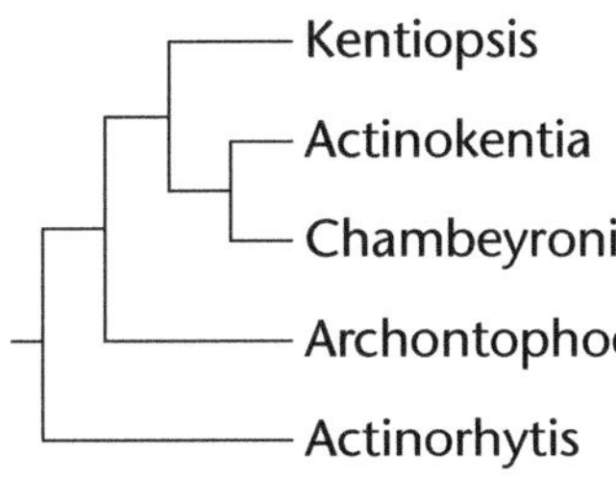

Figure 10.7 Phylogenetic tree summarising the estimated relationships of genera in the Archontophoenicinae. Adapted from Pintaud (1999), Asmussen *et al.* (2006) and Dransfield *et al.* (2008).

subtribe have not been resolved, and it has been provisionally placed as sister to all other genera in the subtribe (Dransfield *et al.* 2008) (Fig. 10.7).

Archontophoenix

Archontophoenix H.Wendl. & Drude, *Linnaea* 39: 182 [211] (1875). *Saguaster* sect. *Archontophoenix* (H.Wendl. & Drude) Kuntze, *Lex. Gen. Phan.*: 495 (1903). Type: *Ptychosperma alexandrae* F.Muell. = *Archontophoenix alexandrae* (F.Muell.) H.Wendl. & Drude.

Ptychosperma sect. *Seaforthia* F.Muell., *Fragm.* 5: 47 (1865) *non* Scheff., *Natuur Tijd. Ned. Indië* 32: 186 (1872). Type: *Ptychosperma alexandrae* F.Muell. = *Archontophoenix alexandrae* (F.Muell.) H.Wendl. & Drude.

Ptychosperma sect. *Alexandra* Scheff., *Natuur Tijd. Ned. Indië* 32: 186 (1872). Type *Ptychosperma alexandrae* F.Muell. = *Archontophoenix alexandrae* (F.Muell.) H.Wendl. & Drude.

Loroma O.F.Cook, *J. Wash. Acad. Sci.* 5: 117 (1915). Type: *Loroma amethystina* O.F.Cook = *Archontophoenix cunninghamiana* (H.Wendl.) H.Wendl. & Drude.

Solitary, tall, erect, pleonanthic, monoecious palms. **Stems** slender to moderate, frequently with a swollen base; leafscars prominent; internodes short to long. **Leaves** paripinnate, cleanly deciduous; leafbases forming a tubular elongate crownshaft, eventually splitting opposite the petiole at abscission, green, brown or purple; petiole lacking to moderately long, channelled on the adaxial surface, rounded on the abaxial surface, glabrous to moderately scaly; rachis long, moderately curved to vertically oriented, angled to flat on the adaxial surface, rounded on the abaxial surface. **Pinnae** in a single plane on the rachis, subopposite to opposite, linear-acute, reduplicate, erect to semi-pendulous, rigid or lax, apex acute to tapered; midrib prominent, parallel veinlets frequently present on the abaxial surface; abaxial surface green or with silver-grey scales; ramenta lacking or present on midrib on the abaxial surface, if present then medifixed. **Inflorescence** infrafoliar, paniculate, branched to 3 or 4 orders, erect to pendulous, branches divaricate, protandrous; prophyll 2-keeled, tubular, fully enclosing the peduncular bract; peduncular bract inserted slightly above attachment of prophyll, tubular; bracts deciduous when flowers still in bud; small to moderate rameal bracts often present; peduncle stout, short; rachis much longer than the peduncle; rachillae erect or pendulous, elongate, flexuous throughout or only toward the apex. **Flowers** unisexual, sessile, in well-spaced or congested triads of a single pistillate flower subtended by a pair of staminate flowers one either side, borne spirally on the rachillae, or only on the proximal portion and then with staminate flowers distally, in pairs or solitary. **Staminate flowers** asymmetric in bud; 3 imbricate sepals; 3 valvate petals, much longer than the sepals; stamens 9–35; filaments curved or reflexed; anthers erect, dorsifixed near the middle, basally bifid, apically pointed, latrorse; pistillode cylindrical, about as long as the stamens, tapered, apically 3-lobed. **Pistillate flowers** smaller than the staminate, symmetric; 3 basally free, imbricate sepals; 3 basally free, imbricate petals, briefly valvate at the apex; staminodes 3, dentiform; gynoecium unilocular, uniovulate; style short; 3 recurved stigmas; ovule hemianatropous, laterally attached. **Pollen** ellipsoidal; aperture monosulcate; exine tectate, finely reticulate. **Fruit** conic-ovoid, ellipsoid, globose to subglobose, 1-seeded; stigmatic remains apical or subapical; epicarp thin, smooth or lightly pebbled; mesocarp thin, with thin or thick longitudinal fibres, terete or flat, unbranched or branched, dissociating or remaining compact in the dried state; endocarp thin, crustaceous or brittle; perianth persistent or not on fruit. **Seed** ovoid, globose to subglobose; hilum lateral; raphe fibres elongate, anastomosing, adherent to the seed; endosperm ruminate; embryo basal; germination adjacent-ligular; eophyll bifid.

Archontophoenix is a genus of six species endemic to Australia. Three of the species are narrow endemics confined to either a single drainage system (*A. myolensis*) or mountain ranges or massifs (*A. maxima*, *A. purpurea*), while the other three have widespread distribution, one confined to Cape York Peninsula (*A. tuckeri*), one to eastern Queensland (*A. alexandrae*) and the other (*A. cunninghamiana*) to central Queensland south to southern New South Wales. Species occur primarily as canopy emergents or subcanopy elements in closed forests in high-rainfall areas on various soil types at low to moderate elevation. *Archontophoenix alexandrae* and *A. cunninghamiana* co-occur in the Eungella

Range, where both grow at high elevation in intermingled populations. There has been no recorded evidence of hybridity between these species in these populations. Characters that distinguish species are presented in Table 10.2.

Genus etymology: '*generi nomen tributum ob regiam staturae folioruuque dispositionis pulchritudinem*' (Wendland and Drude 1875) [The generic name is derived from the stately appearance and the beautiful arrangement of the leaves], from the Greek *archon*, ruler, an Athenian magistrate, and *phoinix*, date palm or palm.

Key to species of Archontophoenix

1 Pinnae green on both adaxial and abaxial surfaces; perianth lilac to purple .... **2. *A. cunninghamiana***

1: Pinnae green on the adaxial surface, silver-grey on the abaxial surface; perianth white-cream or light green to cream-green **2**

2 Mesocarp fibres less than 0.5 mm wide, dissociating from the endocarp in the dried state .... **4. *A. myolensis***

2: Mesocarp fibres greater than 0.5 mm wide, remaining more or less intact around the endocarp in the dried state **3**

3 Fruit greater than 15 mm long **4**

3: Fruit less than 15 mm long 5

4 Pinnae with ramenta on abaxial midrib; stamens 20 or more; crownshaft plum-purple .... **5. *A. purpurea***

4: Pinnae lacking ramenta on abaxial midrib; stamens fewer than 20; crownshaft light green or lime green .... **6. *A. tuckeri***

5 Floral pits with low rounded margins; peduncle to 12 cm wide .... **3. *A. maxima***

5: Floral pits with raised sharp margins; peduncle to 10 cm wide .... **1. *A. alexandrae***

Table 10.2 Distinguishing characters for species of *Archontophoenix*. Characters unique to a single species are highlighted in bold

Species	Pinnae abaxial surface	Flower colour, fruit shape and length	Mesocarp fibres
A. alexandrae	Silver-grey scales, ramenta lacking	Perianth white-cream, fruit ovoid-globose, 8–14 mm	Fibres to 1 mm wide, flat, sparingly branched
A. cunninghamiana	**No silver–grey scales, ramenta present**	**Perianth lilac to purple**, fruit ovoid-globose, 10–15 mm	Fibres to 1 mm wide, flat, branched
A. maxima	Silver-grey scales, ramenta lacking	Perianth white-cream or light green, fruit conic-ovoid, 11–15 mm	Fibres to 1 mm wide, flat, sparingly branched
A. myolensis	Silver-grey scales, ramenta lacking	Perianth white-cream, fruit conic-ovoid–elongate, 13–21 mm	**Fibres terete or flat, less than 0.5 mm wide, branched, forked about midway**
A. purpurea	**Both silver-grey scales and ramenta present**	Perianth white-cream or cream-green, fruit ellipsoid to globose, 20–26 mm	Fibres flat, to 2 mm wide, branched, overlaid and interspersed with thin fibres
A. tuckeri	Silver-grey scales, ramenta lacking	Perianth white-cream, fruit ovoid, 15–25 mm	**Fibres in two layers, outer layer of fibres thin, occasionally branched, inner layer thick, flat, to 3 mm wide, mostly branched; fibres compact in the dried state**

1 Archontophoenix alexandrae

Archontophoenix alexandrae (F.Muell.) H.Wendl. & Drude, *Linnaea* 39: 212 (1875); *Ptychosperma alexandrae* F.Muell., *Fragm.* 5: 47, 213, tab. 43, 44 (1865). Type: Australia. Queensland. Fitzroy R., 'scrub near the upper flats', undated, *E.Bowman*; holotype MEL [7 sheets, including seedlings]; isotype BO (fragment).

Archontophoenix alexandrae var. *schizanthera* H.Wendl. & Drude, *Linnaea* 39: 212, pl. 3, Fig. 6 (1875). Type: Australia. Queensland. Rockingham Bay, July 1866, *E.Bowman*; holotype MEL.

Ptychosperma beatriceae F.Muell., *Chem. and Druggist 4, Austral. Suppl*: 77, Feb. (1882); *Archontophoenix beatriceae* (F.Muell.) F.M.Bailey, *Queensl. Fl.* 5: 1675 (1902); *Archontophoenix alexandrae* var. *beatriceae* (F.Muell.) C.T.White ex L.H.Bailey, *Gentes Herb.* 3: 409 (1935). Type: Australia. Queensland. Mt Elliot, near Port Denison, Nov. 1881, *E.Fitzalan s.n.*; holotype MEL; isotype FI.

Jessenia glazioviana Dammer, *Bot. Jahrb. Syst.* 31: 21 (1902). Type: Cultivation. Brazil. Rio de Janeiro, Quinta de São Christovão, undated, *Glaziou 25537*; holotype B, destroyed; isotype MO.

Stem to 30 m tall, to 30 cm dbh, enlarged at the base to 50 cm diam., becoming slightly fissured vertically with age; leafscars to 20 mm wide; internodes to 15 cm long, green to light grey, often stepped in the lower 2–3 m. **Leaves** 10–12, to 4.5 m long, with a pronounced lateral twist, 60–80 pinnae per side; crownshaft to 130 cm long, elongate, light to mid green or lime green, waxy, swollen, tapered toward the base and apex, irregularly bulging near the attachment of the petiole; petiole to 38 cm long; rachis to 400 cm long. **Pinnae** to 75 cm long, 3–5 cm wide, rigid, dark green on the adaxial surface, silver-grey on the abaxial surface due to presence of persistent dense scales; midrib cream, 10–12 parallel veinlets on abaxial surface; ramenta lacking. **Inflorescence** 50–100 cm long, to 50 cm wide, branched to 4 orders, axes initially white, cream or light green, becoming brown-cream or green in fruit; prophyll to 78 cm long, to 12 cm wide, light to dark green, fibrous, with persistent black scales in the proximal portion; peduncular bract to 44 cm long, to 6 cm wide, beaked, thickly fibrous, with persistent black scales densest distally, white-cream internally; peduncle to 10 cm wide, to 6 cm thick; rachis to 40 cm long, to 2.5 cm wide by 2 cm thick, angular proximally, becoming terete distally; rachillae to 30 cm long, erect to semi-pendulous, becoming flexuous distally; floral pits with raised sharp margins, triangular, to 2 mm high. **Staminate flowers** 6.0–9.5 mm long, 2.0–3.0 mm wide in bud, opening to 10 mm wide at anthesis; sepals to 2 mm long, white to cream; petals 6–7 mm long, 2.0–2.6 mm wide, falcate, pointed, white to cream suffused light brown at the base, light brown on the inner surface distally; stamens 9–16; anthers 3.5–4.0 mm long; filaments curved, to 2 mm long, cream-brown, lacking scales; pistillode acuminate-cylindrical or fusiform. **Pistillate flowers** to 4 mm high, to 3 mm wide; sepals to c. 3 mm long, c. 4 mm wide, imbricate, apex subulate, white-cream; petals to c. 5 mm long, 3–4 mm wide, apex subulate, white-cream; stigmas c. 1 mm long, opening to c. 2 mm wide at receptivity; staminodes c. 1.5 mm long, c. 0.25 mm wide. **Fruit** ovoid-globose, 8–14 mm long, 6–11 mm wide, stigmatic remains apical; epicarp smooth, waxy and glossy, bright red at maturity; mesocarp fibres to 1 mm wide, flat, sparingly branched, remaining tight in the dried state; endocarp brittle, lacking embedded mesocarp fibres; perianth to c. 3 mm long, not persistent on fruit. **Seed** globose, to 8 mm diam. Figures 10.8, 10.9.

Figure 10.8 *Archontophoenix alexandrae* in lowland rainforest, Centenary Lakes, Cairns, north-east Queensland.

Figure 10.9 *Archontophoenix alexandrae.* **Top left:** Inflorescences. **Top right:** Staminate flower at anthesis. **Bottom left:** Pistillate flowers at anthesis. **Bottom right upper:** Mature fruit. **Bottom right lower:** Epicarp removed to reveal mesocarp fibres.

Alexandra palm, King palm

Distribution and ecology: Endemic to Queensland, distributed from Princess Charlotte Bay south to Baffle Creek, near Miriam Vale, most commonly forming large colonies in the coastal lowlands in wet areas, or as a riparian dominant, on various soil types in rainforest, swamp forest and moist sclerophyll forest, 0–1000 m asl (Fig. 10.10). Inland populations are associated with springs or soaks, or are in high-rainfall areas. Flowering all months; fruiting all months. Conservation status – *No present threats.*

Typification: The holotype of *Archontophoenix alexandrae* (basionym *Ptychosperma alexandrae*) is an unnumbered and undated collection by E. Bowman [MEL] from Fitzroy River in Queensland. The protologue includes: *'In virgultis densis silvarum quam maxime umbrossarum ad flumen Fitzroy's River Australiae capricornicae. Bowman'* (Mueller 1865, p. 48)

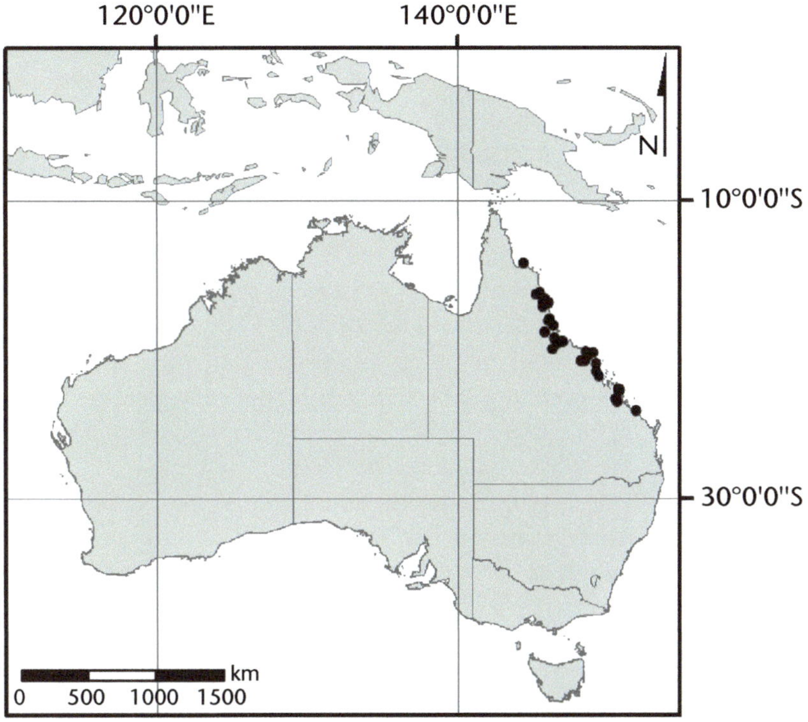

Figure 10.10 *Archontophoenix alexandrae*. **Left:** Distribution based on herbarium specimen data. **Right:** Type specimen. Australia. Queensland. Fitzroy River, 'scrub near the upper flats', undated, *E.Bowman*; holotype MEL [1 of 7 sheets].

[In the dense shady forests of the Fitzroy River, Capricornia region, Australia. Bowman]. The collection consists of seven sheets, including pinnae, inflorescence portions with flowers in bud, a peduncular bract and seedlings. Mueller provided additional descriptive information for *P. alexandrae* on page 213 of the same volume of *Fragm.*, but this is not part of the protologue (Fig. 10.10).

Etymology: '*Palmam inter nostrates eximiam et praepulchram sacravi augustissimae et nobilissimae principi Alexandrae, throni Danici nunc et Britannici gloriae et ornamento*' (Mueller 1865, p. 48): [This palm is the most beautiful in the country and is dedicated to that noble Princess Alexandra, formerly of the Danish throne, now of the British Royal family]. Named for Princess Alexandra of Denmark (b.1844, d.1925), daughter of King Christian IX and Louise of Hessen-Kassel, married to King Edward VII of Great Britain (b.1841, d.1910) in 1863. Chapman (1991) used the spelling '*alexandriae*', but the original spelling given by Mueller is herein maintained.

Notes: *Archontophoenix alexandrae* var. *schizanthera* was described for a collection from Rockingham Bay, and differentiated by thinner inflorescence branches and large prominent bracteoles; *Archontophoenix beatricae* was named for a collection from Mt Elliot, and differentiated by having a lower stem with 'clearly defined steps (at every mark of former yearly foliage) half an inch deep or more, all round up to three to five feet, when these projections cease, and the stem tapers gradually' (Mueller 1882, p. 77). *Jessenia glazioviana* was described for a specimen cultivated in Rio de Janeiro, Brazil.

Archontophoenix alexandrae is the most widespread and prevalent palm in north-east Australia, contributing to a characteristic landscape physiognomy, although most of its former lowland habitat has been cleared for agriculture. Large populations can be seen in the Wet Tropics Bioregion and it is conserved in national parks and conservation reserves. It most closely resembles *A. maxima*, from which it differs by a less robust inflorescence and larger fruits, and *A. myolensis*, which is distinguished by elongate fruit with narrow, rather than wide, fibres in the mesocarp. In these three species the abaxial surface of the pinnae has dense silver-grey scales, and they lack ramenta.

2 Archontophoenix cunninghamiana

Archontophoenix cunninghamiana (H.Wendl.) H.Wendl. & Drude, *Linnaea* 39: 214 (1875); *Ptychosperma cunninghamianum* H.Wendl., *Bot. Zeitung (Berlin)* 16: 346 (1858); *Seaforthia cunninghamii* (H.Wendl.) Hort. ex F.M.Bailey, *Compr. Cat. Queensland Pl.*: 573 (1913); *Loroma cunninghamiana* (H.Wendl.) O.F.Cook, *J. Wash. Acad. Sci.* 5: 118 (1915). Type:

lectotype, figs 1–6 of Tab. 4961 [*Seaforthia elegans* Hook. *non* R.Br.], *in* W.J.Hooker, *Bot. Mag*. 83 (1857) [NB. figs 7–11 are of *Ptychosperma elegans* (R.Br.) Blume], here designated.

Jessenia amazonum Drude, *in* C.F.P. von Martius, *Fl. Bras*. 3: 474, pl. 109 (1882). Type: Cultivation. Brazil. Rio de Janeiro, undated, *Simard (Glaziou?) 8062*; holotype B, destroyed; isotype U.

Loroma amethystina O.F.Cook, *J. Wash. Acad. Sci*. 5: 118 (1915). Type: Cultivation. USA. Santa Barbara, 11 Nov. 1914, *O.F.Cook 694813*; holotype US.

Stem to 30 m tall, to 30 cm dbh, smooth in the distal portion, becoming fissured with age, slightly enlarged at the base; leafscars 1–2 cm wide, raised, stepped in lower portion of stem; internodes 3–24 cm long, dark green to grey. **Leaves** 9–12, to 4.5 m long, arching, with a moderate lateral twist, 70–90 pinnae per side; new leaf occasionally with a bronze tinge; crownshaft 70–140 cm long, narrow, tapered toward the apex, dark green, brown-green or purple-green; petiole to 50 cm long, to 5 cm wide, with rust red-brown scales, densest on the abaxial surface; rachis becoming sharp toward the distal portion on the adaxial surface. **Pinnae** to 100 cm long, 3–10 cm wide, rigid becoming lax and semi-pendulous with age, dark glossy green on the adaxial surface, lighter green on the abaxial surface; midrib brittle; parallel veinlets not prominent, slightly raised in larger pinnae on the adaxial surface; ramenta on midrib on the abaxial surface, to 6 mm long, brown-red, twisted, basi- or medifixed, irregularly spaced, densest in proximal portion of pinnae. **Inflorescence** to 150 cm long, to 60 cm wide, branched to 4 orders, axes initially yellow green, white or cream, becoming light green or mauve in fruit; prophyll 50–60 cm long, to 12 cm wide, dark green with sparse persistent black scales, fibrous, winged, beaked apically; peduncular bract to 58 cm long, thin, papery, mildly fibrous; peduncle to 15 cm long, 5–7 cm wide, yellow-green with red-brown scales, densest on the abaxial surface; rachis to 60 cm long, to 3.5 cm wide by 2 cm thick, tapered to 4 mm thick distally; rachis branches angular, terete distally, white with brown scales in the proximal eighth; rachillae to 55 cm long, to 5 mm wide at attachment of basal triad, to 2 mm wide at distal triad, pendulous, straight in proximal portion, flexuous in extreme distal portion, cream; floral pits hemicupular, with raised sharp margins. **Staminate flowers** 3–6 mm long, 1.5–3 mm wide in bud, opening 7–11 mm wide at anthesis; sepals 1–1.5 mm long, lilac to purple; petals 4–5.5 mm long, 2–3 mm wide, lilac to purple, fibrous, firm, falcate, pointed, gibbous basally, not detaching easily, scattered translucent scales on the outer surface; outer surface with dark dotted longitudinal nerves; stamens 9–18; filaments purple, with light coloured scales in the basal half, reflexed below attachment to anthers; connective elongate; staminal ring dark purple; anthers 1.5–2.5 mm long, lilac; pistillode tapering toward apex, irregularly 2–3-lobed, lilac. **Pistillate flowers** 4–5 mm long, 3–4 mm wide in bud; sepals 2–3 mm long, c. 4 mm wide, lobes triangular, striate, apex acuminate, light green to lilac, glabrous, margins hyaline; petals 4–5 mm long, c. 5 mm wide, triangular, apex subulate, striate, margins smooth to minutely fimbriate, green to dark lilac, glabrous; ovary globose, c. 4 mm diam.; stigmas to 1 mm long, recurved, glistening white. **Fruit** ovoid-globose, 10–15 mm long; stigmatic remains apical; epicarp waxy, bright red at maturity; mesocarp fibres branched, flat, to 1 mm wide, moderately loose in the dried state; endocarp crustaceous, lacking embedded fibres; perianth 1.8–2 mm high, not persistent on fruit. **Seed** globose, to 9 mm diam. Figures 10.11, 10.12.

Figure 10.11 *Archontophoenix cunninghamiana* in moderate-altitude rainforest, Blackall Range, south-east Queensland, c. 600 m altitude.

Figure 10.12 *Archontophoenix cunninghamiana*. **Top left:** Inflorescences. **Top right:** Staminate flowers at anthesis. **Bottom left:** Pistillate flowers at anthesis. **Bottom right:** Fruit, whole, in longitudinal section and with mesocarp fibres exposed.

Bangalow palm, picabbean palm

Distribution and ecology: Endemic to eastern Australia, occurring from Mt Elliot, near Townsville, Queensland, south to Durras Mt, southern New South Wales, usually forming small to large colonies in the coastal lowlands in wet areas but also in hinterland mountains and hills, sometimes up to 120 km inland (Ravensbourne National Park), where either rainfall is high or permanent water occurs. Grows on various soil types in rainforest, vine forest, swamp forest and moist sclerophyll forest, 0–1200 m asl (Fig. 10.13). Flowering all months; fruiting all months. Conservation status – *No present threats*.

Typification: Wendland (1858) did not cite any collections or specimens in the protologue of *Archontophoenix*

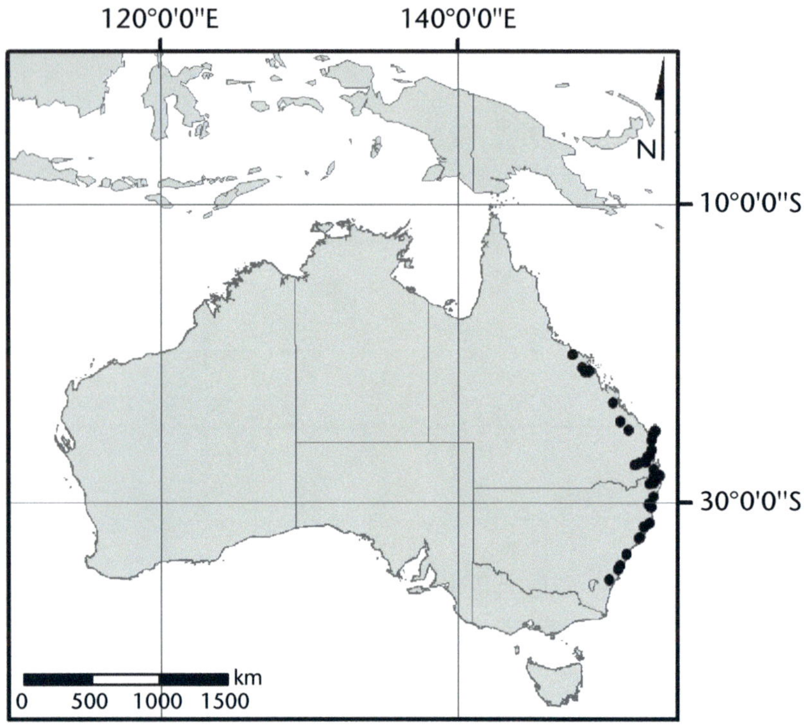

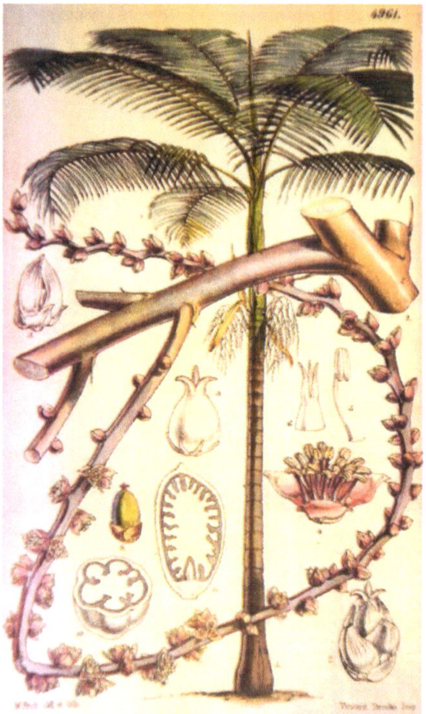

Figure 10.13 *Archontophoenix cunninghamiana*. **Left:** Distribution based on herbarium specimen data. **Right:** Type specimen. Lectotype, Figures 1–6 of Plate 4961 [*Seaforthia elegans* Hook. *non* R.Br.], *in* W.J.Hooker, *Botanical Magazine* 83 (1857).

cunninghamiana (basionym *Ptychosperma cunninghamianum*), but noted collections by Robert Brown and Allan Cunningham. These two distinct taxa had been designated as a single taxon based on mixed specimens and interpreted as representing a single species. In the protologue of *P. cunninghamianum*, Wendland attempted to clarify this situation, and noted that Tab. 4961 in *Curtis's Botanical Magazine* (Hooker 1857) indeed represented two taxa, with Wendland ascertaining Figures 1–8 were of his *P. cunninghamianum* and Figures 9–11 represented *Seaforthia elegans*. In the absence of specimens related to the protologue, Dowe and Hodel (1994) designated Figures 1–7 of Tab. 4961 as the lectotype for *P. cunninghamiana*, while Figures 8–11 represented *S. elegans*. This should be amended with the lectotype of *P. cunninghamianum* being Figures 1–6, and with Figures 7–11 being of *S. elegans*. On close examination of the separate elements within the illustration, and comparing available specimens of both taxa, the Figures representing *A. cunninghamiana* include the habit, inflorescence and staminate flower (Figs 1–6), while the pistillate flower and the fruit (Figs 7–11) represent *S. elegans* (Fig. 10.13). In the text accompanying Tab. 4961, Hooker noted that the plants in Kew Botanic Gardens, from which the illustrations were prepared, did not produce fruit. It is assumed that the illustrated fruits (of *P. elegans*) were of specimens kept in the Kew herbarium, and were mismatched with the live plant of *A. cunninghamiana*.

Etymology: '*Allan Cunningham, von dem der Saamen nach Kew geschickt ist, die Pflanze auf der Ostkuste bis zum 35°S*' (Wendland 1858, p. 347): [Allan Cunningham, who sent seeds to Kew from this plant, found it on the east coast up to 35°S]: Named for Allan Cunningham (b.1791, d.1839), British botanist, who collected extensively in Australia.

Notes: *Jessenia amazonum* was described for a misidentified specimen cultivated in Rio de Janeiro, Brazil. *Loroma amethystina* was a name provided by O.F. Cook, who considered that certain plants cultivated in California represented a new genus allied to *Archontophoenix*.

The relationships of *Archontophoenix cunninghamiana* to others in the genus are not clear; it is characterised by pinnae that lack silver-grey scales, and by the presence of ramenta. Sepals and petals of both staminate and pistillate flowers are characteristically lilac to purple.

3 Archontophoenix maxima

Archontophoenix maxima Dowe, *Austrobaileya* 4: 235, Fig. 1A (1994). Type: Cultivation. Australia. Queensland. Atherton, property of A.K.Irvine, ex Walsh R., 4 May 1993, *J.L.Dowe 106*; holotype BRI [7 sheets].

Stem to 22 m tall, to 30 cm dbh, expanded to 50 cm at the base; leafscars to 40 mm wide; internodes to 15 cm long, green to grey, stepped in the basal portion, becoming fissured with age. **Leaves** 8–11, to 5 m long, ascending, held straight, with a moderate lateral twist, to 90 pinnae per side; crownshaft 74–145 cm long, mid green with sparse black persistent scales, thick textured, heavily fibrous, slightly bulged at the base and at the apex near attachment of the petiole; petiole lacking or to 20 cm long, to 9 cm wide, green, glabrous; rachis to 340 cm long. **Pinnae** to 90 cm long, to 7 cm wide, coriaceous, dark green on the adaxial surface, silver-grey on the abaxial surface due to persistent dense scales; midrib brittle, thick, yellow-green, 7–9 parallel veinlets prominent on the abaxial surface; pinnae remain rigid during the life of the leaf; ramenta lacking. **Inflorescence** 80–140 cm long, to 100 cm wide, branched to 4 orders, smooth, axes cream-green initially, becoming green in fruit; prophyll to 105 cm long, winged with a beak at the apex, thick; peduncular bract to 80 cm long, strongly beaked at the apex, papery; peduncle to 16 cm long, to 12 cm wide, to 4 cm thick, curved downward when fruit mature, proximal portion with prominent lateral 'horns'; rachis with 18–23 main branches, angular, proximal branches to 8 cm wide, to 2 cm thick; rachillae to 90 cm long, rigid, pendulous, proximally straight, 4–7 mm wide near attachment of the basal triad, cream, glabrous; floral pits shallow, with low rounded margins. **Staminate flowers** 5–8 mm long, 2–5 mm wide in bud, strongly asymmetric, opening to 11 mm wide at anthesis; sepals 2–3 mm long, 2–3 mm wide, carinate, glabrous, cream, not readily detaching; petals 5–7 mm long, 2–3 mm wide, irregularly falcate, apically pointed, fleshy, glabrous, white-cream; stamens 11–16, 10–11 mm long; filaments curved, connective short; anthers 2–3 mm long, white; pistillode 8–9 mm long, straight, tapered to the apex, cream-brown toward the apex, shallowly lobed. **Pistillate flowers** 5–7 mm high, 3–4 mm wide in bud; sepals 3–3.5 mm long, 3–4 mm wide, ovate, apex acuminate, cream to light green, glabrous; petals 4–6 mm long, 3–5 mm wide, ovate, apex rounded, cream to light green, glabrous; stigmas to 1.5 mm long, recurved, c. 3 mm wide at receptivity, glistening clear white. **Fruit** conic-ovoid, 11–14 mm long, 7–12 mm wide; stigmatic remains apical, moderately protrusive; epicarp smooth, waxy and glossy, red at maturity; mesocarp fibres to 1 mm wide, flat, sparingly branched; endocarp brittle. **Seed** globose, to 6 mm diam. Figures 10.14, 10.15.

Herberton Range Archontophoenix

Distribution and ecology: Endemic to Atherton Tableland, Queensland, in the Herberton Range and Mt Haig Range, as a riparian element in high-rainfall areas, associated mostly with wet sclerophyll forest and Casuarina-dominated forests on basalt soils, 800–1200 m asl (Fig. 10.16). Flowering June–Aug.; fruiting Dec.–Mar. Conservation status – *No present threats.*

Typification: The holotype of *Archontophoenix maxima* is *Dowe 106* [BRI], collected in May 1993 from a plant cultivated on the property of A.K. Irvine, Atherton,

Figure 10.14 *Archontophoenix maxima* in high-altitude rainforest, Oaky Creek, Walsh River headwaters, Herberton Range, north-east Queensland, c. 1100 m altitude.

Figure 10.15 *Archontophoenix maxima*. **Top left:** Leaf, view of abaxial surface**: Top right:** Staminate flowers at anthesis. **Bottom left:** Pistillate flowers at anthesis. **Bottom right:** Mature fruit.

Queensland. This plant was grown from seed obtained from a population on the upper Walsh River in the Herberton Range. The specimen comprises seven sheets consisting of a complete leaf, inflorescence portions, and staminate and pistillate flowers (Fig. 10.16).

Etymology: 'from the superlative maximus of the Latin *magnus*, "great", in reference to the larger size of the inflorescence, and in particular the greater width of the peduncle as compared to all other species of *Archontophoenix*' (Dowe and Hodel 1994, p. 236).

Notes: *Archontophoenix maxima* is distinguished by the robust peduncle and the leaves which are held more or less flat, not strongly laterally twisted as in other species of *Archontophoenix*. It most closely resembles *A. alexandrae*.

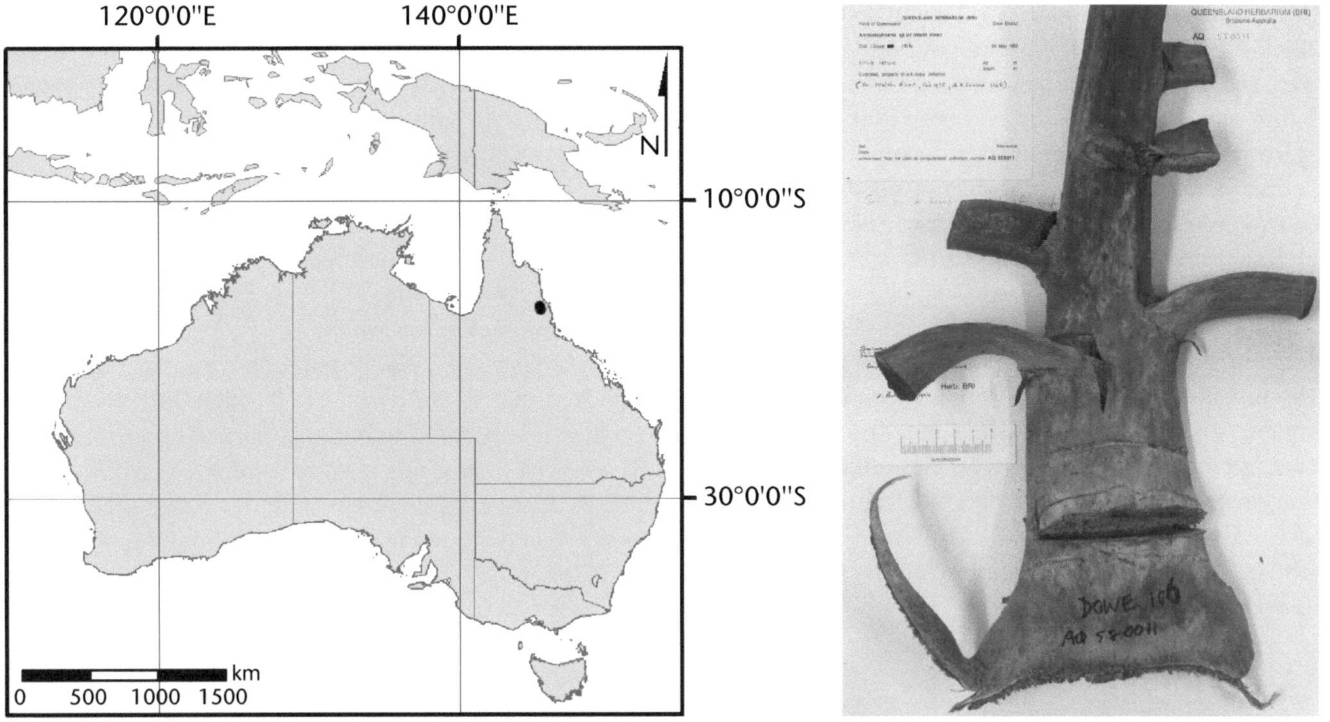

Figure 10.16 *Archontophoenix maxima*. **Left:** Distribution based on herbarium specimen data. **Right:** Type specimen. Australia. Queensland. Atherton, property of A.K.Irvine, ex Walsh River, 4 May 1993, *J.L.Dowe 106*; holotype BRI [1 of 7 sheets].

4 Archontophoenix myolensis

Archontophoenix myolensis Dowe, *Austrobaileya* 4: 237, Fig. 1B (1994). Type: Australia. Queensland. Myola, Warril Ck, 30 May 1976, *A.K.Irvine 1780*; holotype CNS [6 sheets].

Stem to 20 m tall, to 30 cm dbh, expanded to 50 cm at the base, stepped in the lower portion, smooth becoming fissured with age; leafscars to 30 mm wide, slightly raised; internodes to 20 cm long, green to grey. **Leaves** 9–12, to 4 m long, with a prominent lateral twist, 68–71 pinnae per side; crownshaft 74–95 cm long, blue-green, tapered toward the apex, prominent bulges below the attachment of the petiole; petiole to 22 cm long, to 4 cm wide by 2.5 cm thick; rachis to 380 cm long, becoming sharply ridged above in the distal portion. **Pinnae** to 110 cm long, to 6.5 cm wide, coriaceous, dark green on the adaxial surface, with persistent moderately dense silver-grey scales on the abaxial surface; midrib light green, moderately brittle, 9–11 parallel veinlets prominent on abaxial surface; pinnae become lax with age; ramenta lacking. **Inflorescence** 50–157 cm long, to 60 cm wide, branched to 4 orders, axes light cream initially, becoming light green in fruit; prophyll to 45 cm long, thin texture; peduncular bract to 42 cm long, strongly beaked at the apex, papery; peduncle 8.5–23 cm long, 5–7.2 cm wide, by 2.8–3 cm thick, green; rachis to 70 cm long with 21–23 branches; rachillae thin, to 4 mm wide at attachment of basal triad; floral pits cupular with raised sharp margins to 1 mm high. **Staminate flowers** 5.5–9 mm high, 3–3.8 mm wide in bud; sepals 1.5–3 mm high, to 2.5 mm wide, white-cream; petals 7–7.8 mm long, 1.5–3.8 mm wide, white-cream, both surfaces with scattered light brown scales densest toward the base and apex of the inner surface; petals detach easily; stamens 14–24; filaments curved, 3.5 mm long, cream-brown with rusty brown-red scales; anthers to 3 mm long, cream; pistillode tapered, to 4 mm long, sometimes longitudinally grooved, cream with brown-red scales toward the base, 2 or 3 apical lobes. **Pistillate flowers** 4–5 mm long, by 3.5–4 mm wide in bud; sepals 4.5–5 mm long, 3–4 mm long, triangular, apex acuminate, carinate, white-cream, glabrous; petals 5–6 mm long, 4–5 mm wide, triangular, apex acuminate, carinate, white-cream, glabrous; ovule ovoid; stigmas recurved, lobes c. 1 mm long, glistening white.

Fruit conic-ovoid-elongate, 13–21 mm long, 10–12 mm diam.; stigmatic remains prominent, subapical; epicarp waxy, red at maturity; mesocarp fibres terete or flat, to 0.5 mm wide, branched, forked about midway, dissociating from the endocarp in the dried state; endocarp crustaceous; perianth 2–3 mm high, not persistent on fruit. **Seed** ovoid-elongate, 14–20 mm long, 8–9.5 mm wide, dull light brown. Figures 10.17, 10.18.

Myola Archontophoenix

Distribution and ecology: Endemic to the Myola area, Atherton Tableland, Queensland, occurring mostly in remnant riparian forest on Warril Creek and its tributaries, also scattered on nearby slopes on the banks of the Barron River, in rainforest on soils derived from metamorphic rocks, 350–450 m asl (Fig. 10.19). Flowering May–July; fruiting Dec.–Mar. Conservation status – *Endangered* (Bostock and Holland 2007).

Figure 10.17 *Archontophoenix myolensis* in moderate-altitude rainforest, Warril Creek, Myola, Atherton Tableland, north-east Queensland, c. 450 m altitude.

Typification: The holotype of *Archontophoenix myolensis* is *Irvine 1780* [CNS], collected in 1976 from Myola, Queensland. The collection comprises six sheets consisting of leaf and inflorescence portions, staminate and pistillate flowers, and fruit (Fig. 10.19).

Etymology: 'named for the type locality, Myola' (Dowe and Hodel 1994, p. 237): The locality of Myola is located about 5 km west of the township of Kuranda, Atherton Tableland, Queensland.

Notes: *Archontophoenix myolensis* is the rarest species of *Archontophoenix*. The population consists of only a few hundred mature individuals, of which only a small percentage is conserved in national parks or nature reserves. The species' rarity and its population being within an area of urban expansion, places it under considerable threat and it has an Endangered conservation status (Bostock and Holland 2007). Dowe and Hodel (1994) reported that the fruit and seedling morphologies were adaptations to riparian conditions: the fibres in the mesocarp dissociate in the dried state and act as small hooks, which enable the seeds to harness themselves among debris or within sediments. The seedling leaves become pinnate at a much earlier stage than do other species of *Archontophoenix*. This suggests an adaptation to fast-flowing water. The species is distinguished by the elongate fruit and the narrow, rather than wide, fibres in the mesocarp.

5 Archontophoenix purpurea

Archontophoenix purpurea Hodel & Dowe, *Austrobaileya* 4: 238, Fig. 1C (1994). Type: Australia. Queensland. Mt Lewis, 17 Apr. 1993, *M.Ferrero 101*; holotype BRI [8 sheets].

Stem to 25 m tall, to 45 cm dbh, expanded at the base; leafscars to 30 mm wide, slightly sunken; internodes to 15 cm long, green to whitish-grey, initially smooth becoming deeply fissured with age. **Leaves** 7–15, to 6 m long, held flat or with a moderate lateral twist, 55–85 pinnae per side; new leaf frequently with a bronze tinge; crownshaft 105–150 cm long, the surface with a dark blue-grey-green glaucous bloom and moderate to densely flecked with reddish-mahogany scales giving the sheath a plum-purple colour, glossy coppery-brown internally; petiole to 40 cm long, to 12 cm wide, narrowing to 1.5 cm at the rachis, with a light coloured band extending to the sheath; rachis to 4 m long, light green on the adaxial surface and silvery lepidote on the abaxial surface throughout. **Pinnae** to 105 cm long, to 7.5 cm wide, rigid, coriaceous, glossy light green on the adaxial surface, with persistent dense silver-grey scales on the abaxial surface;

Figure 10.18 *Archontophoenix myolensis*. **Top left:** Inflorescence. **Top right:** Staminate flower at anthesis. **Bottom left:** Pistillate flowers at anthesis. **Bottom left upper:** Mature fruit on rachillae. **Bottom left lower:** Fruit, whole, with mesocarp fibres exposed, and in longitudinal section.

midrib brittle, 8–10 parallel veinlets on abaxial surface; ramenta to 5 mm long, medifixed, twisted, red-brown, irregularly positioned, densest toward the base of the pinnae in the proximal half of the leaf, rarely lacking. **Inflorescence** 50–135 cm long, to 65 cm wide, branched to 4 orders, axes initially light green, becoming dark green in fruit; prophyll to 140 cm long, to 25 cm wide, glabrous or covered with reddish-brown scales; peduncular bract attached 5 cm above prophyll, to 78 cm long, to 11 cm wide, glabrous, fibrous; peduncle to 25 cm long, 5.5–9 cm wide, with irregularly placed persistent brown scales, densest on the abaxial surface; rachis to 65 cm long, up to

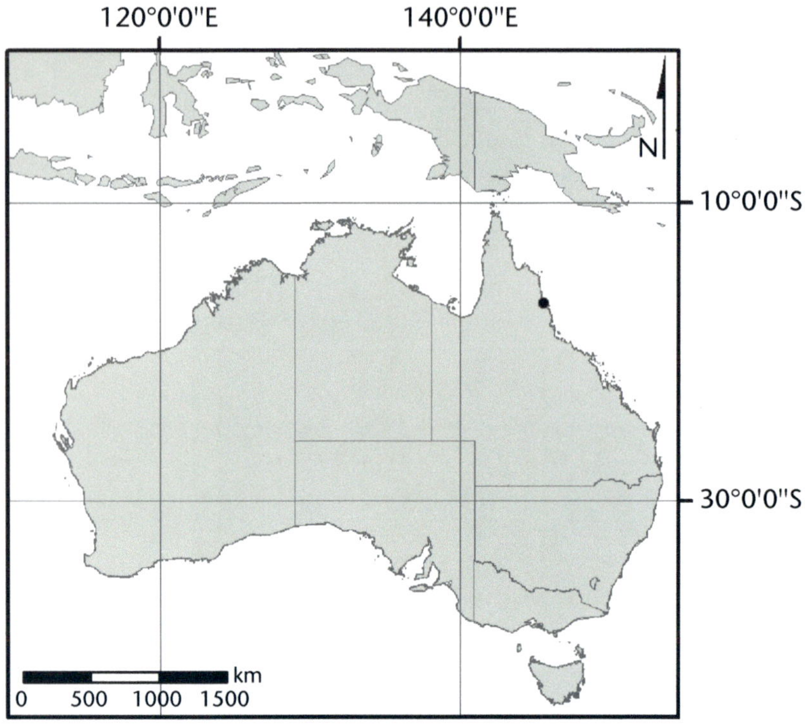

Figure 10.19 *Archontophoenix myolensis*. **Left:** Distribution based on herbarium specimen data. **Right:** Type specimen. Australia. Queensland. Myola, Warril Creek, 30 May 1976, *A.K.Irvine 1780*; holotype CNS [1 of 6 sheets].

12 first-order branches; rachillae to 85 cm long, irregularly flexuous, to 7 mm wide at attachment of the proximal triad, tapering to 3.5 mm wide at distal triad; floral pits with raised rounded to angular margins. **Staminate flowers** 6–7 mm high, 4–5 mm wide in bud, opening to c. 17 mm wide at anthesis; sepals to 4 mm long, to 5 mm wide, white-cream tinged with green; petals 6–11 mm long, 3–5 mm wide, white-cream, small brown scales on both surfaces but densest on the inner surface; stamens 20–35, to 9 mm long; filaments curved, purple-brown with elongate brown-red scales; anthers 2–6 mm long, dorsifixed near the middle, white-cream; pistillode with shallow longitudinal ridges, apex broadly lobed, light brown to cream. **Pistillate flowers** to 10 mm long, to 7 mm wide in bud; sepals 4 mm long, 6 mm wide, green-cream; petals 6.6–9 mm long, to 5 mm wide, cream-green. **Fruit** ellipsoid to globose, 20–26 mm long, 18–22 mm diam.; stigmatic remains apical, light coloured ring to 4 mm diam. surrounds stigmatic remains in the dried state; epicarp glossy, smooth (dries markedly pebbled), red-crimson at maturity; mesocarp fleshy, fibres flat, to 2 mm wide, branched, overlaid and interspersed with thin straight fibres; endocarp thin, crustaceous; perianth 3–9 mm high, persistent on fruit. **Seed** subglobose, to 23 mm long and 18 mm diam., light brown, glossy. Figures 10.20, 10.21.

Figure 10.20 *Archontophoenix purpurea* in high-altitude rainforest, Mt Lewis, north-east Queensland, c. 1100 m altitude.

Figure 10.21 *Archontophoenix purpurea*. **Top left:** Inflorescence. **Top right:** Staminate flower at anthesis. **Bottom left:** Pistillate flowers at anthesis. **Bottom right:** Fruit, whole, with mesocarp fibres exposed, seed, and in longitudinal section.

Mt Lewis Archontophoenix

Distribution and ecology: Endemic to the Mt Finnigan, Mt Spurgeon and Mt Lewis massif, Queensland, on soils derived from weathered granite in rainforest, 400–1200 m asl (Fig. 10.22). Flowering Oct.– April; fruiting Sept.–April. Conservation status – *No present threats*.

Typification: The holotype of *Archontophoenix purpurea* is *Ferrero 101* [BRI] collected in April 1993 from Mt Lewis, Queensland, at about 1200 m elevation. The specimen consists of a complete leaf, a complete inflorescence and mature fruit (Fig. 10.22).

Etymology: 'from the Latin *purpureus*, purple; named for the distinctive colouration of the crownshaft' (Dowe and Hodel 1994, p. 239).

Notes: *Archontophoenix purpurea* is one of the most distinctive species in the genus, having a prominent

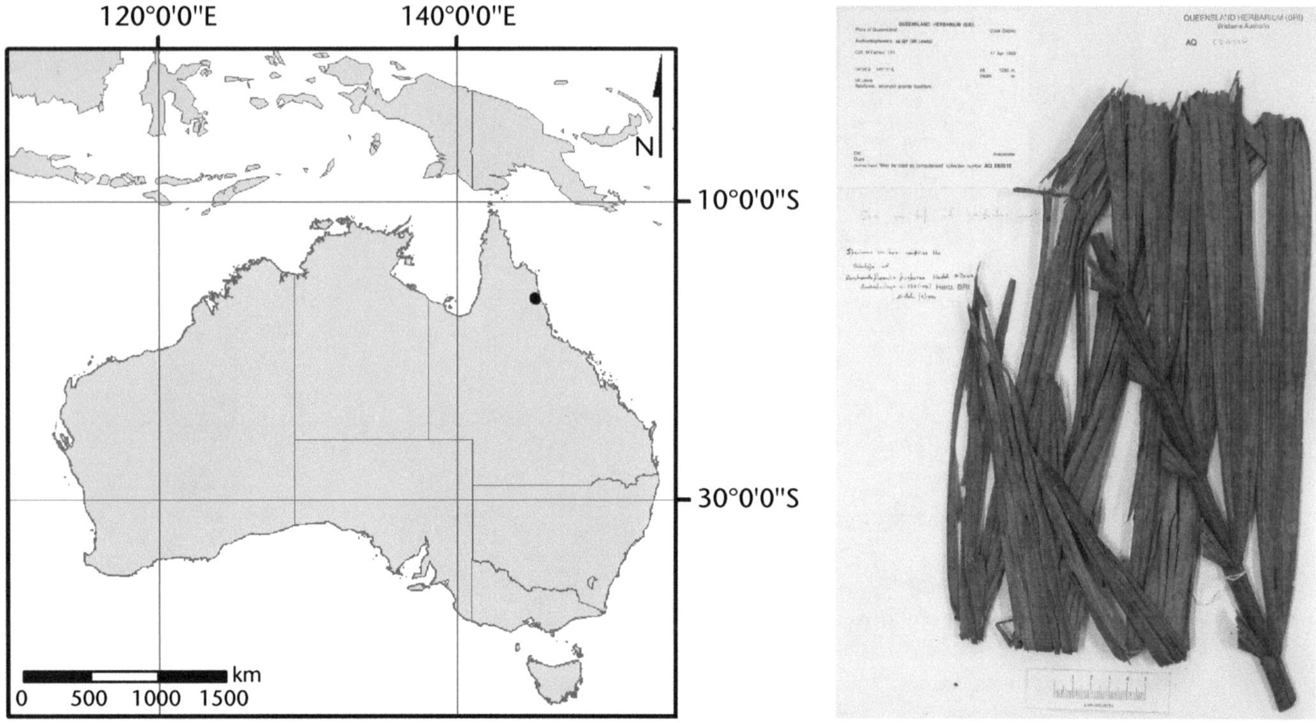

Figure 10.22 *Archontophoenix purpurea*. **Left:** Distribution based on herbarium specimen data. **Right:** Type specimen. Australia. Queensland. Mt Lewis, 17 Apr. 1993, *M.Ferrero 101*; holotype BRI [1 of 8 sheets].

purple crownshaft and large fruit. The colouration of the crownshaft is due to three separate components of surface indumentum. The innermost layer consists of the dark blue-grey surface of the crownshaft, a powdery glaucous bloom and a dense arrangement of reddish-mahogany scales. The three components merge to form the characteristic plum-purple colour. The fruit, at 20–26 mm long, are the largest in the genus. They are red-crimson at maturity, have prominent apical stigmatic remains, and the mesocarp fibres are up to 2 mm wide. In addition, the pinnae have a unique combination of possessing both silver-grey scales and ramenta on the abaxial surface.

6 *Archontophoenix tuckeri*

Archontophoenix tuckeri Dowe, *Austrobaileya* 4: 240, Fig. 1D (1994). Type: Australia. Queensland. Upper Peach R., Bonanza Ck, 12 Aug. 1948, *L.J.Brass 19831*; holotype BRI [2 sheets].

Stem to 20 m tall, to 26 cm dbh, expanded to 60 cm diam. at the base, distinctly tapered to the apex; leafscars to 10 mm wide; internodes to 15 cm long, green to grey, becoming fissured with age. **Leaves** 7–12, to 3.15 m long, held flat or with a moderate lateral twist, 43–68 pinnae per side; new leaves often bronze-red or pink; crownshaft 84–130 cm long, waxy light green or lime green, slightly bulging at the base, not bulged toward the attachment of the petiole, thin textured, tapered toward the apex, light brown and glossy on the adaxial surface; petiole to 44 cm long, to 4 cm wide in the proximal portion; rachis to 175 cm long. **Pinnae** to 93 cm long, to 5 cm wide, slightly falcate, apical third frequently semi-pendulous, dull mid to dark green on the adaxial surface, silver-grey on the abaxial surface due to dense persistent scales; midrib mid green-yellow; 7–8 prominent parallel veinlets on abaxial surface; ramenta lacking. **Inflorescence** 30–60 cm long, to 70 cm wide, axes glabrous, initially light green, becoming dark green in infructescence, branched to 3 orders; prophyll to 45 cm long, 12 cm wide, thin, fibrous, papery, yellow-green, apically beaked, silver-grey on the adaxial surface; peduncular bract to 37 cm long, to 9 cm wide, tubular, thin, yellow-green, silver-yellow on the adaxial surface; peduncle to 12 cm long, to 4.5 cm wide, to 2 cm thick, green, terete, winged at the base; rachis to 50 cm long, angular, with 8 main branches; rachillae stout, to 30 cm long, to 9 mm thick at attachment of the basal triad, flexuous; triads borne throughout rachillae; floral pits cupular with raised sharp margins to 1 mm high. **Staminate flowers** 6.5–7 mm long, 2–3 mm wide in bud, opening to 11 mm wide at anthesis; sepals 1.5–4.2 mm long, to 4 mm wide, white; petals 4.5–7 mm long, 1.8–3.6 mm wide, white, falcate, pointed, with rusty brown scales on both surfaces but densest toward the base of the inner surface; stamens 13–19, cream-brown to light purple; filaments curved, cream-brown, with rusty brown scales at the base; connective short; anthers to 4 mm long, cream; pistillode purple-cream, longitudinally grooved, apex with 2 or 3 lobes, basal portion with rusty red-brown scales. **Pistillate flowers** globose, to 5 mm high in bud; sepals c. 4.5 mm long, c. 4 mm wide, white-cream; petals 5–6 mm long, 4 mm wide, white-cream; stigmas c. 1 mm long, opening 2–3 mm across; ovary globose, c. 4 mm diam. **Fruit** ovoid, 15–25 mm long, 10–15 mm wide; stigmatic remains apical; epicarp c. 0.2 mm thick, glossy, bright red to dark brick red at maturity (drying dark ochre), frequently forms cracks at maturity; mesocarp 2–3 mm thick, fibres in two distinct layers, the outer layer of fibres thin, occasionally branched, the inner layer thick, flat, to 3 mm wide, mostly branched; fibres not usually loosening in the dried state; endocarp crustaceous; perianth to 5–6 mm high, not persistent on fruit. **Seed** globose, to 9 mm diam., dull brown. Figures 10.23, 10.24.

Figure 10.23 *Archontophoenix tuckeri* in lowland rainforest, West Claudie River, Iron Range, north Queensland.

Rocky River Archontophoenix, Iron Range Archontophoenix

Distribution and ecology: Endemic to Cape York Peninsula, Queensland, occurring from Lockerbie Scrub south to McIlwraith Range, where it is uncommon, forming small populations, mostly as a riparian element in rainforest on various soil types, 0–500 m asl (Fig. 10.25). Flowering June–Feb.; fruiting Aug.–Nov. Conservation status – *No present threats.*

Typification: The holotype of *Archontophoenix tuckeri* is *Brass 19831* [BRI], collected in August 1948 from Bonanza Creek, a tributary of Upper Peach Creek in the McIlwraith Range, Queensland. The collection consists of leaf and inflorescence portions, staminate flowers and fruit, comprising two sheets (Fig. 10.25).

Etymology: 'named for Robert Tucker (1955–92) who first published notes on the difference of this species' (Dowe and Hodel 1994, p. 241). Robert James Thomas Tucker, horticulturist, botanical illustrator, taxonomist (Pandanaceae) and landscape designer for Townsville Botanic Gardens, 1982–92, and designer of The Palmetum, Townsville, a botanic garden devoted to palms. Tucker collected palms and pandanus throughout Cape York Peninsula (deposited in BRI and CNS). He was the author of *The Palms of Subequatorial Queensland* (Tucker 1988) and taxonomic works on the Pandanaceae (Tucker 1994).

Notes: *Archontophoenix tuckeri* is distinguished by the narrow elongate light green crownshaft and large bright red fruit 15–25 mm long.

■ BASSELINIINAE

Relationships

Relationships of the Basseliniinae to other subtribes and of its six genera to each other are not fully resolved, although there are unique morphological characters, such as an incompletely encircling prophyll at attachment, that define the group (Norup *et al.* 2006; Pintaud and Baker 2008; Dransfield *et al.* 2008) (Figs 10.6, 10.26). The subtribe

Figure 10.24 *Archontophoenix tuckeri*. **Top left:** Inflorescence with bracts attached. **Top right:** Staminate flowers at anthesis. **Bottom left:** Pistillate flowers at anthesis. **Bottom right upper:** Mature fruit. **Bottom right lower:** Fruit, whole and in longitudinal section.

occurs in the south-west Pacific archipelagoes and the relationships of its only Australian representative, *Lepidorrhachis*, are unclear.

Lepidorrhachis

Lepidorrhachis (H.Wendl. & Drude) O.F.Cook, *J. Heredity* 18: 408 (1927); *Clinostigma* sect. *Lepidorrhachis* H.Wendl. & Drude, *Linnaea* 39: 186 [218] (1875). Type: *Kentia mooreana* F.Muell. = *Lepidorrhachis mooreana* (F.Muell.) O.F.Cook.

Solitary, moderate, erect, pleonanthic, monoecious palms. **Stems** moderate, frequently with a swollen base; leafscars prominent; internodes short. **Leaves** paripinnate, cleanly deciduous; leafbases not forming a tubular crownshaft, split opposite the petiole, green and brown-scurfy

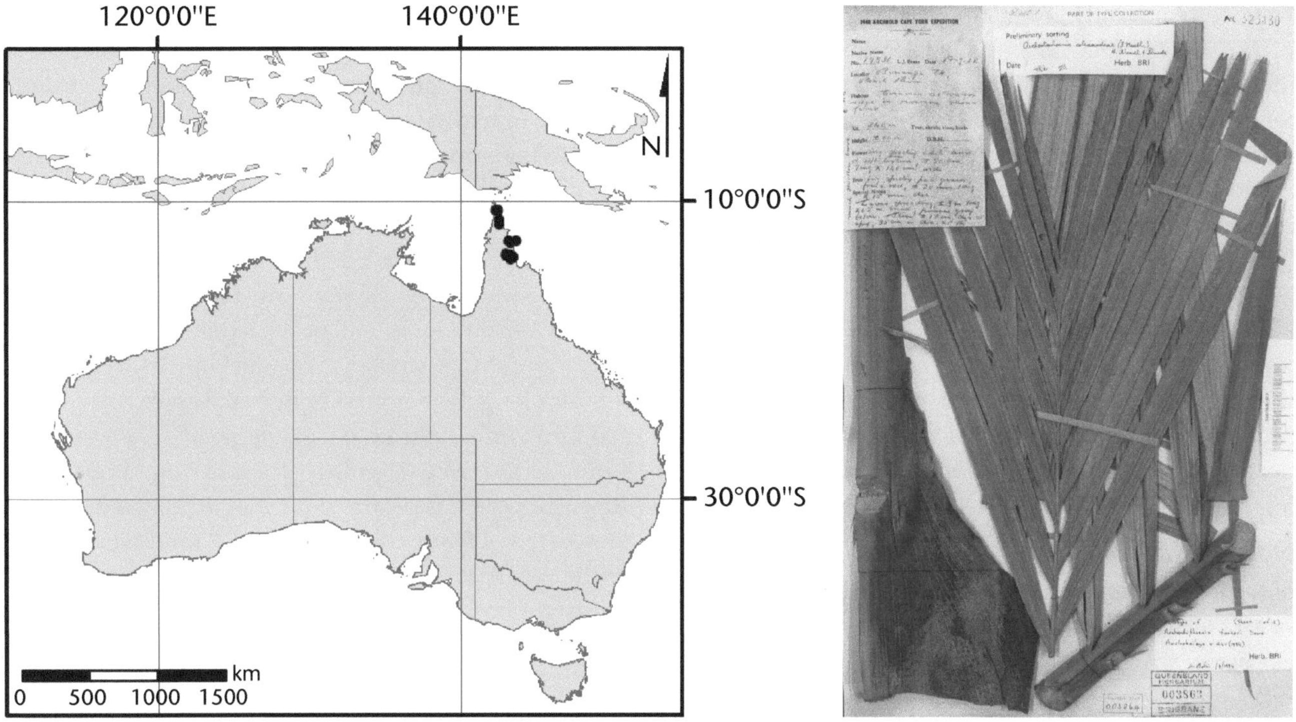

Figure 10.25 *Archontophoenix tuckeri*. **Left:** Distribution based on herbarium specimen data. **Right:** Type specimen. Australia. Queensland. Upper Peach River, Bonanza Creek, 12 Aug. 1948, *L.J.Brass 19831*; holotype BRI [1 of 2 sheets].

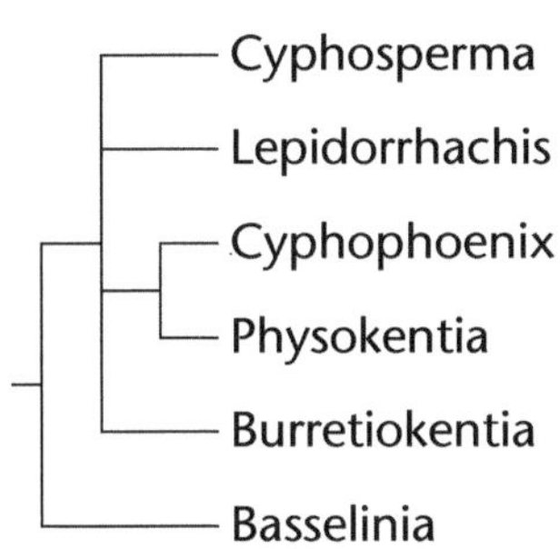

Figure 10.26 Phylogenetic tree summarising the estimated relationships of genera in the Basseliniinae. Adapted from Asmussen *et al*. (2006) and Dransfield *et al*. (2008).

toward the petiole; petiole short, channelled on the adaxial surface, rounded on the abaxial surface, densely brown-scurfy; rachis arching. **Pinnae** in a single plane on the rachis, subopposite to opposite, linear-acute, reduplicate, erect, rigid; midrib prominent, parallel veinlets prominent, margins thickened, abaxial surface green, adaxial surface green, scales on midrib and veinlets on both surfaces; ramenta lacking. **Inflorescence** infrafoliar at maturity, paniculate, branched to 3 orders, erect, branches divaricate, protandrous; prophyll tubular, 2-keeled, fully or incompletely enclosing the peduncular bract at point of attachment; peduncular bract inserted slightly above attachment of prophyll, tubular, apically exserted from the prophyll, deciduous immediately prior to floral anthesis; short and wide rameal bracts present; peduncle short, densely covered with brown membranous scales; rachis longer than the peduncle, densely covered with brown membranous scales; rachillae erect, short. **Flowers** unisexual, sessile, in triads of a single pistillate flower subtended by a pair of staminate flowers one either side, borne spirally throughout the rachillae, or only on the proximal portion and then with staminate flowers distally, in pairs or solitary; triads subtended by an acute bract. **Staminate flowers** asymmetric in bud; 3 imbricate keeled sepals; 3 asymmetric valvate petals, about twice as long as sepals; stamens 6; filaments inflexed; anthers oblong, dorsifixed, apically rounded, latrorse; pistillode cylindrical, slightly longer than the stamens, tapered, apically expanded. **Pistillate flowers** smaller than the staminate, symmetric; 3 rounded imbricate sepals; 3 imbricate petals briefly valvate at the apex; staminodes 3, dentiform; gynoecium unilocular, uniovulate; style short, stigmas indistinct; ovule pendulous, hemianatropous. **Pollen** ellipsoidal; aperture monosulcate; exine tectate, finely punctuate. **Fruit** globose to subglobose, 1-seeded; stigmatic remains subapical to lateral; epicarp thin, smooth or lightly granular, red at maturity; mesocarp thin, fleshy, with thin longitudinal terete unbranched or branched fibres; endocarp thin, brittle, operculum at the base; perianth not persistent on fruit. **Seed** globose to subglobose, hilum apical, raphe fibres anastomising, adherent to the seed; endosperm homogeneous; embryo subbasal; germination adjacent-ligular; eophyll bifid.

Etymology: *'costa media et segmentorum nervi marginesque subtus squamulis dense inspersis fuscis leprosis quasi villosa'* (Wendland and Drude 1875): [midribs and veinlets on the segment margins have dark scattered leprose scales and hair below]: With reference to the scales on the midrib and parallel veinlets of the pinnae; from the Greek *lepis*, a scale, and *rachis*, spine or backbone.

Lepidorrhachis mooreana

Lepidorrhachis mooreana (F.Muell.) O.F.Cook, *J. Heredity* 18: 408 (1927); *Kentia mooreana* F.Muell., *Fragm.* 7: 101 (1870); *Clinostigma mooreana* (F.Muell.) F.Muell., *Fragm.* 8: 235 (1874); *Clinostigma mooreanum* (F.Muell.) H.Wendl. & Drude, *Linnaea* 39: 218 (June 1875); *Clinostigma moorei* (F.Muell.) F.Muell., *Fragm.* 9: 78 (June 1875) *ortho. var.* Type. Australia. New South Wales. Lord Howe Island, summit of Mt Gower, undated, *C.Moore & W.Carron s.n.*; holotype MEL [4 sheets].

Stem to 2 m tall, erect to leaning, 15–20 cm dbh, base swollen; leafscars prominent, raised, 1–2 cm wide; internodes compacted, indistinct or to 3 cm long, green. **Leaves** 8–10 in the crown, rigid, arcuate, 1–1.5 m long, with a moderated lateral twist, 40–60 pinnae per side; petiole base split opposite the petiole, 15–40 cm long, swollen, loose, mid green to dark green, brown-scurfy toward the petiole; petiole stout, 20–30 cm long, deeply concave above, convex below; rachis to 120 cm long, ridged above, convex below. **Pinnae** 25–40 cm long, 2–4 cm wide, rigid, lanceolate, acuminate, obliquely erect, dark green on the adaxial surface, lighter green on the abaxial surface; midrib prominent on both surfaces, a prominent veinlet either side of the midrib with 2–3 less prominent parallel veinlets on the adaxial surface. **Inflorescence** 30–50 cm long, much branched and spreading rigidly, to 40 cm wide, branched to 3 orders, axes initially light green, becoming dark green in infructescence; prophyll to 30 cm long, tubular, 2-keeled, green, partially or fully enclosing the peduncle at attachment, with persistent black scales in the proximal portion; peduncular bract tubular, to 40 cm long, to 10 cm wide, exserted from the prophyll, beaked, thickly fibrous; peduncle to 8 cm long; rachis indistinct from other branches, to 30 cm long, to 2.5 cm wide by 2 cm thick, angular proximally, becoming terete distally; rachillae to 15 cm long, erect to semipendulous, becoming flexuous distally; floral pits with raised acute margins. **Flowers** subsessile, embedded in the rachillae in bud. **Staminate flowers** 3–6 mm long, 2–5 mm wide in bud, opening to c. 10 mm wide at anthesis; sepals 1–2 mm long, 1.5–3 mm wide, cream-light yellow; petals 3–5 mm long, 3–5 mm wide, cream-light yellow, suffused light brown at the base, light brown on the inner surface distally; stamens to c. 2 mm long; filaments 1–2 mm long; anthers 1–2 mm long, dull orange-yellow; pistillode cylindrical, apically expanded, c. 2.5 mm long. **Pistillate flowers** 4–8 mm long, 5–9 mm wide in bud; sepals c. 3 mm long, c. 4 mm wide, light green-cream; petals c. 4 mm long, c. 4 mm wide, cream; stigmas c. 0.8 mm long. **Fruit** globose, 10–12 mm diam.; epicarp to 0.1 mm thick, red at maturity; mesocarp c. 1.5 mm thick, fibres to 0.2 mm thick; endocarp brittle; operculum circular, c. 1 mm diam.; perianth not persistent on fruit. **Seed** globose to subglobose, to 7 mm diam. Figures 10.27, 10.28.

Figure 10.27 *Lepidorrhachis mooreana* in subtropical montane rainforest, Lord Howe Island. Photo by William J. Baker, Royal Botanic Gardens Kew.

Figure 10.28 *Lepidorrhachis mooreana*. **Top left:** Staminate flowers at anthesis. Photo by William J. Baker, Royal Botanic Gardens Kew. **Top right:** Pistillate flowers. Photo by Ian Hutton. **Bottom left:** Infructescence with mature fruit. Photo by Jeanne Price. **Bottom right:** Mature fruit. Photo by William J. Baker, Royal Botanic Gardens Kew.

Little mountain palm

Distribution and ecology: Endemic to Lord Howe Island, on the summits of Mt Lidgbird and Mt Gower, in subtropical montane rainforest, occurring in scattered colonies in open or shaded positions in stunted moss forest, which is frequently in cloud, 750–875 m asl (Fig. 10.29) (Baker and Hutton 2006). Pickard (1983) described the habitat as 'Gnarled mossy forest' subformation in 'Evergreen closed forest' formation. Soils are deeply weathered basalt. Flowering June–Dec.; fruiting Aug.–Feb. Conservation status – *Vulnerable* (Johnson 1996), *Endangered*, here designated.

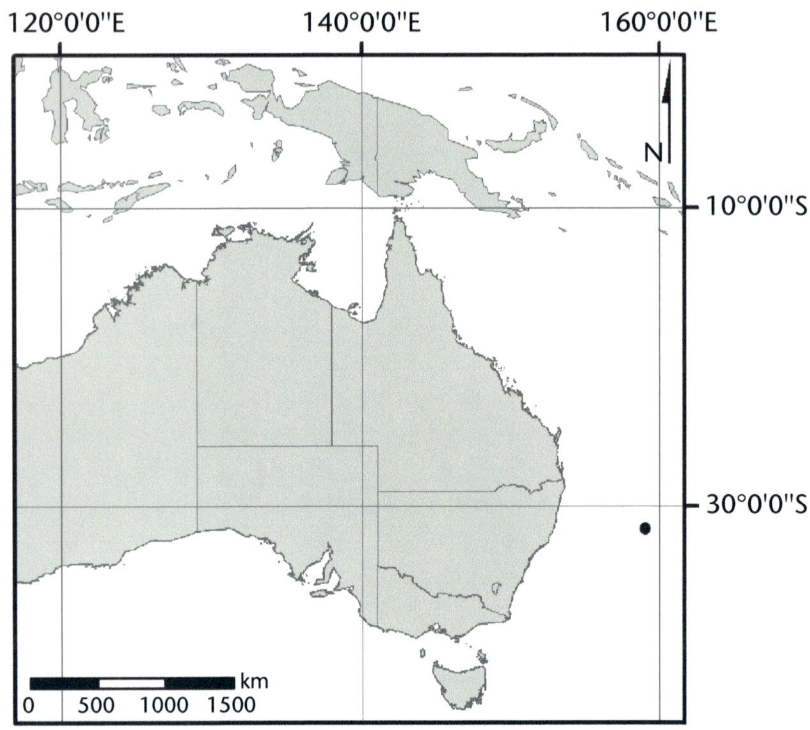

Figure 10.29 *Lepidorrhachis mooreana*. **Left:** Distribution based on herbarium specimen data. **Right:** Type specimen. Australia. Lord Howe Island, summit of Mt Gower, undated, *C.Moore and W.Carron s.n.*; holotype MEL [1 of 4 sheets].

Typification: The holotype of *Lepidorrhachis mooreana* (basionym *Kentia mooreana*) is an unnumbered and undated collection that does not include the collector's name [MEL] collected from the summit of Mt Gower, Lord Howe Island. However, the handwriting is that of Charles Moore. Moore was accompanied by William Carron to Lord Howe Island. The collection consists of a leaf portion, an inflorescence and fruits (Fig. 10.29). The label data described it as a 'dwarf palm' and the number '42' is provided, but this cannot be related to any documentation. Mueller (1870) did not provide a collector's name in the protologue. Wendland and Drude (1875), in their transferral to *Clinostigma*, included verbatim Mueller's specimen citation but cited Moore and Carron as the collectors, which is followed here. Moore was cited as the sole collector by Green (1994).

Etymology: There was no explicit explanation of the derivation of the specific epithet provided in the protologue, but presumably named for Charles Moore (b.1820, d.1905), Director of Sydney Botanic Gardens 1848–96, having visited Lord Howe Island in 1869 when he collected the type specimen (Moore 1871, 1884; Moore and Betche 1893).

Notes: Essig *et al.* (1999) noted the distinctiveness of the exocarp in *L. mooreana*: subepidermal layer of tanniniferous-pigment tissues, an inner layer of brachysclereids and the whole structure crossed by oblique fibrous bundles. In contrast to other species in the Basseliniinae in which the prophyll only partially encircles the peduncle, *L. mooreana* has a prophyll that either fully or only partially encircles the peduncle.

Lepidorrhachis mooreana is distinguished by a solitary habit with a short stem, arching pinnate leaves and a short, deeply split crownshaft, a much-branched inflorescence, staminate flowers with six stamens, fruit globose to 12 mm diameter and homogeneous endosperm.

■ LINOSPADICINAE

Relationships

The monophyly of the Linospadicinae is not fully supported with the inclusion of *Calyptrocalyx*, although the remaining genera, *Howea*, *Laccospadix* and *Linospadix* are resolved as monophyletic in most studies (Asmussen *et al.* 2006; Norup *et al.* 2006; Dransfield *et al.* 2008). The position of the subtribe within the Areceae is uncertain (Lewis and Doyle 2002) (Fig. 10.30). Each genus can be distinguished by a combination of characters (Table 10.3).

Table 10.3 Distinguishing characters of the genera in the Linospadicinae

Genus	Habit	Leaves	Inflorescence	Peduncular bract attachment	Anthers	Endosperm
Calyptrocalyx	Solitary or clustering	Irregularly to regularly pinnate	1-spicate or multispicate	Adjacent to the prophyll attachment, inflorescence becomes exserted	Dorsifixed, versatile	Homogeneous or ruminate
Linospadix	Solitary or clustering	Irregularly to regularly pinnate	1-spicate	Distal end of peduncle, inflorescence not exserted	Subbasifixed to basifixed, not versatile	Homogeneous
Laccospadix	Solitary or clustering	Regularly pinnate	1-spicate	Midway between prophyll attachment and lowest flowers, inflorescence not exserted	Basifixed, not versatile	Ruminate
Howea	Solitary	Regularly pinnate	1-spicate or multispicate	Midway between prophyll attachment and lowest flowers, inflorescence not exserted	Basifixed, not versatile	Homogeneous

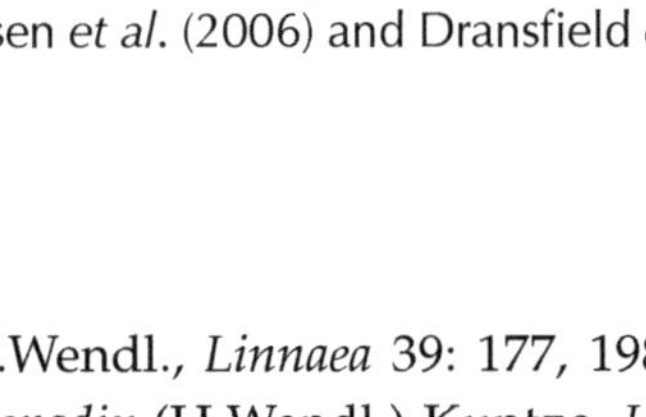

Figure 10.30 Phylogenetic tree summarising the estimated relationships of genera in the Linospadicinae. Adapted from Hahn (2002a), Asmussen *et al.* (2006) and Dransfield *et al.* (2008).

Linospadix

Linospadix H.Wendl., *Linnaea* 39: 177, 198 (1875); *Bacularia* sect. *Linospadix* (H.Wendl.) Kuntze, *Lex. Gen. Phan.*: 57 (1903). Type: *Areca monostachya* Mart. = *Linospadix monostachyos* (Mart.) H.Wendl.

Bacularia F.Muell., *Fragm.* 7: 103 (1870) *nom illeg.*; *Linospadix* subg. *Bacularia* (F.Muell.) Drude, *in* H.G.A.Engler & K.A.E.Prantl, *Nat. Pflanzenf.* 2(3): 67 (1887). *Bacularia* F.Muell. sect. *Bacularia*, *Lex. Gen. Phan.*: 57 (1903). *Bacularia* sect. *Eubacularia* Kuntze, *Lex. Gen. Phan.*: 57 (1903). Type: *Areca monostachya* Mart. = *Linospadix monostachyos* (Mart.) H.Wendl.

Linospadix Becc. *non* H.Wendl., *Malesia* 1: 62 (1877); Hook.f., *in* G.Bentham & J.D.Hooker, *Gen. Pl.* 3: 903 (1883); *Linospadix* subg. *Eulinospadix* Drude, *in* H.G.A.Engler & K.A.E.Prantl, *Nat. Pflanzenf.* 2(3): 67 (1887) = *Paralinospadix* Burret.

Solitary or clustering, small to moderate, pleonanthic, monoecious palms. **Stems** slender, leafscars conspicuous and raised; internodes long. **Leaves** paripinnate, irregularly or evenly segmented, entire and bifid; leafbases clasping, not forming a tubular crownshaft, with scattered scales, splitting opposite the petiole, margins fibrous or smooth; petiole long or lacking, channelled on the adaxial surface, rounded on the abaxial surface, glabrous or scaly. Pinnae in a single plane on the rachis, single and subopposite to opposite, linear, straight or curved, acute to acuminate, or united; apices praemorse when segments are united, acuminate when not united; ramenta on abaxial midrib and veinlets. **Inflorescence** interfoliar, solitary, spicate, erect to pendulous, protandrous; prophyll flattened, tubular, attached at base of peduncle, enclosing inflorescence in bud, persistent; peduncular bract tubular, attached at the apex of peduncle just below the most basal flowers with portion of peduncle exposed, deciduous; peduncle elongate, erect; rachilla about the same length as the peduncle to much less. **Flowers** unisexual, spirally arranged in shallow pits; in congested triads of a single pistillate flower subtended by a pair of staminate flowers one either side, borne spirally on the rachillae, or only on the proximal portion and then with staminate flowers distally, in pairs or solitary. **Staminate flowers** sessile, symmetric in bud; 3 broadly imbricate sepals; 3 thick, apically valvate petals, 2–3 times longer than sepals; stamens 4–20, erect, filaments short, anthers sagittate, subbasifixed to approaching basifixed, not versatile, latrorse; pistillode lacking or small. **Pistillate flowers** same size or larger than staminate flowers; 3 broadly imbricate sepals; 3 thick petals, apically valvate, longer than the sepals; staminodes 3–6, dentiform; gynoecium unilocular, uniovulate; 3 recurved stigmas; ovule laterally attached, campylotropous. **Pollen** ellipsoidal to circular; aperture monosulcate; exine tectate, finely to coarsely reticulate. **Fruit** ellipsoid, ovoid, cylindrical, globose or pyriform; stigmatic remains apical; epicarp smooth, striate or rugose, yellow or red at maturity; mesocarp fleshy, thin, with longitudinal fibres appressed to the endocarp; endocarp thin, adhering to the seed. **Seed** globose to ellipsoid to club-shaped, subbasally attached, raphe attached for less than half the length of seed; endosperm homogeneous; embryo basal; germination adjacent-ligular; eophyll bifid.

Linospadix is a genus of seven species distributed in Australia, where there are five species all of which are endemic, and New Guinea, where there are two species. Although each species is readily distinguishable, relationships among them are unclear. *Linospadix* resolves as sister to *Laccospadix* and *Howea* in most studies (Fig. 10.30).

Etymology: The generic description was provided in two parts. First mention was by Wendland and Drude (1875, p. 177): '*Spadices intra frondes simplices elongati graciles, longissime pedunculati*' [inflorescences interfoliar, unbranched long and thin, the peduncle very long]; the second mention (p. 198) was: '*Spadices simplices, tenues et penduli*' [inflorescences unbranched, thin and pendulous]: Named with reference to the thin inflorescence, from the Latin *line*, a now disused measurement of one-twelfth of an inch, and *spadix*, inflorescence.

Key to species of Linospadix in Australia

1 Stem solitary .... **4. *L. monostachyos***
1: Stems clustering 2
2 Fruit globose to turbinate; yellow-orange or pink to red .... **2. *L. microcaryus***
2: Fruit elongate to cylindrical; yellow or red 3
3 Petiole mostly lacking or rarely to 2 cm long; leaf most often bifid, infrequently partially segmented or pinnate .... **1. *L. apetiolatus***
3: Petiole 2–80 cm long; leaf variously pinnate but not bifid 4
4 Leaf lamina rigid; veinlets prominent on both adaxial and abaxial surfaces; staminate flowers globose, not widely opening at anthesis .... **5. *L. palmerianus***
4: Leaf lamina lax, or chartaceous; veinlets prominent only on adaxial surface; staminate flowers elongate, widely opening at anthesis .... **3. *L. minor***

1 Linospadix apetiolatus

Linospadix apetiolatus Dowe & A.K.Irvine, *Principes* 41: 215, figs. 10, 11 (1997). Type: Australia. Queensland. Mt Lewis, 27 km from Rex Hwy, on forestry road, 24 Aug. 1996, *J.L.Dowe 369 with M.D.Ferrero & L.T.Smith*; holotype BRI; isotypes CNS, K.

Clustering, small palm. **Stems** 2–6, erect or leaning to almost horizontal, 2–5 m tall, 15–25 mm diam.; leafscars 3–5 mm wide; internodes 2–40 mm long, green; vegetative aerial growths from nodes are common. **Leaves** 6–14, 30–90 cm long, 10–20 cm wide, simply bifid or with broad bilobed terminals and evenly segmented laterals on both sides of midrib and broader pinnae or with one lobe entire and running the length of the rachis and the other lobe segmented into pinnae; dull or semi-glossy light green above, lighter green below, with a metallic sheen on both surfaces; midrib prominent on both surfaces; veinlets prominent on both surfaces; petiole lacking or rarely to 2 cm long; lamina thick. **Inflorescence** to 80 cm long, rachillae green in flower and in fruit. **Staminate flowers** elongate to bullet-shaped in bud, 2.8–4 mm long, opening 2–3 mm wide at anthesis; sepals 0.8–1.2 mm long, 0.7–1.1 mm wide, light green, glabrous, margins entire; petals 1.5–3 mm long, 1–1.2 mm wide, apex shortly acuminate to rounded, cream, glabrous, margins entire; stamens 4–7, 1.5–2 mm long; connective not extending beyond the anthers; anthers subbasifixed, 0.8–1.2 mm long, cream; pistillode lacking. **Pistillate flowers** globose, 2–3 mm wide in bud; sepals 0.5–1 mm long, 0.5–1 mm wide, broadly imbricate, rounded, light green, glabrous, margins entire; petals 1.5–2.5 mm long, 1.5–1.7 mm wide, apex acuminate, cream to green or dull mauve; stigmas slightly exserted, recurved, each c. 0.5 mm long, transluscent glistening cream. **Fruit** elongate-cylindrical, 10–15 mm long, 5–6 mm wide; epicarp moderately rugose when fully mature, yellow or red at maturity; mesocarp to 2 mm thick. **Seed** elongate-ellipsoid, 9–10 mm long, 3–4 mm wide. Figures 10.31, 10.32.

Mt Lewis walking-stick palm

Distribution and ecology: Endemic to north-east Queensland on Mt Spurgeon and Mt Lewis, in rainforest in small to large populations as an understorey element on granitic soils, 400–1300 m asl (Fig. 10.33). Flowering Apr.–May; fruiting Nov.–Mar. Conservation status – *No present threats*.

Typification: The holotype of *Linospadix apetiolatus* is *Dowe 369 with Ferrero & Smith* [BRI], collected in 1996 from Mt Lewis, north-east Queensland. The collection consists of stem portions, leaves, inflorescences and both staminate and pistillate flowers (Fig. 10.33).

Figure 10.31 *Linospadix apetiolatus* in high-altitude rainforest, Mt Lewis, north-east Queensland, c. 1200 m altitude.

Etymology: 'in reference to the lack of a petiole on the leaf' (Dowe and Irvine 1997). From the Latin *a*, without, and *petiolus*, petiole.

Notes: *Linospadix apetiolatus* is distinguished by a clustering habit, thin stems up to 5 m tall, and petiole lacking or short, leaf usually undivided (although there are partially segmented forms) and ellipsoidal fruit.

2 *Linospadix microcaryus*

Linospadix microcaryus (Domin) Burret, *Notizbl. Bot. Gard. Berlin-Dahlem* 12: 331 (1935); *Bacularia microcarya* Domin, *Biblioth. Bot.* 85: 499 (1915). Type: Australia. Queensland. Harveys Ck, Jan. 1910, *K.Domin 2083, 2085, 2086, 2089, 2090*; holotype PR [5 sheets].

Linospadix sessilifolius (Becc.) R.W.Johnson, *Queensl. Naturalist* 22: 19 (1981); *Bacularia sessilifolia* Becc., *Atti Soc. Tosc. Sci. Nat. Pisa Processi Verbali* 44: 133 (1934). Type: Australia. Queensland. Russell R., 1886, *W.A.Sayer s.n.*; holotype FI.

Bacularia sessilifolia var. *multisecta* Becc. ex Martelli, *Nuovo Giorn. Bot. Ital.* n.s. 42: 30 (1935). Type: not designated.

Clustering, small palm. **Stems** 1–6, erect to leaning, to 3 m tall, 7–25 mm diam.; leafscars to 5 mm wide; internodes to 60 mm long, green. **Leaves** 5–9, 28–70 cm long, irregularly segmented with united pinnae, segments broadly adnate to the rachis, or regularly pinnate; petiole 1–23 cm long, 3–6 mm wide. **Pinnae** 3–23 per leaf, 11–36.5 cm long, 0.9–7.4 cm wide, semi-glossy, lettuce to mid green above, sometimes dark green when in deep shade, lighter green below; midrib and veinlets prominent on both surfaces; terminal pair broader than adjacent laterals and often basal pair broader than laterals. **Inflorescence** to 90 cm long. **Staminate flowers** globose in bud, 2–5 mm long, not widely opening at anthesis; sepals c. 1 mm long, c. 1 mm wide, dark green; petals c. 4 mm long, c. 3 mm wide, cupular, orange-yellow, apically rounded; stamens 8–12, c. 2 mm long; filaments 0.2 mm long; anthers 1 mm long, light brown. **Pistillate flowers** to 2 mm long, to 1.5 mm wide in bud; sepals to 1 mm long, imbricate, green, glabrous; petals basally connate, segments to 2 mm long, apices valvate, curved inward, green, glabrous; stigmas not exserted, held within the petals. **Fruit** globose-turbinate to subglobose, 5–9 mm long, 5–8 mm wide; epicarp smooth, c. 0.1 mm thick, yellow-orange, or pink to red at maturity; mesocarp c. 0.8 mm thick, succulent. **Seed** globose to subglobose, to c. 4.5 mm long, to c. 4 mm wide. Figures 10.34, 10.35.

Figure 10.32 *Linospadix apetiolatus*. **Top left:** Typical undivided leaf. **Top right:** Staminate flowers at anthesis. **Bottom left:** Pistillate flowers at anthesis. **Bottom right:** Immature fruit, c. half-size.

Clustered walking-stick palm, small-fruited walking-stick palm

Distribution and ecology: Endemic to north-east Queensland from about Mt Spurgeon in the north, to about the Johnson River and Palmerston Range, occurring as an understorey element in rainforest, usually in small groups or scattered individuals on soils predominantly of basaltic origin, less common on soils derived from granite, 20–1200 m asl (Fig. 10.36). Flowering all months; fruiting all months. Conservation status – *Rare* (Bostock and Holland 2007).

Typification: The holotype [PR] of *Linospadix microcaryus* (basionym *Bacularia microcarya*) was collected by Karel Domin from Harveys Creek, a stream with its headwaters in Bellenden Ker Range, north-east Queensland.

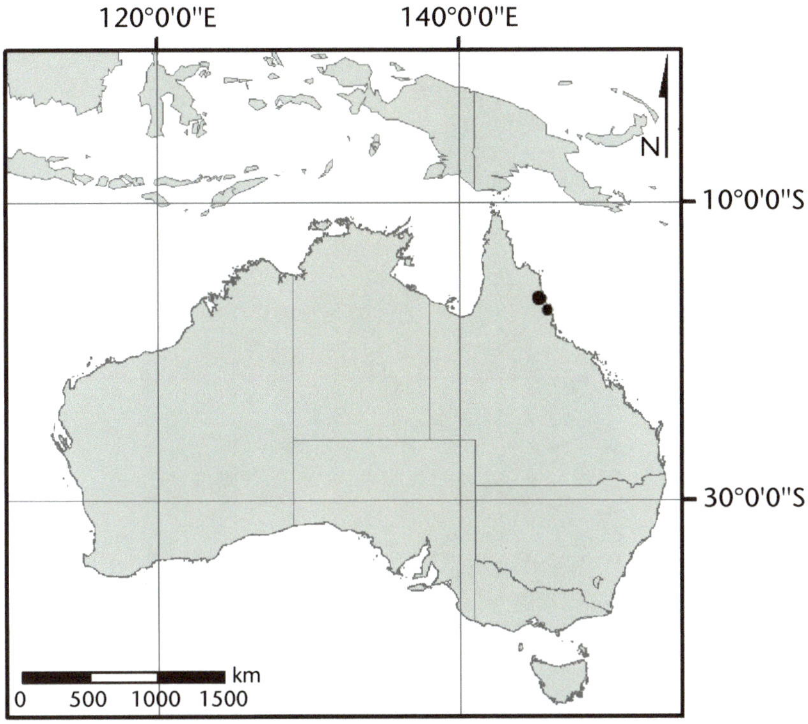

Figure 10.33 *Linospadix apetiolatus*. **Left:** Distribution based on herbarium specimen data. **Right:** Type specimen. Australia. Queensland. Mt Lewis, 27 km from Rex Hwy, on forestry road, 24 Aug. 1996, *J.L.Dowe 369 with M.D.Ferrero and L.T.Smith*; holotype BRI [1 of 4 sheets].

Figure 10.34 *Linospadix microcaryus* in lowland rainforest, Majuba Creek, eastern footslopes of Bartle Frere, Bellenden Ker Range, north-east Queensland.

The collection consists of a number of entire leaves, inflorescences with flowers and an infructescence with fruit (Fig. 10.36). The designation of the holotype as '*Harvey's Ck., 1889, Bailey s.n.* (holotype: BRI [AQ754431])', proposed by Dowe and Irvine (1997), was incorrect and is hereby rescinded.

Etymology: '*fructus 7–8 mm longus*' (Domin 1915) [fruit 7–8 mm long]: With reference to the relatively small fruit, from the Greek *micros*, small, and *caryon*, fruit.

Notes: *Bacularia sessilifolia* was described with a short petiole, based on a collection from Russell River, but the description of fruit places it as *L. microcaryus*. *Bacularia sessilifolia* var. *multisecta* was listed by Martelli, and later as *Linospadix microcaryus* var. *multisecta* by Burret.

Linospadix microcaryus is distinguished by the clustering habit, thin stems up to 3 m tall, a short to long petiole, an irregularly segmented leaf and small turbinate or globose fruit.

3 Linospadix minor

Linospadix minor (W.Hill) Burret, *Notizbl. Bot. Gard. Berlin-Dahlem* 12: 330 (1935); *Areca minor* W.Hill, *Rep. Brisbane Bot. Gard. 1874*: 6 (1874); *Kentia minor* (W.Hill) F.Muell., *Fragm.* 8: 235 (1874); *Bacularia minor* (W.Hill) F.Muell., *Fragm.* 11:

Figure 10.35 *Linospadix microcaryus*. **Top left:** Pinnate leaf form. **Top right:** Staminate flowers at anthesis. **Bottom left:** Pistillate flowers at anthesis. **Bottom right:** Mature fruit.

58 (1878). Type: Australia. Queensland. Bellenden Ker Range, Nov. 1873, *W.Hill s.n.*; lectotype MEL [3 sheets].

Linospadix intermedia (C.T.White) R.W.Johnson, *Queensl. Naturalist* 22: 19 (1981); *Bacularia intermedia* C.T.White, *Proc. Royal Soc. Queensland* 47: 83 (1936). Type: Australia. Queensland. Mowbray R., 21 Jan. 1932, *L.J.Brass 1975*; lectotype BRI.

Clustering, small palm. **Stems** 2–15, erect or leaning, 1–5 m tall, 7–20 mm diam.; leafscars raised, to 5 mm wide; internodes to 60 mm long, green. **Leaves** 7–12, to 110 cm long, irregularly segmented with united pinnae, segments broadly adnate to the rachis, or regularly pinnate with narrow pinnae; petiole 4–51 cm long, 1–4 mm wide. **Pinnae** 3–24 per leaf, semi-glossy

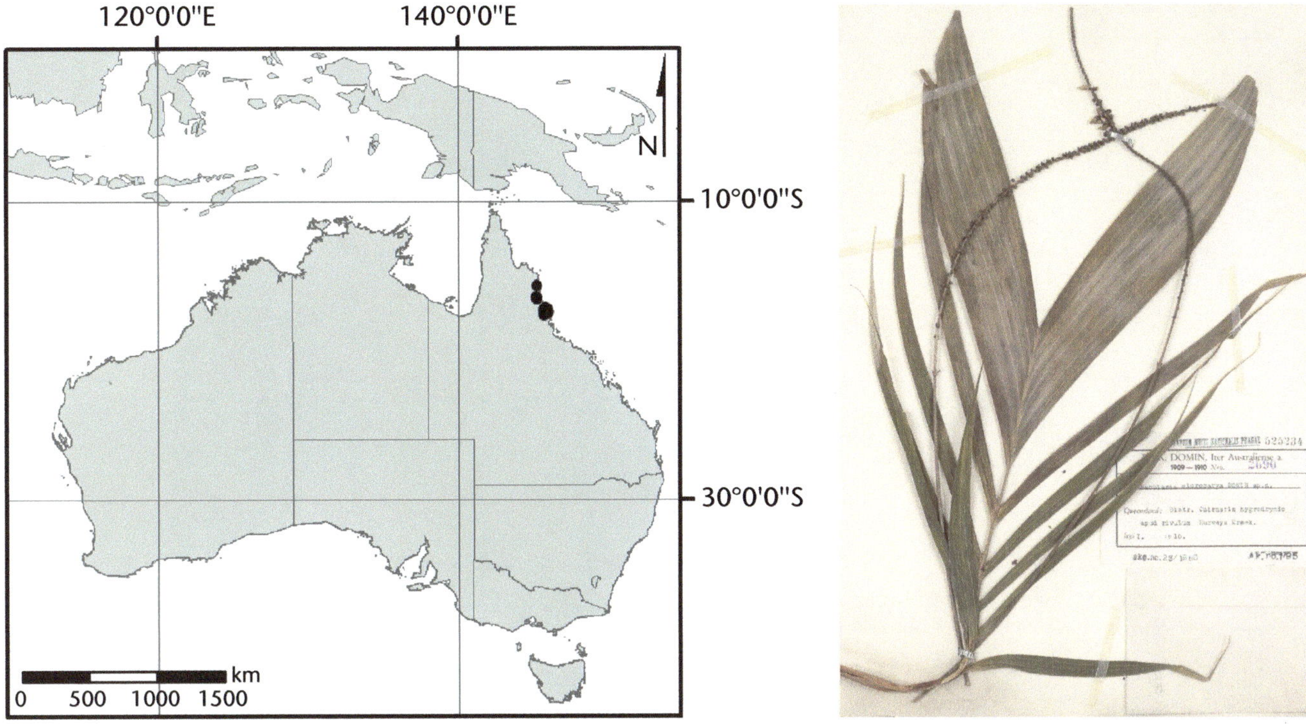

Figure 10.36 *Linospadix microcaryus*. **Left:** Distribution based on herbarium specimen data. **Right:** Type specimen. Australia. Queensland. Harveys Creek, Jan. 1910, *K.Domin 2083, 2085, 2086, 2089, 2090*; holotype PR [1 of 5 sheets].

dark green above, lighter green below, midrib prominent on both surfaces, veinlets not prominent on lower surface; lamina chartaceous, lax, irregularly corrugated to flat. **Inflorescence** to 80 cm long; floral pits shallow, margins scarcely raised, smooth. **Staminate flowers** bullet-shaped in bud, 3 mm long, to 2 mm wide, widely opening at anthesis; sepals to 2 mm long, green, glabrous; petals to 3 mm long, apically rounded, with conspicuous longitudinal striations, glabrous, green at anthesis, opening to 5 mm wide at anthesis; stamens 7–20, attached at different levels in the staminal cluster; connective not extending beyond the anther; anther lobes irregular. **Pistillate flowers** ovate, 2–3 mm long; sepals 0.5–1.2 mm long, gibbous, cream; petals 2.5–3 mm long, c. 1.5 mm wide, imbricate, cream to light yellow; stigmas exserted c. 0.5 mm. **Fruit** elongate-cylindrical, 8–18 mm long, 3–8 mm diam.; epicarp irregularly rugose when fruit is fully mature, to 0.1 mm thick, yellow, orange or red at maturity; mesocarp 1.5–2 mm thick; endocarp 0.2–0.4 mm thick. **Seed** elongate-ellipsoid to club-shaped, to 15 mm long, 3.5–4 mm wide in the middle, widening toward the apex; embryo basal, c. 1.5 mm long. Figures 10.37, 10.38.

Figure 10.37 *Linospadix minor* in lowland rainforest, Oliver Creek, Daintree National Park, north-east Queensland.

Figure 10.38 *Linospadix minor*. **Top left:** Inflorescence in bud with withered peduncular bract. **Top right:** Staminate flowers at anthesis. **Bottom left:** Pistillate flowers at anthesis. **Bottom right upper:** Mature fruit, red form. **Bottom right lower:** Mature fruit, orange form.

Mission Beach walking-stick palm

Distribution and ecology: Endemic to north-east Queensland, uncommon at McIlwraith Range on Cape York Peninsula, then with a disjunction to just south of Cooktown, then south commonly to Mission Beach, in rainforest as an understorey element in small groups or scattered individuals on basaltic and granitic soils, 0–1300 m asl (Fig. 10.39). Flowering all months; fruiting all months. Conservation status – *No present threats*.

Typification: Walter Hill did not provide details of any collections in the protologue of *Linospadix minor* (basionym *Areca minor*), but mentioned distribution in Moresby and

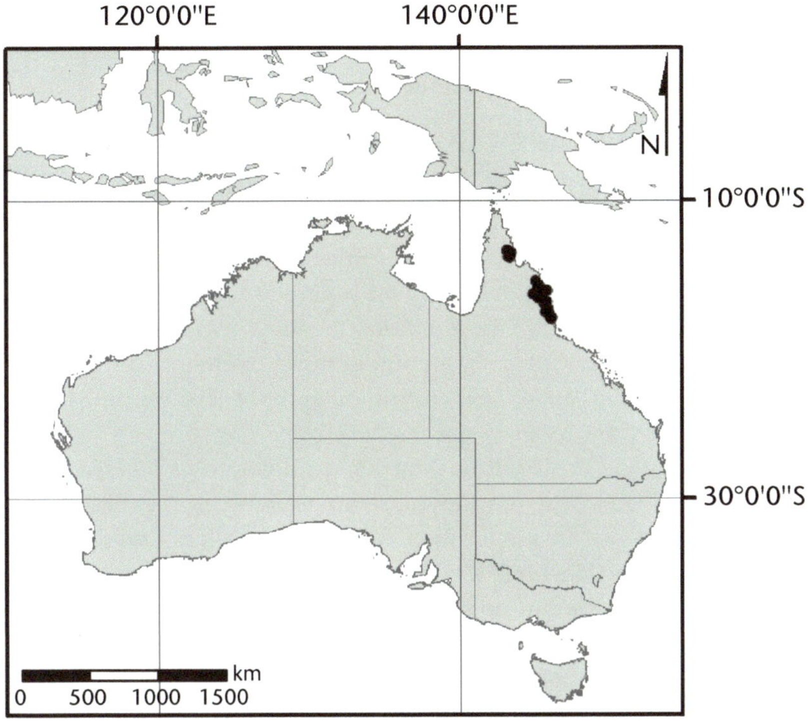

Figure 10.39 *Linospadix minor*. **Left:** Distribution based on herbarium specimen data. **Right:** Type specimen. Australia. Queensland. Bellenden Ker Range, Nov. 1873, *W.Hill s.n.*; lectotype MEL [1 of 3 sheets].

Russell Rivers and Bellenden Ker Range. The specimen, conserved in MEL, labelled as '*Kentia minor, Bellenden Ker Ranges, W. Hill*', was chosen as the lectotype (Dowe and Irvine 1997), and those labelled '*Kentia minor, Moresby's & Russell's Rivers & Bellenden Ker Ranges, W. Hill*' and '*Areca minor W.H. No 1, Bellenden Kerr Ranges & Ranges near Morelian Harbour 1873*' as syntypes. The lectotype consists of a full leaf and a single inflorescence (Fig. 10.39).

Etymology: 'caudex 2 to 5 feet high, about 1/2 an inch in diameter' (Hill 1874): There was no explanation provided in the protologue by the author, but probably referring to the lesser stature of this species compared to *L. monostachyos*, which Hill would have known from southern Queensland: From the Latin *minor*, smaller or less.

Notes: *Linospadix intermedia* was described as intermediate between *L. palmerianus* and *L. monostachyos*, with leaves with 2–4 opposite or alternate segments on either side of the rachis, and the intermediate segments and sometimes the lowest pair much narrower than the uppermost. This leaf type falls within the variation in *L. minor*.

Although Mueller (1878) was the first to cite the name *Linospadix minor*, he did so in the context of a possible alternative name, but still indicated that *Bacularia minor* was the preferred and correct name. Burret (1935), in his revision of *Linospadix*, provided the first nomenclaturally legal combination.

Linospadix minor is the most variable of the Australian species. Plants may be sparsely or densely clustered and leaves may be small to large with few to many segments. The fruit is ellipsoidal and may be yellow, orange or red at maturity. Coloured fruits may occur in the same population and this character appears to be variable, although an individual plant appears to maintain the same colour during its life.

4 *Linospadix monostachyos*

Linospadix monostachyos (Mart.) H.Wendl., *Linnaea* 39: 199 (1875); *Areca monostachya* Mart., *Hist. Nat. Palm. Edn 1*, 3: 178, tab. 155, Fig. 4 sub. 1–19 (1838); *Kentia monostachya* (Mart.) F.Muell., *Fragm.* 7: 82 (1870); *Bacularia monostachya* (Mart.) F.Muell., *Fragm.* 11: 58 (1878). Type: lectotype, illustration of *Areca monostachya in* C.F.P. von Martius, *Hist. Nat Palm.* 3: tab. 155. Fig. 4 sub. 1–19 (1838) *fide* Dowe and Irvine (1997).

Solitary, small to moderate palm. **Stem** erect, 1.3–6 m tall, 1.2–5 cm diam.; leafscars raised, 2–4 mm wide; internodes 15–40 mm long, green, grey with age; leafbases persistent immediately below crown, caducous. **Leaves** 5–10, to 130 cm long, irregularly segmented with united or single pinnae, or regularly pinnate with single pinnae; petiole 24–80 cm long, 5.5–8 mm wide. **Pinnae** 10–30 per leaf, to 30 cm long, 5–200 mm wide, glossy mid to dark green above, lighter green below; midrib and veinlets prominent on both surfaces. **Inflorescence** to 100 cm long; peduncle 30–60 cm long; prophyll 20–30 cm long; rachilla to 50 cm long. **Staminate flowers** pointed in bud, angular-pyramidal, 6–13 mm long, not widely opening at anthesis; sepals c. 1.5 mm long, connate, green-brown; petals rigid, 5–13 mm long, 7–10 mm wide, with deep longitudinal striations in the dried state, apically pointed, acutely heeled, straw-coloured at anthesis; stamens 8–12; connective apiculate and extending beyond attachment of anthers; anthers c. 2 mm long. **Pistillate flowers** ovoid, 5–6 mm long, 2–4 mm wide; sepals c. 1.2 mm long, carinate, green; petals 3–5 mm long, light green; stigmas not exserted, enclosed within the petals; ovary c. 4 mm long. **Fruit** globose to subglobose, or narrowly ovoid, 9–15 mm long, 4–10 mm wide; epicarp smooth, c. 0.5 mm thick, red at maturity; mesocarp 2–5 mm thick, succulent; endocarp 0.3 mm thick, with fibres compressed against it. **Seed** elongate-ellipsoid, 7–11 mm long, 2–5 mm wide. Figures 10.40, 10.41.

Figure 10.40 *Linospadix monostachyos* in moderate-altitude rainforest, Mt Warning, New South Wales, pinnate leaf form, c. 600 m altitude.

Walking-stick palm, Midgin-bil

Distribution and ecology: Endemic to eastern Australia in southern Queensland just north of Gympie, south into northern New South Wales to John's Mt, near Taree, in rainforest primarily on basaltic soils, occurring in moderate-sized populations or small groups and scattered individuals as an understorey element, 0–1200 m asl (Fig. 10.42); Flowering all months; fruiting all months. Conservation status – *No present threats*.

Typification: Martius' protologue of *Linospadix monostachyos* (basionym *Areca monostachya*) included reference to a specimen collected by Allan Cunningham in the Hastings River area, near Port Macquarie, New South Wales, but no specimen relating to this location has been found. The protologue included an illustration of a leaf, an inflorescence and flowers [Martius, *Hist. Nat Palm.* 3: tab. 155. Fig. 4 sub. 1–19 (1838)], and this was chosen as the lectotype (Fig. 10.42) (Dowe and Irvine 1997).

Etymology: *'spadice simplicissimo filiformis'* (Martius 1838) [inflorescence unbranched and thin]: With reference to the inflorescence, which is a single spike, from the Greek *mono*, single, and *stachyus*, inflorescence.

Notes: *Linospadix monostachyos* is distinguished by the solitary habit with stems to 6 m tall, leaves with variously united segments, or infrequently regularly pinnate (Mt Warning population) and long pendulous infructescences with globose to subglobose red fruit.

5 Linospadix palmerianus

Linospadix palmerianus (F.M.Bailey) Burret, *Notizbl. Bot. Gard. Berlin-Dahlem* 12: 331 (1935); *Bacularia palmeriana* F.M.Bailey, *Rep. Meston Bellenden-Ker Range* 24 (16 October 1889). Type: Australia. Queensland. Bellenden Ker Range, 'South Peak', 1889, *F.M.Bailey*; holotype BRI [3 sheets] (*fide* Dowe and Broughton 2007).

Linospadix aequisegmentosa (Domin) Burret, *Notizbl. Bot. Gard. Berlin-Dahlem* 12: 331 (1935); *Bacularia aequisegmentosa* Domin, *Biblioth. Bot.* 85: 500, pl.18, figs 1–8. (1915). Type: Australia. Queensland. Bellenden Ker Range, 800 m alt., Dec. 1909, *K.Domin 2091–2097*; holotype PR [6 sheets].

Stems clustering, erect to leaning, 1–6, to 3 m tall, 5–20 mm diam.; leafscars 1–4 mm wide; internodes 1–4 cm long, green. **Leaves** 8–12, to 60 cm long, regularly segmented with united pinnae, most often with

Figure 10.41 *Linospadix monostachyos*. **Top left:** Typical leaf form. **Top right:** Staminate flowers at anthesis. **Bottom left:** Pistillate flowers at anthesis. **Bottom right upper:** Mature fruit. **Bottom right lower:** Fruit, whole, with epicarp removed to reveal mesocarp fibres, and in longitudinal section.

two segments either side of rachis, segments broadly adnate to the rachis, or regularly and finely pinnate; petiole 6–13 cm long; rachis to 50 cm long; lamina rigid. **Pinnae** 4–24 per leaf, dull to semi-glossy dark green above, lighter green below; midrib prominent on both surfaces, veinlets numerous and prominent on both surfaces, terminal pinnae broader than laterals. **Inflorescence** to 60 cm long. **Staminate flowers** globose in bud, 2.5–3 mm long, not widely opening at anthesis; sepals c. 1 mm long, c. 1 mm wide, dark green; petals 2.5 mm long, 2.5 mm wide, orange, cupular, apically rounded, orange-yellow; stamens 6–9, c. 1.5 mm long; filaments c. 0.2 mm

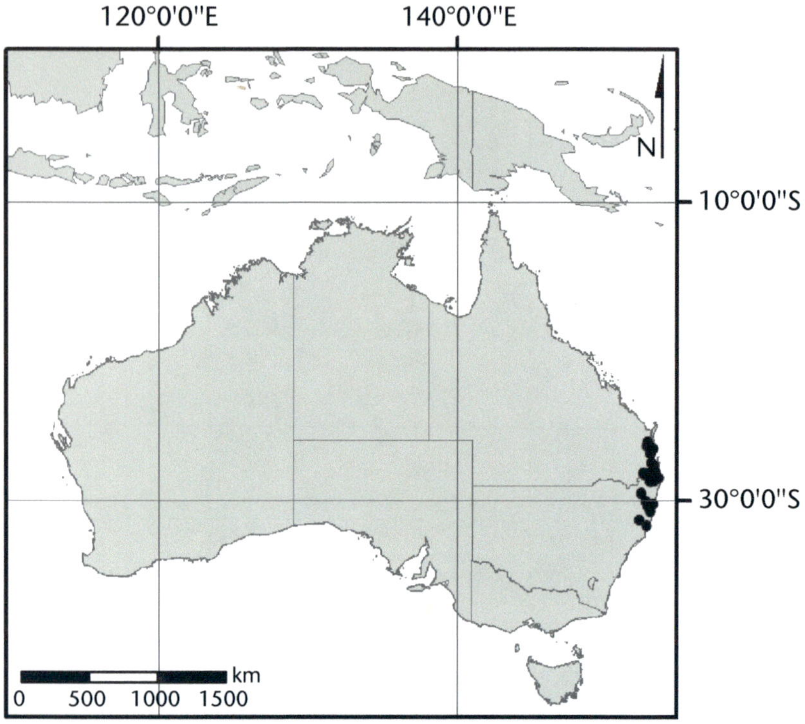

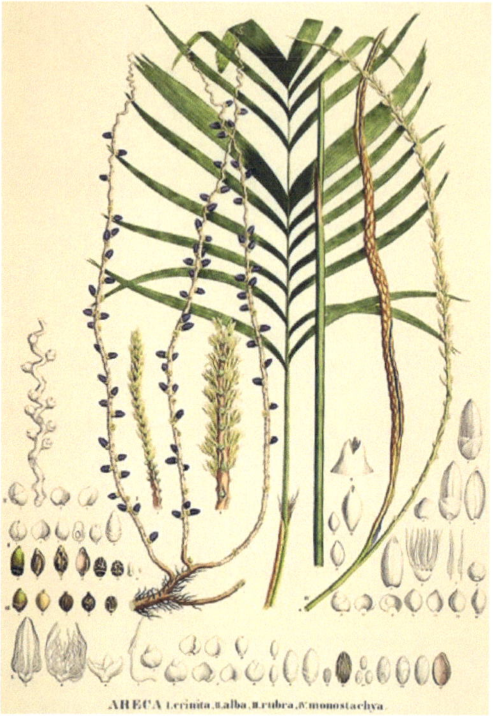

Figure 10.42 *Linospadix monostachyos*. **Left:** Distribution based on herbarium specimen data. **Right:** Type specimen. Lectotype, illustration of *Areca monostachya in* C.F.P. von Martius, *Historia Naturalis Palmarum* 3: Plate 155, figure 4, sub. 1–19 (1838).

long; anthers oblong, c. 1 mm long, light brown; connective not extending beyond the anthers. **Pistillate flowers** globose, to 3 mm long in bud; sepals to 1 mm long, triangular, apex rounded, whitish-cream, apically minutely fringed; petals to 3 mm long, triangular, apically acuminate, cream; stigmas c. 0.5 mm long, recurved. **Fruit** elongate-cylindrical, 10–15 mm long, 5–7 mm wide; epicarp smooth or irregularly rugose, yellow or red at maturity; mesocarp to 1.5 mm thick. **Seed** elongate-ellipsoid, 8–12 mm long, 3–4 mm wide. Figures 10.43, 10.44.

Bartle Frere walking-stick palm

Distribution and ecology: *Linospadix palmerianus* is endemic to north-east Queensland in the Bellenden Ker Range, Murray Prior Range and upland areas to the immediate west of Bartle Frere on Atherton Tableland, in rainforest, most common on basalt and granitic soils, occurring as small groups or scattered individuals, 300–1600 m asl (Fig. 10.45). Flowering all months; fruiting all months. Conservation status – *Rare* (Bostock and Holland 2007).

Typification: In the protologue of *Linospadix palmerianus* (basionym *Bacularia palmeriana*), Bailey (1889, p. 24) noted 'My specimens were gathered at the base of the

Figure 10.43 *Linospadix palmerianus* in high-altitude rainforest, Bartle Frere western slopes, Bellenden Ker Range, north-east Queensland, c. 1000 m altitude.

Figure 10.44 *Linospadix palmerianus*. **Top left:** Epiphylls on a leaf, common at high altitudes. **Top right:** Staminate flowers at anthesis. **Bottom left:** Immature fruit. **Bottom right upper:** Mature fruit. **Bottom right lower:** Fruit, whole and in longitudinal section.

leading spur at about 2,000 feet, and then from that to the summit of the south peak, an altitude of over 5,000 feet'. The specimen chosen as the lectotype (Dowe and Irvine 1997) is labelled '*Bacularia palmeriana,* this is from the base of the mountain. I have but ample specimens of the others'. Included on the same sheet are two fruits labelled 'unripe fruit off plant on top of mountain'. This specimen therefore represents two collections, of which the fruit is superfluous to typification; the lectotype is the portion consisting of a full crown with a portion of stem and a number of inflorescences (Dowe and Irvine 1997) (Fig. 10.45).

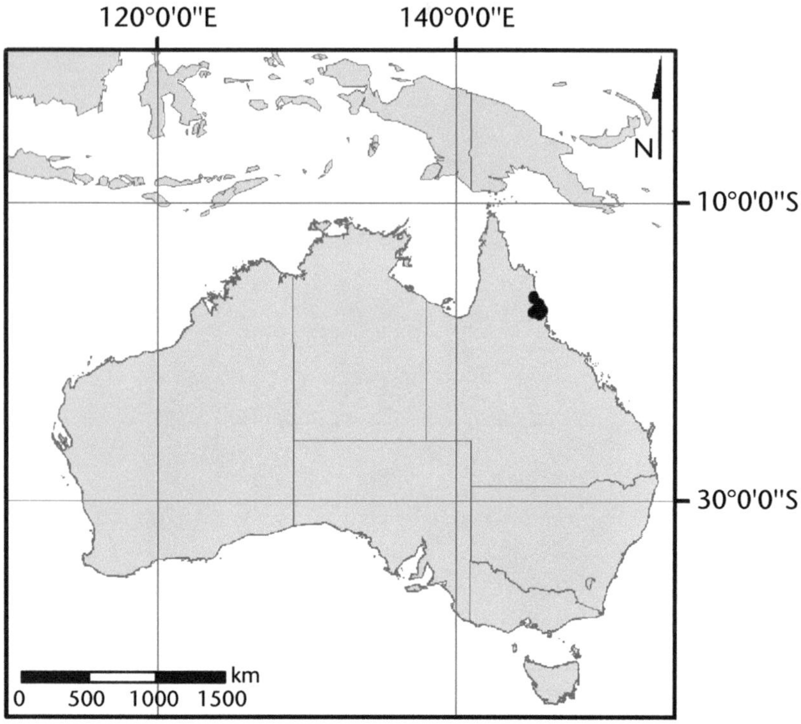

Figure 10.45 *Linospadix palmerianus*. **Left:** Distribution based on herbarium specimen data. **Right:** Type specimen. Australia. Queensland. Bellenden Ker Range, 'South Peak', 1889, *F.M.Bailey*; holotype BRI [1 of 3 sheets].

Etymology: 'After Edward Palmer, M.L.A., to whom we are indebted for much useful information as to the uses made by the aboriginal of our indigenous plants' Bailey 1889, p. 24): Edward Palmer (b. c. 1840, d.1899), author, pastoralist, amateur anthropologist, Member of the Legislative Assembly of Queensland for Burke District 1883–88 and Carpentaria District 1888–93. Author of *Early Days in North Queensland* (Palmer 1903) and a paper on the anthropology of the Aboriginal tribes of north Queensland in which he provided ethnobotanical information (Palmer 1884). A few Palmer plant specimens were cited in publications on the Queensland flora (Bailey 1890).

Notes: *Linospadix aequisegmentosa* (basionym *Bacularia aequisegmentosa*) was described with regularly pinnate leaves, but is referable to *L. palmerianus* as this variation occurs within the species. Dowe and Irvine (1997) designated the illustration included with Domin's protologue [*Biblioth. Bot.* 85: tab. 18, figs 1–8 (1915)] as the lectotype for this name (Dowe and Irvine 1997), but this has been rescinded as the holotype was subsequently located at PR. The holotype consists of seven sheets, with complete leaves, stem portions, inflorescences and flowers.

Linospadix palmerianus is distinguished by its usually small stature but with stems to 3 m tall, leaves regularly segmented with united pinnae, most often with two segments either side of rachis, infrequently simply bifid, segments broadly adnate to the rachis, or regularly and finely pinnate, and elongate-cylindrical red or yellow fruit.

Laccospadix

Laccospadix H.Wendl. & Drude, *Linnaea* 39: 178 [205] (June 1875); *Calyptrocalyx* subg. *Laccospadix* (H.Wendl. & Drude) Drude, *in* H.G.A.Engler & K.A.E.Prantl, *Nat. Pflanzenfam.* 2(3): 69 (1887); *Saguaster* sect. *Laccospadix* (H.Wendl. & Drude) Kuntze, *in* T.E.Post & O.Kuntze, *Lex. Gen. Phan.*: 495 (1903). Type: *Laccospadix australasicus* H.Wendl. & Drude.

[*Calyptrocalyx* Blume, *in* R.H.C.C.Scheffer, *Ann. Jard. Bot. Buitenzorg* 1: 140 (1876)].

[*Laccospadix* Drude & H.Wendl., *Nachr. Königl. Ges. Wiss. Georg-Augusts-Univ. 1875*: 59 (February, 1875) *nom. illeg.*].

The first publication of *Laccospadix* has been cited as Drude & H.Wendl. in *Nachr. Königl. Ges. Wiss. Georg-Augusts-Univ. 1875*: 59 (1875) by many authors, including Chapman (1991), Uhl and Dransfield (1987) and Dransfield *et al.* (2008). The use of the name was not validated in that publication as it was a summary dealing with generic differences of certain palm genera dealt with in Wendland and Drude's *Palmae Australasicae* in *Linnaea* 39 (1875), but without direct reference. Subsequently, Hooker (1883) cited *Linnaea* 39 as the valid place of publication of *Laccospadix*, as did Burret (1928). The publication of *Nachr. Königl. Ges. Wiss. Georg-Augusts-Univ.* 1875 occurred in February 1875, while the publication of *Linnaea* occurred in June 1875. Drude and Wendland did not include a species name or a generic diagnosis in the former, thus invalidating it as the publication place according to Articles in the ICBN. Therefore, the first valid publication of *Laccospadix* was in *Linnaea* 39, in which was included a protologue and species name. Therefore, the correct author citation should be *Laccospadix* H.Wendl. & Drude.

Solitary or clustering, small to moderate, pleonanthic, monoecious palms. **Stems** slender; leafscars conspicuous, internodes short to long. **Leaves** paripinnate; leafbase clasping, not forming a tubular crownshaft, caducous, deeply split opposite the petiole, margins fibrous, glabrous to lightly scaly; petiole long, channelled on the adaxial surface, rounded on the abaxial surface, glabrous to moderately scaly; rachis long, moderately curved, angled to flat on the adaxial surface, rounded on the abaxial surface. **Pinnae** in a single plane on the rachis, subopposite to opposite, linear acuminate, reduplicate, rigid, apex acute; midrib prominent, parallel veinlets present on abaxial surface; ramenta on abaxial midrib, medifixed. **Inflorescence** interfoliar, spicate, initially erect, pendulous in fruit, protandrous; prophyll short, within leafbases, flattened, tubular, 2-keeled; peduncular bract exserted from prophyll, attached midway between attachment of prophyll and lowest flowers on the rachilla, enclosing the rachilla in bud to almost floral maturity, persistent and withering to a fibrous mass or fully deciduous and leaving a prominent raised scar; peduncle short, elliptical in cross-section; rachilla much longer than the peduncle, slightly thicker. **Flowers** unisexual, sessile to subpedicellate, spirally arranged in pits with raised margins, in triads of a single pistillate flower subtended by a pair of staminate flowers one either side in proximal portion, paired or solitary staminate flowers in distal portion. **Staminate flowers** symmetric in bud; 3 broadly imbricate, keeled sepals; 3 thick, apically valvate petals c. twice as long as sepals; stamens 9–12, erect; filaments short, basally connate; anthers elongate, basifixed, not versatile, latrorse; connective broad or narrow, sometimes extending beyond the anthers and then apicular; pistillode small, trifid. **Pistillate flowers** globular, same size or slightly smaller than the staminate flowers; 3 broadly imbricate sepals; 3 thick valvate petals, longer than the sepals; staminodes 3–6, dentate; gynoecium unilocular, uniovulate; 3 recurved stigmas; ovule campylotropous, laterally attached. **Pollen** ellipsoidal or circular; aperture monosulcate; exine tectate, scabrate or finely reticulate. **Fruit** ellipsoidal to ovoid, 1-seeded, stigmatic remains apical; epicarp smooth, red at maturity; mesocarp fleshy, longitudinal fibres throughout the mesocarp, densest toward the endocarp; endocarp thin, adhering to the seed; perianth not persistent on fruit. **Seed** ovoid, laterally attached, hilum short, raphe extending full length of the seed, branches anastomosing; endosperm ruminate; embryo basal; germination adjacent-ligular; eophyll bifid.

Laccospadix is a monotypic genus endemic to north-east Australia.

Relationships

Laccospadix is placed within the Linospadicinae, which also includes the genera of *Howea*, *Linospadix* and *Calyptrocalyx*. The latter genus does not occur in Australia. The Linospadicinae is distinguished by spicate inflorescences, with flowers most often borne in pits with an enlarged proximal lip. Flower morphology is variable, and seeds can be with either ruminate or homogeneous endosperm. *Laccospadix* resolves as sister to *Howea* in most studies (Fig. 10.30).

Etymology: '*Glomeruli triflori foveis e spadicis substantiâ bracteatis in orthostichos dispositis immersi*' (Wendland and Drude 1875, p. 178). [Triads immersed in orthostichously arranged pits with prominent bracts on the spadix]: With reference to the flowers being borne in pits on the inflorescence; from the Greek *lakkos*, reservoir or pit, and *spadix*, inflorescence.

Laccospadix australasicus

Laccospadix australasicus H.Wendl. & Drude, *Linnaea* 39: 206, pl. 2, Fig. 3 (1875); *Ptychosperma laccospadix* Benth., *Fl. Austral.* 7: 140 (1878) *nom. illeg.*; *Calyptrocalyx australasicus* (H.Wendl. & Drude) Hook.f., *in* G.Bentham and J.D.Hooker, *Gen. Pl.* 3: 903 (1883); *Calyptrocalyx laccospadix* (Benth.) F.M.Bailey, *Catal. Plants Queensl.*: 50 (1890). Type: Australia. Queensland. Rockingham Bay, 8 Feb. 1866, *J.Dallachy 27*; holotype MEL; isotype BO, K.

Stems 1–8 m tall, to 10 cm dbh, with 1–4 dominant, erect or leaning; leafscars irregular, raised, 5–20 mm wide, white to grey; internodes 5–50 mm long, green. **Leaves** 6–8, to 2.5 m long, moderately arching to horizontal, new leaf emerges bronze, 10–30 pinnae per side; leafbases 30–50 cm long, green, glabrous, margins moderately lacerate-fibrous, fibres coarse; petiole 30–100 cm long, 2–3.5 cm wide; rachis to 1.5 m long. **Pinnae** 30–40 cm long, 2–4 cm wide, dark green on the adaxial surface, lighter green on the abaxial surface; midrib prominent on both surfaces; ramenta to 4 mm long, medifixed. **Inflorescence** 50–150 cm long; peduncle 20–40 cm long, 6–8 mm wide, subterete, green; rachilla 35–110 cm long, 6–10 mm wide, terete, green; prophyll 15–20 cm long, 3 cm wide, fibrous; peduncular bract 60–70 cm long, persistent after withering. **Staminate flowers** 4–9 mm long in bud, opening to c. 8 mm at anthesis; sepals 1–3 mm long, green; petals 4–9 mm long, with deep impressions of the stamens on the inner surface, green to straw yellow, glabrous; stamens unequal length, 5–8 mm long; filaments 0.5–1.0 mm long; anthers 2–4 mm long, cream to orange. **Pistillate flowers** 3–4 mm high in bud; sepals c. 1 mm long, dark reddish-green; petals 3–4 mm long, dull yellow-orange to green; stigmas to c. 0.5 mm long, spreading to c. 1 mm wide, cream to reddish-yellow. **Fruit** 9–18 mm long, 5–11 mm wide; epicarp c. 0.6 mm thick, red at maturity; mesocarp c. 2 mm thick, fleshy, finely fibrous; endocarp fibrous, adherent to the seed; perianth persistent on fruit. **Seed** 5–9 mm long, 4–5 mm wide. Figures 10.46, 10.47.

Figure 10.46 *Laccospadix australasicus* in high-altitude rainforest, Cloudy Creek, Paluma Range, north-east Queensland, c. 900 m altitude.

Misty mountain palm, Atherton palm

Distribution and ecology: Endemic to north-east Queensland, from Mt Spurgeon south to Paluma Range, most common at moderate to high altitudes, infrequently at low altitudes in wetter areas, in rainforest, 100–1600 m asl (Fig. 10.48). Flowering all months; fruiting all months. Conservation status – *No present threats.*

Typification: The holotype of *Laccospadix australasicus* is a collection by John Dallachy [MEL] from Rockingham Bay, Queensland. The collection consists of leaf portions, a segment of infructescence and fruits (Fig. 10.48).

Domin (1915) proposed that the author of the combination *C. australasicus* was Scheffer. Scheffer (1876) indicated that he considered *Laccospadix* and *Calyptrocalyx* to be a single taxon, but did not apply the new combination, and Domin's proposal is invalid. The first application of the combination was provided by Hooker (1883), although he also noted there, incorrectly, that the combination had been applied by Scheffer. Bailey (1890) attributed the novel combination *Calyptrocalyx laccospadix* to Bentham, an apparent incorrect transcription as Bentham did not apply that particular combination, but subsequently maintained placement in *Calyptrocalyx* as *C. australasicus* (Bailey 1902, 1912). Transference to *Laccospadix* was provided by Burret (1928) and followed by Martelli (1935) and Beccari and Pichi-Sermolli (1955).

Etymology: *'Australasiae ora tropica orientalis'* (Wendland and Drude 1875): In the eastern tropics of Australia; from the Latin *australis*, southern.

Notes: *Laccospadix australasicus* is distinguished by an either solitary or clustering habit, regularly pinnate leaves, a 1-spicate inflorescence, and seed with ruminate endosperm.

Howea

Howea Becc., *Malesia* 1: 41 [66] (1877) [as *Howeia*]. Type: *Kentia belmoreana* C.Moore & F.Muell. = *Howea belmoreana* (C.Moore & F.Muell.) Becc.

Figure 10.47 *Laccospadix australasicus*. **Top left:** Inflorescence with flowers in bud and persistent peduncular bract. **Top right:** Staminate flowers at anthesis. **Bottom left:** Pistillate flowers at anthesis. **Bottom right upper:** Mature fruit. **Bottom right lower:** Fruit, whole and in longitudinal section.

Grisebachia Drude & H.Wendl., *Nachr. Königl. Ges. Wiss. Georg-Augusts-Univ. 1875*: 55 (1875) *nom. illeg.*, *non* Klotzsch (1838). Type: *Grisebachia belmoreana* (C.Moore & F.Muell.) Drude & H.Wendl. = *Howea belmoreana* (C.Moore & F.Muell.) Becc.

Denea O.F.Cook, *J. Wash. Acad. Sci.* 16: 395 (1926). Type: *Denea forsteriana* (C.Moore & F.Muell.) O.F.Cook = *Howea forsteriana* (C.Moore & F.Muell.) Becc.

Solitary, moderate to tall, erect, emergent, pleonanthic, monoecious palms. **Stems** slender, frequently with a flared base; leafscars prominent, horizontal to oblique; internodes short to long. **Leaves** paripinnate; sheaths not forming a distinct crownshaft but splitting opposite the petiole producing a mass of interconnecting fibres; petiole short to long, concave on the adaxial surface, rounded to angled on the abaxial surface, scaly; rachis long, ridged

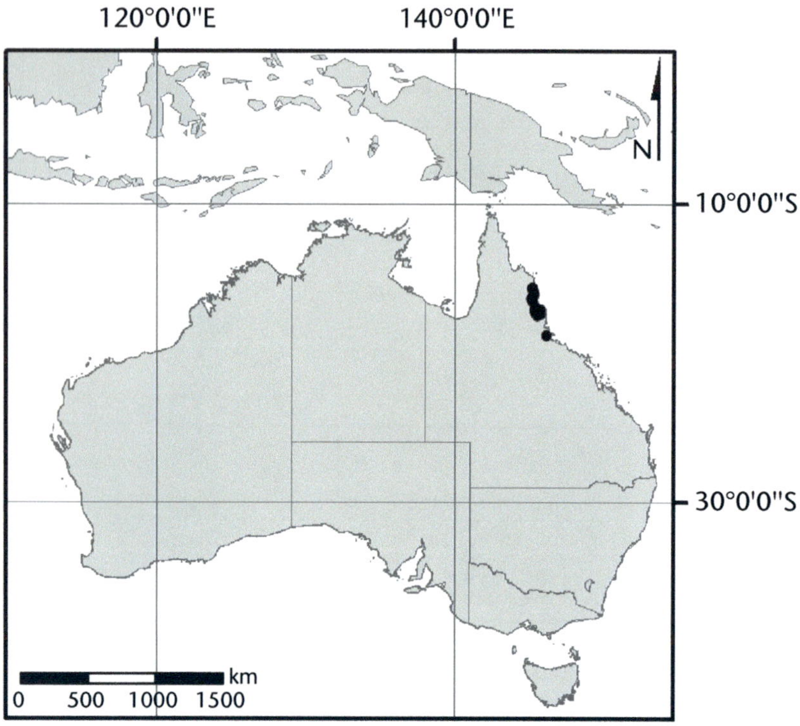

Figure 10.48 *Laccospadix australasicus*. **Left:** Distribution based on herbarium specimen data. **Right:** Type specimen. Australia. Queensland. Rockingham Bay, 8 Feb. 1866, *J.Dallachy 27*; holotype MEL.

and angled on the adaxial surface, rounded to angled on the abaxial surface, scaly. **Pinnae** inserted in a single plane along rachis, linear-acute, reduplicate, subopposite, erectly curved to semi-pendulous, firm, apex acuminate or briefly cleft; midrib prominent, parallel veinlets present on abaxial surface; abaxial surface glabrous or dotted with scales and floccose indumentum; ramenta on the abaxial surface of the midrib, medifixed. **Inflorescence** at first interfoliar, becoming infrafoliar after leaf abscission, usually slightly shorter than the leaves, spicate, or compound with 3–8 spikes sharing a common base, at first erect on emergence, becoming pendulous with maturity of fruit, protandrous; prophyll tubular, bicarinate, membranous; peduncular bract attached near the prophyll or at a distance along the peduncle, fully enclosing the inflorescence until anthesis, tubular, membranous to papery fibrous, splitting with maturity along almost its full length, becoming disintegrated and deciduous and leaving a prominent annular scar where previously attached; peduncle elliptic in cross-section, shorter than or same length at the rachis, scaly; rachis terete, densely scaly; floral pits deep, margins prominent on distal side, raised, sharp, less raised on proximal side, rounded, each pit with a partially sunken triad, or distally with paired staminate flowers; floral bracteoles sepal-like. **Flowers** unisexual, spirally arranged in triads of a single pistillate flower subtended by a pair of staminate flowers each side. **Staminate flowers** singly exserted at anthesis, on a stalk-like base as long as the sepals; 3 imbricate keeled sepals, margins dentate; 3 valvate petals, much longer than the sepals; stamens 30–70; filaments elongate, connate at the base for much of their length; connective pointed; anthers basifixed, oblong, not versatile, latrorse; pistillode lacking. **Pistillate flowers** smaller than the staminate, globular; 3 imbricate rounded sepals, margins dentate; 3 imbricate petals briefly valvate at the apex; staminodes 3–6, triangular, forming a membranous ring or irregular; gynoecium unilocular, uniovulate; style short; 3 stigmas, short, recurved; ovule laterally attached, campylotropus. **Pollen** ellipsoidal; aperture monosulcate; exine tectate, scabrate. **Fruit** ovoid, slightly ridged on one side, 1-seeded; stigmatic remains apical; epicarp thin, smooth, shiny, dark green to yellowish-green or reddish-brown; mesocarp thin, moderately fleshy with numerous longitudinal fibres; endocarp thin, cartilaginous, separating from the seed; perianth not persistent on fruit. **Seed** ovoid to subglobose; hilum lateral; raphe fibres extending one-third or less the length of the seed;

endosperm homogeneous; embryo basal; germination adjacent-ligular; eophyll bifid.

A genus of two species, endemic to Lord Howe Island and morphologically most similar to *Laccospadix*. The monophyly of *Howea* is strongly supported (Savolainen *et al.* 2006) and there is a sister relationship with *Laccospadix* (Norup *et al.* 2006; Dransfield *et al.* 2008). Molecular dating methods have placed divergence of *Howea* from *Laccospadix* at 4.6–5.5 Mya, and subsequent sympatric speciation of the two *Howea* species at 1.9–0.5 Mya (Savolainen *et al.* 2006).

Etymology: '*Abita. – Isola Lord Howe.*' [Distribution – Lord Howe Island]: With reference to Lord Howe Island, discovered and named by Lt Henry Lidgbird Ball, in 1778, in honour of Lord (Richard) Howe (b.1726, d.1799), Admiral of the British Fleet.

Key to species of Howea

1 Leaf rachis arching; pinnae erect on the rachis; inflorescence a solitary spike .... **1. *H. belmoreana***

1: Leaf rachis ± straight; pinnae pendulous on the rachis; inflorescences with multiple spikes sharing a common base and prophyll .... **2. *H. forsteriana***

1 Howea belmoreana

Howea belmoreana (C.Moore & F.Muell.) Becc., *Malesia* 1: 66 (1877); *Kentia belmoreana* C.Moore & F.Muell., *Fragm.* 7: 99 (1870); *Grisebachia belmoreana* (C.Moore & F.Muell.) Drude & H.Wendl., *Nachrich. K. Gesell. Wissensch. Götting. 1875*: 58 (1875). Type: Australia. New South Wales. Lord Howe Island, undated, *C.Moore & W.Carron s.n.*; holotype MEL [1 sheet]; isotype BO.

Stem 3–12 m tall, 8–15 cm dbh, flared at the base; leafscars prominent, irregular, 1–3 cm wide; internodes 2–30 cm long, green, becoming grey with age. **Leaves** 10–18, 2–5 m long, ascending, 30–40 pinnae per side; leafbase 25–30 cm long, green, glabrous, margin moderately lacerate-fibrous, fibres thick and meshed; petiole 20–100 cm long, 1.5–3 cm wide, deeply channelled above, glabrous; rachis to 4 m long, arching. **Pinnae** 30–60 cm long, 2.5–4 cm wide, erect, subopposite, lanceolate, linear acute, dull green on the adaxial surface and on the abaxial surface; midrib prominent on upper surface only; ramenta dark brown, strongly twisted and undulate. **Inflorescence** 1-spiked, 1–2 m long, to 4 cm thick; peduncle 15–20 cm long, subterete, green; rachilla 80–190 cm long, terete, green; floral pits widely spaced; prophyll 15–20 cm long, 3 cm wide, woody; peduncular bract 1–2 m long, papery, withering, caducous or persistent. **Staminate flowers** ovate, 11–14 mm long, 4–5 mm wide in bud, opening to c. 8 mm wide at anthesis; sepals to 3 mm long, carinate, imbricate, apically rounded, green; petals 8–10 mm long, oblong, curved inward, cupular, longitudinally striately nerved, woody, green; stamens 30–70, 6–8 mm long; filaments c. 1 mm long; connective broad, extending beyond the anthers, apicular; anthers linear, 6–7 mm long. **Pistillate flowers** globose to elongate, c. 10 mm long, c. 8 mm wide in bud; sepals 5–7 mm long, 4–6 mm wide, margins ciliate, green; petals 8–10 mm long, 4–6 mm wide, green; staminodes dentate. **Fruit** ovoid-ellipsoidal, 3–4 cm long, 2–3 cm diam., slightly beaked; perianth not persistent on fruit; epicarp c. 0.2 mm thick, glossy, brownish-red at maturity; mesocarp 5–8 mm thick, fibrous, dry; endocarp thin, papery. **Seed** ovate, 15–20 mm long, 8–12 mm wide. Figures 10.49, 10.50.

Curly palm, Kentia palm

Distribution and ecology: Endemic to Lord Howe Island, in evergreen closed forest on alluvium derived from

Figure 10.49 *Howea belmoreana* in moderate-altitude forest, Lord Howe Island. Photo by William J. Baker, Royal Botanic Gardens Kew.

Figure 10.50 *Howea belmoreana*. **Top left upper:** Crown with inflorescences. Photo by Jeanne Price. **Top left lower:** Abaxial surface of pinnae with ramenta. **Top right:** Staminate flowers at anthesis. **Bottom left:** Pistillate flowers at anthesis. **Bottom right:** Infructescence with mature fruit. Photos by William J. Baker, Royal Botanic Gardens Kew.

basalt, calcarenite and coral sand, in rocky soils or scree, forming large and small colonies, often in association with *H. forsteriana*, 0–400 m asl (Fig. 10.51). Flowering all months; fruiting all months. Conservation status – *Vulnerable* (IUCN 2008).

Typification: The holotype of *Howea belmoreana* (basionym *Kentia belmoreana*) is an undated and unnumbered collection by Charles Moore and William Carron [MEL] collected in 1869 from Lord Howe Island. The collection consists of a number of inflorescence portions and flowers (Fig. 10.51). In the protologue, Mueller stated he had not seen any leaves. The protologue consists only of descriptions of the inflorescence and fruit.

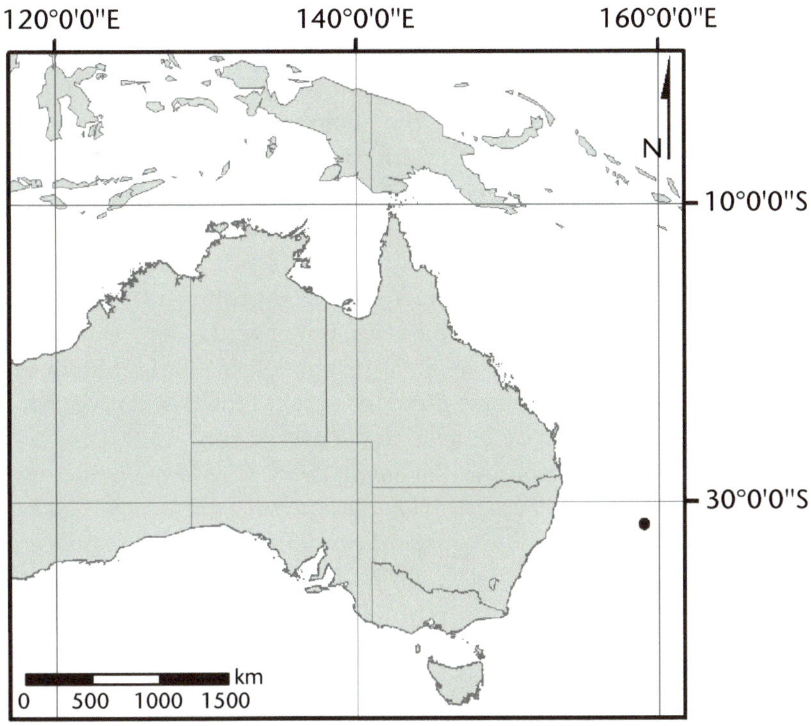

Figure 10.51 *Howea belmoreana*. **Left:** Distribution based on herbarium specimen data. **Right:** Type specimen. Australia. Lord Howe Island, undated, *C.Moore and W.Carron s.n.*; holotype MEL.

Etymology: *'Hancce praenobilem plantam Australiae tribuimus excellentissimo docto et praehonorabili Comiti S.R.L.C. de Belmore, serenissimae Britaniorum Reginae Consiliario, territorii Novae Austro-Cambriae elato Gubernatori'* (Mueller 1870). [This stately Australian plant is named as a tribute to Earl S.R.L.C. Belmore, British Queens Counsellor, Governor of New South Wales]: Named for Sir Somerset Richard Lowry-Corry (b.1835, d.1913), 4th Earl of Belmore, Governor of New South Wales 1868–72.

Notes: *Howea belmoreana* is distinguished by arching leaves with erect pinnae, and a 1-spicate inflorescence.

2 Howea forsteriana

Howea forsteriana (C.Moore & F.Muell.) Becc., *Malesia* 1: 66 (1877); *Kentia forsteriana* C.Moore & F.Muell., *Fragm.* 7: 100 (1870); *Grisebachia forsteriana* (C.Moore & F.Muell.) Drude & H.Wendl., *Nachrich. K. Gesell. Wissensch. Götting. 1875*: 58 (1875); *Denea forsteriana* (C.Moore & F.Muell.) O.F.Cook, *Journ. Wash. Acad.* 16: 397 (1926). Type: Australia. New South Wales. Lord Howe Island, undated, *C.Moore s.n.*; holotype MEL [1 sheet]; isotype BO.

Moderate to large palms. **Stem** 3–25 m tall, 12–15 cm dbh, flared at the base; leafscars prominent, irregular, 1–3 cm wide; internodes 2–30 cm long, green, becoming grey with age. **Leaves** 15–18, 3–5 m long, horizontal or descending, 60–80 pinnae per side; leafbase 35–45 cm long, green, glabrous, margin moderately lacerate-fibrous, fibres thick and meshed; petiole 30–190 cm long, 3–6 cm wide; rachis to 4 m long, ± straight. **Pinnae** 30–60 cm long, 3–4 cm wide, linear acute, horizontal to semi-pendulous, dark green on the adaxial surface and on the abaxial surface; midrib prominent on both sides; a prominent parallel vein each side of midrib c. half-way to margin; c. 10 minor parallel veinlets each side of midrib; margins thickened; ramenta on abaxial midrib, sparse, confined to proximal portion of midrib, medifixed, twisted, fibrous, rusty brown, to 6 mm long. **Inflorescence** 3–8-spiked, 100–180 cm long, sharing a common base and prophyll; peduncle 3–5 cm long, subterete, green; rachilla 70–125 cm long, terete, green; floral pits widely spaced; prophyll 20–30 cm long, woody, enclosing all spikes; peduncular bract 100–170 cm long, papery, fibrous, withering, persistent or deciduous, present on each spike. **Staminate flowers** ovate, 11–12 mm long, 5–7 mm wide in bud, opening to c. 10 mm at maturity; sepals to 3 mm long, carinate, imbricate, apically rounded, green; petals 8–10 mm long, oblong, cupular, longitudinally striately nerved, woody, green at anthesis drying to light brown; stamens 65–100, 4–5 mm long; filaments c. 0.5 mm long, thick; connective broad, extending beyond the anthers, apicular; anthers linear, to 2 mm long. **Pistillate flowers** globose, 5–7 mm diam.; sepals c. 2 mm long, carinate, apically rounded, margins lacerate, dark green; petals 3–6 mm long, cupular, imbricate, light green to straw-coloured; stigma c. 2 mm long; staminodes dentiform. **Fruit** ellipsoidal, 30–50 mm long, 20–30 mm wide, apically and basally tapered; perianth not persistent on the fruit; epicarp c. 2 mm thick, bright or dull orange to red at maturity; mesocarp 3–9 mm thick; endocarp thin, papery. **Seed** ellipsoid-ovate, 18–20 mm long, 12–15 mm wide. Figures 10.52. 10.53.

Kentia palm, Howea palm, sentry palm

Distribution and ecology: Endemic to Lord Howe Island, most common at low elevation, in evergreen closed forest, closed evergreen scrub and boulder beaches, on alluvium derived from basalt and calcarenite, but most common on calcarenite and coral sand, in stony soils and scree, forming large and dense monospecific or small populations, 0–400 m asl (Fig. 10.54). Flowering all months; fruiting all months. Conservation status – *Vulnerable* (IUCN 2008).

Figure 10.52 *Howea forsteriana*, in lowland forest, Lord Howe Island. Photo by William J. Baker, Royal Botanic Gardens Kew.

Typification: The holotype of *Howea forsteriana* (basionym *Kentia forsteriana*) is an undated and unnumbered collection by Charles Moore [MEL] collected from Lord Howe Island. The collection consists solely of a packet of intact fruits (Fig. 10.54). In the protologue, Mueller had difficulty finding any difference between this species and *H. belmoreana*, but nevertheless created a new species primarily at the instigation of Charles Moore, who appears to have provided most of the descriptive characters used by Mueller in the description.

Etymology: '*Voluntati amici C. Moore libenter nunc impono huic palmae nomen viri permagno aestimandi Guilielmo Forster, Neo-Cambriae digni Senatoris ibique disciplinarum fautoris*' (Mueller 1870). [With respect to their friendship, C. Moore now wishes to pay tribute by naming this palm after William Forster, dignified Senator of New South Wales]: Named for William Forster (b.1818, d.1882), politician in New South Wales, Member of the Legislative Assembly 1856–82, Premier and Colonial Secretary 1859–69, Colonial Secretary 1863–65, Secretary for Lands 1868–70 and Colonial Treasurer 1875–76.

Notes: *Howea forsteriana* is distinguished by the horizontal to descending leaves with horizontal to semi-pendulous pinnae, and the multispicate inflorescence.

■ PTYCHOSPERMATINAE

Relationships

The Ptychospermatinae is monophyletic but its relationship to other subtribes is presently unresolved (Lewis and Doyle 2002; Asmussen *et al.* 2006; Norup *et al.* 2006; Dransfield *et al.* 2008). Of the 12 genera in the subtribe, four occur in Australia and represent considerable evolutionary variation within the subtribe (Fig. 10.55). Three of the Australian genera, *Carpentaria*, *Normanbya* and *Wodyetia* are monotypic; the fourth, *Ptychosperma*, is represented by two species of which one is endemic.

Ptychosperma

Ptychosperma Labill., *Mém. Cl. Sci. Math. Inst. Natl. France* 9: 252 (1809); *Saguaster* sect. *Ptychosperma* (Labill.) Kuntze, *Lex. Gen. Phan.*: 495 (1903). Type: *Ptychosperma gracile* Labill.

Seaforthia R.Br., *Prodr.*: 267 (1810). Type: *Seaforthia elegans* R.Br. = *Ptychosperma elegans* (R.Br.) Blume.

Figure 10.53 *Howea forsteriana*. **Top left upper:** In habitat, Lord Howe Island. Photo by William J. Baker, Royal Botanic Gardens Kew. **Top left lower:** Abaxial surface of pinnae with ramenta. Photo by William J. Baker, Royal Botanic Gardens Kew. **Top right:** Staminate flowers at anthesis. **Bottom left:** Pistillate flowers at anthesis. **Bottom right:** Infructescence with mature fruit.

Seaforthia sect. *Spiranthe* Mart., *Hist. Nat. Palm. Edn 1,* 3: 181 (1838); *Ptychosperma* sect. *Spiranthe* (Mart.) H.Wendl. & Drude, *Linnaea* 39: 215 (1875). Type: not designated.

Drymophloeus subg. *Actinophloeus* Becc., *Malesia* 1: 42 (1877); *Actinophloeus* (Becc.) Becc., *Ann. Jard. Bot. Buitenzorg* 2: 126 (1885); *Ptychosperma* subg. *Actinophloeus* (Becc.) Becc., *Nuovo Giorn. Bot. Ital.* n.s. 42: 18 (1935). Type: *Drymophloeus ambiguus* Becc. = *Ptychosperma ambiguum* (Becc.) Becc. ex Martelli.

Ptychosperma sect. *Seaforthia* Scheff., *Natuur Tijd. Ned. Indië* 32: 186 (1872) *non* F.Muell., *Fragm.* 5: 47 (1865). Type: *Ptychosperma seaforthia* Miq. = *Ptychosperma elegans* (R.Br.) Blume.

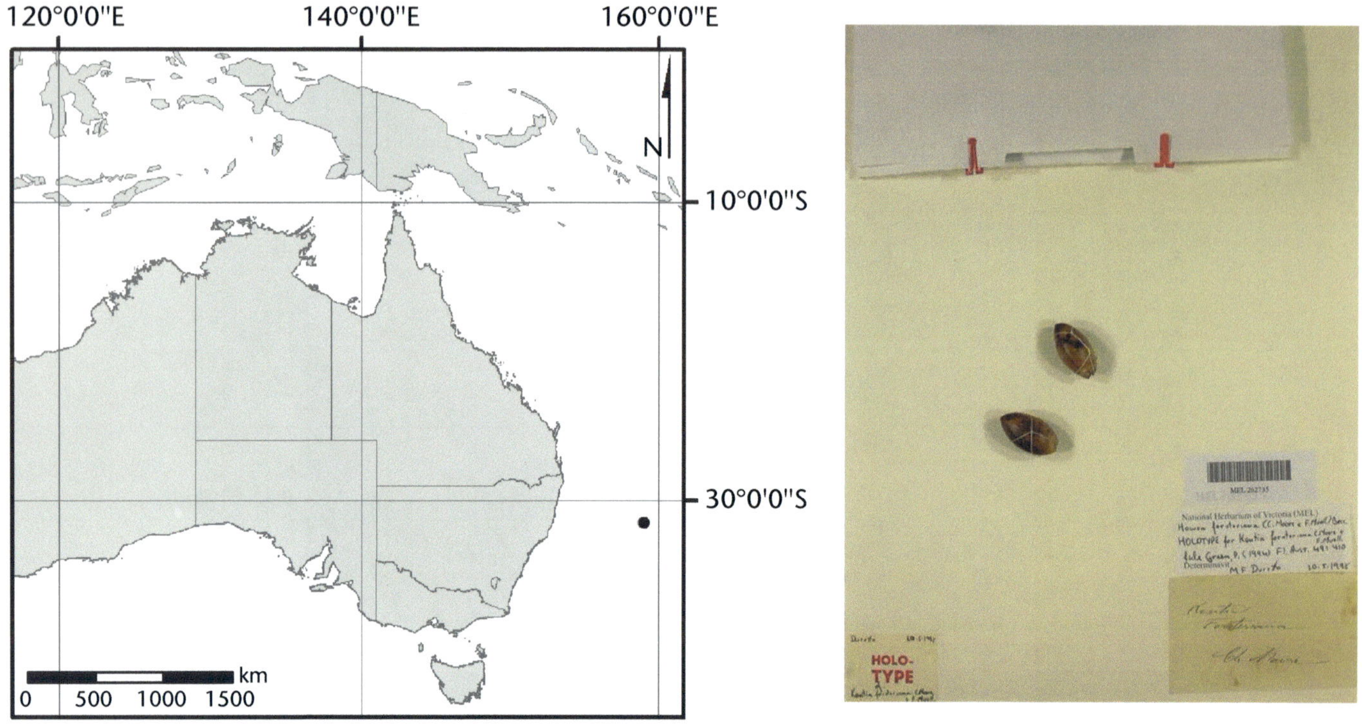

Figure 10.54 *Howea forsteriana*. **Left:** Distribution based on herbarium specimen data. **Right:** Type specimen. Australia. Lord Howe Island, undated, *C.Moore s.n.*; holotype MEL.

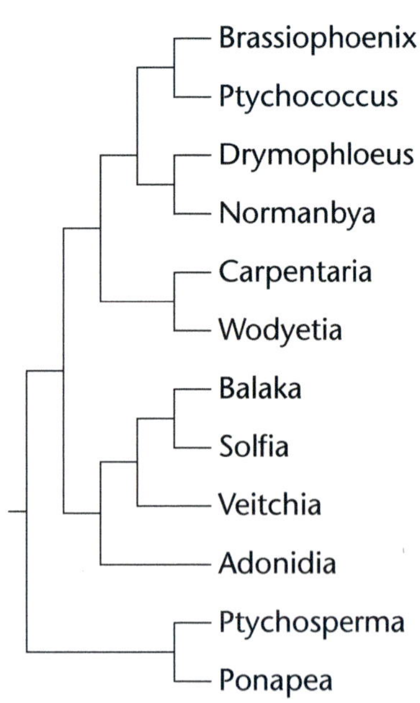

Figure 10.55 Phylogenetic tree summarising the estimated relationships of genera in the Ptychospermatinae. Adapted from Zona (1999), Lewis and Doyle (2002), Asmussen *et al.* (2006), Norup *et al.* (2006) and Dransfield *et al.* (2008).

NB: Generic synonyms which do not have a taxonomic association with Australian taxa are not included in this treatment. A full list of generic synonyms is provided in Dransfield *et al.* (2008).

Solitary or clustering, small to moderate, pleonanthic, monoecious palms. **Stems** slender to robust, leaning to erect, unarmed; leafscars prominent to obscure; internodes short to long. **Leaves** paripinnate, cleanly deciduous; leafbases forming a tubular elongate crownshaft, variously scaly, tomentose or glabrous, eventually splitting opposite the petiole at abscission; ligulate extension in the upper part; petiole short to long, channelled on the adaxial surface, rounded on the abaxial surface, glabrous to densely scaly or tomentose; rachis short to long, arching to straight, ridged on the adaxial surface, rounded on the abaxial surface, scaly. **Pinnae** single or variously united, subopposite to opposite or grouped, linear to cuneate or medianly wide, reduplicate, erect to horizontal; apex truncate and praemorse or notched, terminal pair basally united; midrib prominent; glabrous or densely scaly on the midrib and parallel veinlets; ramenta on abaxial midrib, or lacking. **Inflorescence** infrafoliar, paniculate, erect to spreading, branched 2–6 orders, protandrous; prophyll tubular, bicarinate, fully enclosing the peduncular bract; peduncular bract tubular, beaked, inserted slightly above attachment of prophyll, or distant; an incomplete second peduncular bract often present; rachis bracts triangular, short to extended; peduncle short to long, angled, glabrous, scaly or tomentose; rachis short when peduncle is long, or long; branches and rachillae green to strongly coloured; rachillae long, straight to flexuous, fleshy to fibrous. **Flowers** unisexual, sessile, in spirally arranged triads of a single pistillate flower subtended by a pair of staminate flowers one either side, with solitary or paired staminate flowers distally. **Staminate flowers** bullet-shaped to ovoid in bud; 3 broadly imbricate, rounded and gibbous sepals with fimbriate margins; 3 hard valvate petals, much longer than the sepals; stamens 9–100+, in antesepalous whorls of 3, and multiple antepetalous whorls; filaments short; anthers long, basally sagittate, dorsifixed, versatile; pistillode lageniform, longer than the stamens, apically irregularly lobed, trifid or papillate. **Pistillate flowers** shorter than the staminate, conic-ovoid; 3 broadly imbricate, rounded, sometimes slightly gibbous sepals with fimbriate margins; 3 broadly imbricate, pointed petals, valvate at the apex; staminodes 1–6, dentiform, linear or fused; gynoecium conic-ovoid, 1–3 locular; 3 short recurved stigmas; ovule hemianatropous, 5-angled, arillate. **Pollen** ellipsoidal or subtriangular; aperture monosulcate or trichotomosulcate; exine tectate, finely reticulate. **Fruit** globose to ellipsoid; stigmatic remains apical; epicarp granular-colliculose, red, orange or black-purple at maturity; mesocarp fleshy, mucilaginous or tanniniferous, with irritant crystals; endocarp 5-lobed, fibrous, adherent to the seed; perianth persistent or not. **Seed** 5-lobed; hilum lateral, raphe branches few; endosperm homogeneous or ruminate; the embryo basal or subbasal, germination adjacent-ligular; eophyll bifid.

Ptychosperma is a genus of c. 29 species, with greatest diversity in New Guinea, and with outliers in the Moluccas, Solomon Islands and Australia (2 spp. with one endemic). *Ptychosperma* resolves in molecular phylogenetic studies to be monophyletic and to have various relationships, most closely with *Ponapea*, *Drymophloeus* or *Normanbya* (Norup *et al.* 2006; Zona 1999; Hahn 2002a; Dransfield *et al.* 2008).

Etymology: *'Je l'ai appelé ptychosperma, dénomination tirée de la forme de l'amande . . . il renferme une amande (albumen) ovale, amincie au sommet, marquée dans sa longueur de cinq stries profondes'* Labillardière (1809, p. 252). [I have called it *Ptychosperma*, based on the form of the seed . . . it contains an oval seed (endosperm), pointed at the top, marked along its length by five deep furrows]: With reference to the lobed endocarp and seed, when viewed in cross-section, from the Greek *ptyche*, folded, and *sperma*, seed.

Key to species of Ptychosperma in Australia

1 Stem solitary; pinnae 30–60 each side of the rachis, to 84 cm long; endosperm deeply ruminate .... **1. *P. elegans***

1: Stems clustered; pinnae 15–40 each side of the rachis, to 56 cm long; endosperm homogeneous .... **2. *P. macarthurii***

1 Ptychosperma elegans

Ptychosperma elegans (R.Br.) Blume, *Rumphia* 2: 118 (1843); *Seaforthia elegans* R.Br., *Prodr.*: 267 (1810); *Ptychosperma seaforthia* Miq., *Fl. Ned. Ind.* 3: 21 (1855) *nom. illeg.*; *Archontophoenix elegans* (R.Br.) F.M.Bailey, *Catal. Plants Queensl.*: 50 (1890) *nom. illeg.*; *Saguaster elegans* (R.Br.) Kuntze, *Revis. Gen. Pl.* 2: 735 (1891). Type: Australia. Queensland. Northumberland Is., 1802, *R.Brown s.n.* (Bennett 5794); holotype BM [3 sheets]; isotype K.

Archontophoenix veitchii H.Wendl. & Drude, *Linnaea* 39: 213, pl. 2, Fig. 8 (1875); *Ptychosperma veitchii* H.Wendl., *Linnaea* 39: 213 (1875). Type: Australia. Queensland. Gould Island, Rockingham Bay, Aug. 1868, *P.O'Shanesy*; holotype MEL.

Ptychosperma capitis-yorkii H.Wendl. & Drude, *Linnaea* 39: 217 (1875); *Saguaster capitis-yorkii* (H.Wendl. & Drude) Kuntze, *Revis. Gen. Pl.* 2: 735 (1891); *Actinophloeus capitis-yorkii* (H.Wendl. & Drude) Burret, *Repert. Spec. Nov. Regni Veg.* 24: 266 (1928). Type: Australia. Queensland. Somerset, undated, *Veitch s.n.*; holotype GOET [presumed lost or destroyed, *pers. comm.* J. Heinricks), *fide* Essig (1978)]; isotypes BM, MEL.

Pinanga smithii W.Hill, *Catalog. Brisbane Bot. Gard.*: 20 (1875). Type: not designated.

Ptychosperma elegans var. *sphaerocarpum* Becc., *Ann. Jard. Bot. Buitenzorg* 2: 88, tab. 3, 4 (1885); *Ptychosperma wendlandianum* var. *sphaerocarpum* (Becc.) Burret, *Notizbl. Bot. Gart. Berlin-Dahlem* 10: 205 (1927). Type: lectotype, illustration of *Ptychosperma elegans* var. *sphaerocarpum* in O. Beccari, *Ann. Jard. Bot. Buitenzorg* 2: tab. 3, 4 (1885) (*fide* Essig 1978).

Ptychosperma jardinei (F.M.Bailey) F.M.Bailey, *Queensland Agric. J.* 23: 35 (1909); *Archontophoenix jardinei*

F.M.Bailey, *Queensland Agric. J.* 2: 129 (1898). Type: Australia. Queensland. Somerset, *F.Jardine s.n.*; holotype MEL; isotype BH.

Ptychosperma wendlandiana Burret, *Notizbl. Bot. Gart. Berlin-Dahlem* 10: 205 (1927). Type: Australia. Queensland. Cape York, 1877, *E.Daemel s.n.*; holotype K; isotype BO, MEL.

Stems solitary, to 12 m tall, to 8 cm dbh, slightly enlarged at the base; leafscars raised, to 20 mm wide; internodes to 15 cm long, green to grey. **Leaves** 7–11, to 3.3 m long, 30–60 pinnae each side of rachis; crown spreading; crownshaft to c. 65 cm long, densely white-woolly and minutely brown-lepidote with punctiform to lacerate-peltate scales, ligule lacking, or to c. 3 cm long, triangular; petiole 19–35 cm long, minutely lepidote, with brown punctiform scales, more sparsely lepidote above than below; rachis 150–230 cm long, lepidote like petiole. **Pinnae** regular to subopposite, to 84 cm long, to 8 cm wide, dark green on the adaxial surface, lighter green on the abaxial surface; basal pinnae reduced; ramenta scattered along midrib and parallel veinlets on abaxial surface, to c. 7 mm long, pale brown, basifixed or medifixed. **Inflorescence** 65–70 cm long, 100–125 cm broad, branched to 4 or 5 orders, axes pale green, becoming orange-yellow in fruit, thickly but finely brown lepidote-tomentose in the lower parts, sparsely so in the extremities, white-woolly when young; prophyll to 55 cm long, to 20 cm wide, glabrous with scattered light and dark scales densest toward the apex; peduncular bract to 50 cm long, to 18 cm wide, with scattered light and dark scales toward the apex; rameal bracts reduced to inconspicuous stubs in horizontal scar-like grooves; peduncle c. 7 cm long; rachillae to 30 cm long, to 2.5 mm thick; floral pits shallow, margins lipped, triangular, sharp. **Staminate flowers** 4.5–6 mm long, to 2 mm wide in bud, opening 12–15 mm wide at anthesis; sepals 1.5–2 mm long, to 2 mm wide, imbricate, gibbous, margins smooth, light green, glabrous; petals to 6 mm long, to 2 mm wide, thick, apex curved inward, light green, glabrous; stamens 12–22, to 6 mm long; filaments 4–5 mm long, straight; anthers c. 1.5 mm long, medifixed, versatile, white; pistillode 7–10 mm long, straight in lower portion, flexuous in distal portion, widening slightly to the apex, apex bulbous, 3-lobed. **Pistillate flowers** 3.5–5 mm long; sepals to 2.5 mm long, to 2.5 mm wide, imbricate, green, glabrous, margins fimbriate; petals to 5 mm long, to 2.5 mm wide, green, glabrous; staminodes c. 6, narrow-dentiform or fused into double segments; stigmas c. 1 mm long, recurved, white. **Fruit** globose to ellipsoid, 9–15 mm long, 8–10 mm diam.; epicarp c. 0.1 mm thick, smooth, semi-glossy, red at maturity; mesocarp to 2 mm thick, succulent; endocarp to 0.5 mm thick; tawny fibre layer compacted against the endocarp; perianth not persistent on fruit. **Seed** 10–11 mm long, 7–7.5 mm diam., lobes obtusely angled; endosperm deeply ruminate. Figures 10.56, 10.57.

Solitare palm

Distribution and ecology: Endemic to Queensland, where distributed from larger islands in western Torres Strait to near Yeppoon and Great Keppel Island, primarily in coastal forest, within metres of the shore-line and at low elevation on windward slopes, to high elevation in high-rainfall areas, in mesophyll vine forest, moist sclerophyll forest and littoral forest, sheltered gullies within drier and seasonal sclerophyll forest, and less frequent in wetter rainforest, 0–900 m asl (Fig. 10.58). Flowering all months; fruiting all month. Conservation status – *No present threats*.

Figure 10.56 *Ptychosperma elegans* in lowland rainforest, West Claudie River, Iron Range, north Queensland.

Figure 10.57 *Ptychosperma elegans*. **Top left:** Crownshaft. **Top right:** Staminate flowers at anthesis. **Bottom left:** Pistillate flowers at anthesis. **Bottom right upper:** Infructescence with mature fruit. **Bottom right lower:** Fruit, whole and with epicarp removed to show furrows, and seed in cross-section.

Typification: The holotype [BM] of *Ptychosperma elegans* (basionym *Seaforthia elegans*) is an unnumbered collection made by Robert Brown in 1802 from the Northumberland Islands off the north-east coast of Queensland. The collection consists of two complete leaves, a complete inflorescence and a portion of an inflorescence (Fig. 10.58).

Etymology: Brown (1810) provided no explanation for the species name in the protologue, but probably with reference to the elegant appearance of the species, from the Latin *elegans*, elegant or refined.

Notes: *Ptychosperma elegans* is characterd by a solitary habit and ruminate endosperm. The species is widely distributed along the Queensland coast, and Shapcott (1998b)

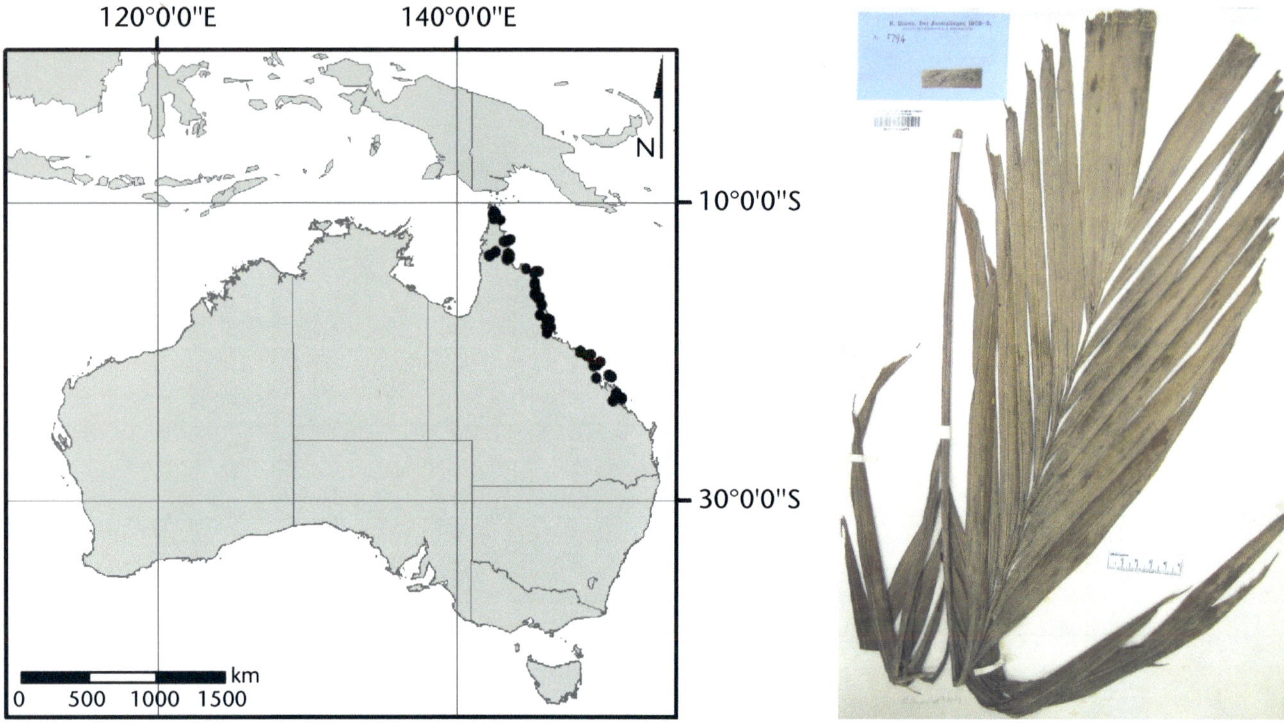

Figure 10.58 *Ptychosperma elegans*. **Left:** Distribution based on herbarium specimen data. **Right:** Type specimen. Australia. Queensland. Northumberland Island, 1802, *R.Brown s.n.* (Bennett 5794); holotype BM [1 of 3 sheets].

found considerable genetic variation within the population. Closest relationships appear to be with solitary-stemmed species in New Guinea and Solomon Islands.

2 Ptychosperma macarthurii

Ptychosperma macarthurii (H.Wendl. ex H.J.Veitch) H.Wendl. ex Hook.f., *Rep. Progr. Condition Roy. Bot. Gard. Kew 1882*: 55 (1884); *Kentia macarthurii* H.Wendl. ex H.J.Veitch, *Cat. Pl. 1879*: 26, pl. p.15 (1879); *Saguaster macarthurii* (H.Wendl. ex H.J.Veitch) Kuntze, *Revis Gen. Pl.* 2: 735 (1891); *Actinophloeus macarthurii* (H.Wendl. ex H.J.Veitch) Becc. ex Wigman, *Bull. Dép. Agric. Indes Néerl.* 31: 12 (1909) . Type: lectotype. Illustration of *'Kentia macarthuri'*, p. 15, *James Veitch & Sons' Cat. Pl. 1879* (1879) (Dowe 2007a), and rescinded the neotype as designated by Essig (1978).

[*Actinophloeus macarthurii* (H.Wendl. ex H.J.Veitch) Becc. ex Raderm., *Ann. Jard. Bot. Buitenzorg* 35: 12 (1925), sometimes incorrectly noted as the place of publication of the new combination].

Ptychosperma bleeseri Burret, *Repert. Sp. Nov. Regni Veg.* 24: 266 (1928). Type: Australia. Northern Territory. Bankers Jungle, 27 Aug. 1925, *A.K.Bleeser 430*: holotype B, destroyed.

Actinophloeus hospitus Burret, *Notizbl. Bot. Gart. Berlin-Dahlem* 11: 206 (1931); *Ptychosperma hospitum* (Burret) Burret, *Notizbl. Bot. Gart. Berlin-Dahlem* 12: 596 (1935); *Actinophloeus macarthurii* var. *hospitus* L.H.Bailey, *Fairchild Trop. Gard. Occas. Paper* 7: 4 (1940). Type: Cultivation. Indonesia. Bogor Botanic Gardens, 1929, *V-H.17*; holotype B, destroyed.

Ptychosperma julianettii Becc., *Atti Soc. Tosc. Sci. Pisa Processi Verbali* 44: 143 (1934). Type: Australia. Queensland. Torres Strait, Hammond Is. [Keriri Island], Mar. 1891, *A.Guilianetti s.n.*; holotype FI.

['Kentia MacArthuri' Hort., *Belgique Hort.* 27: 241 (1877) *nom. nud.*].

Stems clustered, 2–8 dominant, to 10 m tall, 3–7 cm dbh, slightly expanded at the base, occasionally leaning to prostrate with crown erect; leafscars raised, 15 mm wide, grey; internodes to 16 cm long, at first green, ageing to grey. **Leaves** 4–10, to 3 m long, arched, 15–40 pinnae each side of rachis; crownshaft 30–60 cm long, deciduously white-woolly and often with coarse dark scales in patches in the upper part; ligule to 4 cm long; petiole 13–58 cm long, finely brown-lepidote with tiny punctiform to lacerate-peltate scales; rachis about 100–200 cm long, lepidote like petiole or more coarsely so above. **Pinnae** regular to subopposite, or grouped; basal pinnae crowded and reduced; midleaf pinnae 20–56 cm long, 2–5.7 cm wide, apex obliquely praemorse or shallowly praemorse-notched; apical pinnae basally joined, with 1–4 ribs; ramenta basifixed, narrow, dark, twisted, to about 8 mm long, or sometimes lacking. **Inflorescence** 20–45 cm long, 45–60 cm wide, branched to 3 orders; axes yellow-green to green, sparsely to moderately brown-lepidote-tomentose; prophyll 30–50 cm long, 5–10 cm wide, prominently winged, glabrous to sparsely woolly and pale lepidote with punctiform to lacerate-peltate scales; peduncular bract exserted from the apex of the prophyll, 32–45 cm long, 4–10 cm wide, gradually acuminate, glabrous to sparsely woolly and pale lepidote with punctiform to lacerate-peltate scales; rameal bracts to 6 cm long; peduncle 2.5–6.5 cm long; rachis to 45 cm long; rachillae 17–37 cm long, 2–3.5 mm diam., erect to arching pendulous; floral pits shallow, margins raised, triangular, sharp. **Staminate flowers** 6–8 mm long and 2.5–3.5 mm wide in bud, opening to 15 mm wide at anthesis; sepals 1.5–2.5 mm high, to c. 2.5 mm wide, glabrous, weakly white-fimbriate, yellow-green to light green; petals to 7 mm long, to 3.5 mm wide, thick, apex curved inward, glabrous, yellow-green to light green; stamens 23–40, to 7 mm long; filaments to 6 mm long; anthers to 1.8 mm long, white; pistillode to c. 10 mm long, straight to apically coiled, white. **Pistillate flowers** 3–4 mm long, 3–4 mm wide in bud; sepals 2–3 mm long, 2–3 mm wide, imbricate, margins entire to finely fimbriate, cream-green; petals 3–4 mm long, 2–3.5 mm wide, imbricate, margins entire to finely fimbriate, cream-green, occasionally infused with pink toward the apex; stigmas c. 0.5 mm long, recurved, glistening clear white; ovary turbinate; staminodes 3–6, narrow-dentiform, truncate, sometimes fused into broader segments. **Fruit** irregular-ovoid, 12–18 mm long, 8–12 mm diam.; stigmatic remains c. 1.5 mm high; epicarp 0.1–0.2 mm thick, bright red at maturity; mesocarp succulent and watery, 3–4 mm thick, tawny fibres thin; endocarp 0.2–0.4 mm thick; perianth c. 6 mm high, not persistent on fruit. **Seed** ovoid, 9–12 mm long, 5–7 mm diam.; endosperm homogeneous. Figures 10.59, 10.60.

Macarthur palm

Distribution and ecology: In New Guinea and Australia, in Queensland and Northern Territory, in the first state from Torres Strait and Cape York Peninsula to southern McIlwraith Range, and in the latter around Darwin. Primarily in lowland and hillslope rainforests, but also in monsoon forest, littoral closed forest and brackish mangroves, 0–400 m asl (Fig. 10.61). Flowering all year; fruiting all year. Conservation status – *No present threats* (Queensland), *Endangered* (Northern Territory) (NRETAS 2009).

Figure 10.59 *Ptychosperma macarthurii* in seasonally dry monsoonal forest, Pajinka, Cape York Peninsula, north Queensland.

Typification: Although Essig (1978) proposed a neotype for *P. macarthurii*, in the apparent absence of any existing typifying materials, the rules of priority provide for the illustration accompanying the protologue to be designated as the lectotype (Dowe 2007a) (Fig. 10.61). The use of the name *Kentia macarthurii* for cultivated plants preceded its taxonomic formalisation by a number of years. At the 1877 Liege Exposition in Belgium, plants named as *K. macarthurii* and supposedly from New Caledonia were exhibited as a 'new and rare plant' (Anon. 1877). Soon after, plants named *K. macarthurii* were advertised in the Veitch's Nursery Catalogue of Plants (Veitch 1879). The name of the species was apparently suggested by Hermann Wendland, with Harry James Veitch

Figure 10.60 *Ptychosperma macarthurii*. **Top left:** Inflorescence with flowers in bud. **Top right:** Staminate flower at anthesis. **Bottom left:** Pistillate flowers at anthesis. **Bottom right:** Mature fruit.

providing the description. This publication formalised the name. Some years later, Hooker (1884) provided a list of palms growing in Kew Gardens, and the transfer to *Ptychosperma macarthurii* was provided.

Etymology: '*Kentia macarthurii* . . . It was named by Mr. Wendland, after our valued correspondent, Sir William Macarthur, near Sydney, N.S.W., to whom we are indebted for its introduction': Named for Sir William MacArthur (b.1800, d.1882), horticulturist and sheep breeder, whose gardener J. Reedy first collected seed of this species obtained during the Chevert Expedition to New Guinea organised by William John Macleay, May–September 1875 (Dowe 2007a). In the protologue, H.J. Veitch, of Veitch Nursery, Chelsea, made reference to seeds being sent by Macarthur (1855–86).

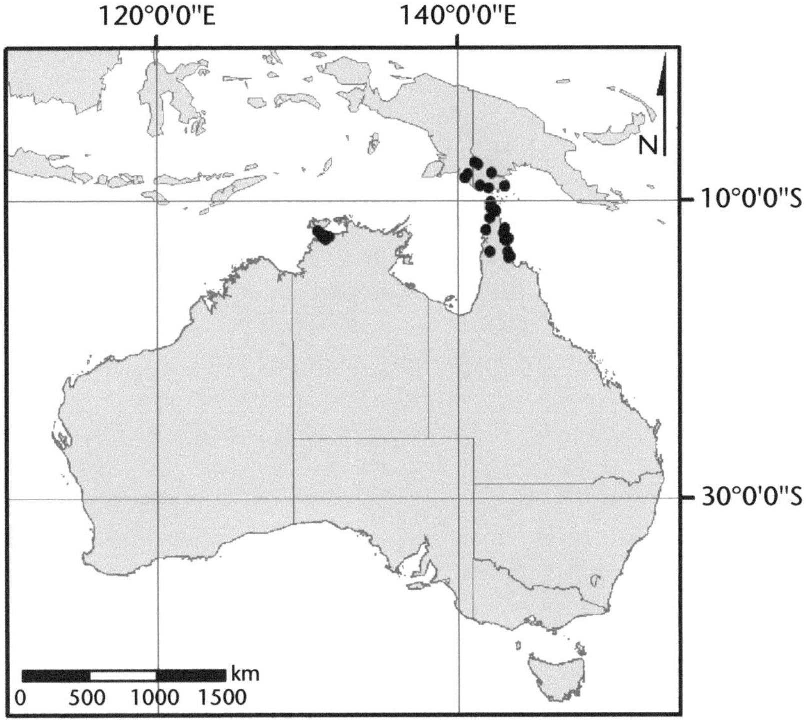

Figure 10.61 *Ptychosperma macarthurii*. **Left:** Distribution based on herbarium specimen data. **Right:** Type specimen. Lectotype, illustration of '*Kentia macarthuri*', p. 15, *James Veitch and Sons' Catalogue of Plants including Novelties for 1879* (1879).

Notes: *Ptychosperma bleeseri* was named for a collection from near Darwin in the Northern Territory. Dixon *et al.* (2003) assessed the taxonomic validity of this taxon and proposed that it be placed as a synonym of *P. macarthurii*, a proposal that is followed here. *Ptychosperma julianettii* was named for a collection from Hammond Island, Torres Strait.

Ptychosperma macarthurii is distinguished by a clustering habit and homogeneous endosperm. There appears to be close relationships with a number of species in New Guinea. Shapcott (1998b) found little genetic variation within the Northern Territory populations, thus suggesting a possible single introduction, whereas the Queensland populations had moderate variation among populations.

Carpentaria

Carpentaria Becc., *Ann. Jard. Bot. Buitenzorg* 2: 128 (1885). Type: *Kentia acuminata* H.Wendl. & Drude = *Carpentaria acuminata* (H.Wendl. & Drude) Becc.

Tall, solitary, erect, pleonanthic, monoecious palms. **Stem** slender, sometimes moderately ventricose; leafscars prominent or obscure; internodes moderate to long. **Leaves** paripinnate, cleanly deciduous; leafsheaths forming a tubular crownshaft, moderately to densely scaly; petiole short, channelled on the adaxial surface, rounded on the abaxial surface, moderately scaly; rachis strongly arching, concave on the adaxial surface, distally ridged, glabrous, rounded on the abaxial surface, scattered dark scales on the abaxial surface. **Pinnae** in a single plane on the rachis, clustered basally, subopposite to opposite, linear to lanceolate, reduplicate, erect to semipendulous, rigid or lax, apically truncate and praemorse; midrib prominent; ramenta sparse on midrib on the abaxial surface, medifixed, or lacking. **Inflorescence** infrafoliar, divaricate panicle branched to 4 orders, erect to pendulous, protandrous; prophyll tubular, prominently 2-keeled, peduncular bract tubular, apically exserted from prophyll, beaked, inserted slightly above attachment of prophyll; bracts deciduous prior to floral anthesis; small to moderate rameal bracts often present; peduncle short; rachis elongate, first-order branches subtended by small ridged bract; rachillae short, thin to moderate, slightly flexuous. **Flowers** unisexual, sessile, in well-spaced triads of a single pistillate flower subtended by a pair of staminate flowers one either side, borne spirally throughout the rachillae, or only on the proximal portion and then with staminate flowers distally, in pairs or solitary. **Staminate flowers** symmetrical in bud; sepals 3, broadly imbricate, gibbous; petals 3, broadly ovate, thick, valvate; stamens 30–40; filaments erect, short; anthers oblong to elliptical, dorsifixed, latrorse, connective broad; pistillode as long as or longer than the stamens, lageniform. **Pistillate flowers** ovoid; sepals 3, imbricate, rounded, margins irregularly fimbriate; petals 3, broadly ovate and imbricate, briefly valvate at the apex; staminodes 3, dentiform, bifid; gynoecium ovoid, unilocular, uniovulate; stigmas 3, fleshy, recurved; ovule laterally attached. **Pollen** ellipsoidal; aperture monosulcate; exine tectate, finely reticulate. **Fruit** ovoid to globose; stigmatic remains apical; epicarp smooth; mesocarp fleshy to succulent, irritative, fibrous toward endocarp, black fibres anastomosing; endocarp thin, glassy; perianth persistent on fruit. **Seed** ovoid, laterally attached, terete in cross-section; endosperm homogeneous; embryo basal; eophyll bifid; germination adjacent-ligular.

Carpentaria is a monotypic genus endemic to the Top End of the Northern Territory, Australia. Within the Ptychospermatinae, closest relationship is with *Wodyetia* (Asmussen *et al.* 2006; Norup *et al.* 2006; Dransfield *et al.* 2008).

Etymology: '*Ritengo che costituisca anch'essa un genere autonomo, che chiamo Carpentaria (C. acuminata), caratterizzato dal fruit con mesocarpio carnoso, dal seme non ruminato, dale foglie coi segmenti inferiori acuminati ed i superiori premorsi*' (Beccari 1885, p. 128) [I consider that this constitutes a new genus, that I call *Carpentaria* (*C. acuminata*), characterised by fruit with a fleshy mesocarp, by the non-ruminate seed, and with leaves in which the lower segments are acuminate and the apices praemorse]: The etymological derivation was not provided in the protologue, but possibly with reference to the locality in which the palm was collected, around the Gulf of Carpentaria, northern Australia.

Carpentaria acuminata

Carpentaria acuminata (H.Wendl. & Drude) Becc., *Ann. Jard. Bot. Buitenzorg* 2: 128 (1885); *Kentia acuminata* H.Wendl. & Drude, *Linnaea* 39: 207, tab. 2, Fig. 4 (1875). Type: Australia. Northern Territory, Escape Cliffs, undated, *C.Hulls*; holotype MEL.

Stem to 30 m tall, to 25 cm dbh, grey, enlarged at the base and moderately ventricose or with regular width; leafscars to 20 mm wide; internodes to 30 cm long, smooth or marked with light to dark hard excrescences. **Leaves** 7–14, to 4 m long, with a recurved apex, 55–70 pinnae per side; crownshaft to 120 cm long, whitish to grey-green, dark rough scales toward the apex and into the petiole; petiole to 60 cm long, with scattered dark scales; rachis to 340 cm long, glabrous. **Pinnae** 30–60 cm long, 2–4 cm wide, dark green on adaxial surface, lighter green on the abaxial surface; midrib raised on adaxial surface, raised less on abaxial surface; parallel veinlets numerous either side of midrib, light green, not raised; ramenta sparse, medifixed, or lacking. **Inflorescence** to 1.5 m long, to 60 cm wide, branched to 4 orders; axes cream-green; prophyll to 40 cm long, to 10 cm wide, green, with a thin pubescence particularly on the abaxial surface, prominently winged; peduncular bract exserted 10–15 cm from the prophyll apex, to 50 cm long, to 8 cm wide, apically pointed, green, with a thin pubescence densest in the distal portion; peduncle to 12 cm long, to 6 cm wide, to 2 cm thick; rachis to 80 cm long, 1–3 cm wide; rachillae to 16 cm long, semi-pendulous, moderately flexuous. **Staminate flowers** to 10 mm long in bud, bullet-shaped, opening to c. 12 mm wide at anthesis; sepals 4–5 mm long, cream-green; petals 7–9 mm long, to 4–6 mm wide, white or cream-green to ivory-green; stamens 6–7 mm long; filaments 2–3 mm long; anthers 3–3.5 mm long, white-cream; pistillode c. 10 mm long. **Pistillate flowers** 5–6 mm high, 4–5 mm wide; sepals 3 mm long, c. 3 mm wide, cream-green; petals 4–5 mm long, c. 3 mm wide, cream-green; stigmas opening to c. 2 mm wide. **Fruit** to 20 mm long, to 18 mm wide; epicarp c. 1.5 mm thick, red at maturity; mesocarp to c. 3.5 mm thick, fibres to 0.7 mm thick; perianth not persistent on fruit. **Seed** globose, to 15 mm diam. Figures 10.62, 10.63.

Carpentaria palm, Darwin palm

Distribution and ecology: Endemic to the Northern Territory where it occurs from Fitzmaurice River in the west to Harris Creek in the east and inland to about Pine Creek, in monsoon forest or vine forest, as a canopy or subcanopy element on various soil types, often associated with permanent springs and watercourses, 0–200 m asl (Fig. 10.64). Some populations may be temporarily inundated during the wet season. Flowering Sept.–Jan.; fruiting Dec.–Mar. Conservation status – *No present threats*.

Typification: The holotype of *Carpentaria acuminata* (basionym *Kentia acuminata*) is an unnumbered and undated collection by *C. Hulls* [MEL] from Escape Cliffs, to the north-east of Darwin, Northern Territory. The

Figure 10.62 *Carpentaria acuminata* in lowland monsoon forest, Green Ant Creek, Litchfield National Park, Northern Territory.

collection consists of a single pinna, fruit and a seedling (Fig. 10.64).

Etymology: '*Fructus [vivi succulenti] in statu sicco plicati et runcinate, oblique acuminati stigmatum residues validis in vertice obliquis*' (Wendland and Drude 1875). [Fruit [succulent when fresh] in the dry state folded and sharply incised, stigmatic remains acuminate and turned to one side and twisted]: With reference to the obliquely oriented and pointed stigmatic remains of the fruit in the dried state, from the Latin *acumen*, point.

Notes: *Carpentaria acuminata* is a tall solitary-stemmed palm distinguished by the strongly arching leaves, proximal pinnae being grouped, the obscurely obliquely truncate leaflet apices, bright red globose fruit and a slightly ventricose stem. Genetic variation within the total population suggests that fragmentation has occurred and there was little evidence of recent gene flow among discrete subpopulations (Shapcott 1998a).

Wodyetia

Wodyetia A.K.Irvine, *Principes* 27: 161 (1983). Type: *Wodyetia bifurcata* A.K.Irvine.

Solitary, moderate, erect, pleonanthic, monoecious palms. **Stems** columnar, slightly ventricose; leafscars congested; internodes short. **Leaves** paripinnate, plumose, cleanly deciduous; leafbases forming a tubular elongate crownshaft, eventually splitting opposite the petiole at abscission, white-grey-green toward the petiole; petiole short, stout, flat to concave on the adaxial surface, rounded on the abaxial surface, densely scaly, with whitish tomentum and scattered brown scales; rachis long, angled on the adaxial surface, rounded on the abaxial surface, densely scaly, grey to brown. **Pinnae** in a single plane on the rachis, subopposite to opposite or clustered, reduplicate, divided longitudinally to base into 2–17 linear segments, apically truncate and praemorse or dentate, some with 1–4 midribs, some lacking midribs, outer pinnae segments with thickened margins, densely scaly; ramenta lacking. **Inflorescence** infrafoliar, paniculate, branched to 4 orders, erect to pendulous in fruit, branches divaricate, protandrous; prophyll tubular, 2-keeled, bluntly beaked, completely enclosing peduncular bracts; peduncular bract tubular; incomplete peduncular bracts present; peduncle short, flattened; rachis long, branches stout, angled; rachis bracts low, subtending lower branches; rachillae short, terete, rigid. **Flowers** unisexual, sessile, spirally arranged in well-spaced triads of a single pistillate flower subtended by a pair of staminate flowers one either side, or only on the proximal portion and then with staminate flowers distally, in pairs or solitary. **Staminate flowers** symmetrical, bullet-shaped in bud; sepals 3, imbricate, rounded, inflated, margins dentate; petals 3, thick, rigid, ovate, valvate; stamens 58–72; filaments short; anthers long, versatile, apically bifid, basally sagittate, dorsifixed, latrorse; pistillode lageniform, longer than stamens, apically lobed or pappilate. **Pistillate flower** ovoid; sepals 3, broad, imbricate, margins tattered; petals 3, broad, imbricate, hooded, apically valvate; staminodes 3–6, dentiform, small; gynoecium conic-ovoid, unilocular, uniovulate; style short; 3 large recurved stigmas. **Pollen** ellipsoidal; aperture monosulcate; exine tectate, punctate. **Fruit** globose-ovoid, 1-seeded, beaked; stigmatic remains apical; epicarp smooth, thin, orange-red at maturity; mesocarp fleshy, moderately fibrous; endocarp with thick, flat, black fibres branched 1–4 times, inner layer of horizontal fibres; perianth not persistent on fruit. **Seed** ellipsoidal, slightly pointed, laterally attached, terete in cross-section; raphe

Figure 10.63 *Carpentaria acuminata*. **Top left:** Staminate flower at anthesis. **Top right:** Pistillate flowers at anthesis. **Bottom left:** Mature fruit. **Bottom right:** Fruit, whole, in longitudinal section and with epicarp removed to reveal mesocarp fibres.

branches sparse, anastomosing; hilum lateral; endosperm homogeneous; embryo basal; germination adjacent-ligular; eophyll simple or bilobed, apices oblique-acute.

Wodyetia is a monotypic genus endemic to north-east Queensland. Within the Ptychospermatinae, it is most consistently placed nearest to *Carpentaria* in molecular phylogenies, rather than the morphologically similar and geographically closer *Normanbya* (Norup *et al.* 2006; Dransfield *et al.* 2008).

Etymology: 'Wodyeti (Wad-yeti) was their name clearly intended for Johnny Flinders, the last surviving, male Aboriginal with traditional knowledge of the area, who died in 1978 at about 78 years old. Wodyeti acted as an anthropological and linguistic informant for

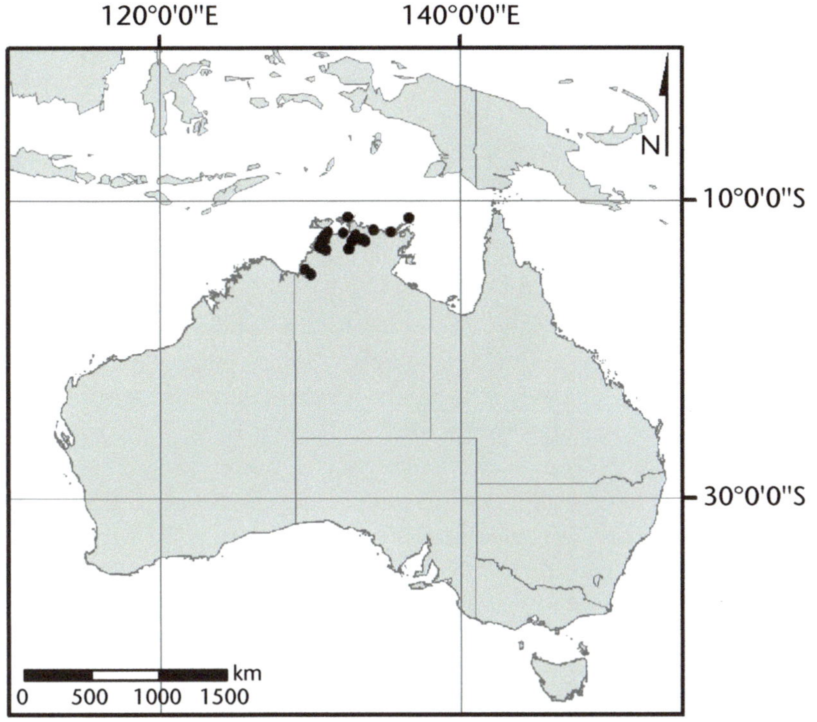

Figure 10.64 *Carpentaria acuminata*. **Left:** Distribution based on herbarium specimen data. **Right:** Type specimen. Australia. Northern Territory. Escape Cliffs, undated, *C.Hulls*; holotype MEL.

researchers such as Chase and Sutton' (Irvine 1983). Named for an Aboriginal elder known as Wad-yeti.

Wodyetia bifurcata

Wodyetia bifurcata A.K.Irvine, *Principes* 27: 165, figs pp. 162–164 s.n. (1983). Type: Australia. Queensland. Melville Ra., 4 km N of Abbey Peak, 23 Nov. 1981, *A.K.Irvine 2184*; holotype CNS [10 sheets].

Stem 6–15 m tall, 20–25 cm dbh, slightly to strongly ventricose, enlarged at the base; leafscars to 3 cm wide; internodes 20 cm long, grey. **Leaves** 6–10, 2.6–3.2 m long, arching, 90–110 pinnae each side of rachis; crownshaft 80–120 cm long, light green with greyish-white bloom; petiole 25–90 cm long, 5.0–5.6 cm wide, greenish-greyish-white on the adaxial surface, with brownish lacerate-peltate scales, with fringed and lacerate-peltate scales on the abaxial surface; rachis to 230 cm long, greenish-greyish-white on the adaxial surface, with brownish lacerate-peltate scales, with fringed and lacerate-peltate scales on the abaxial surface. **Pinnae** to 70 cm long, to 4 cm wide; proximal 1–4 pinnae sometimes entire or divided into 1–7 segments; midleaf pinnae with 11–17 segments; distal and near-terminal pinnae having 1–7 segments; apical pinnae undivided, basally united; pinnae glossy light green above, paler flat green with faint whitish sheen below; all segments twisted; ramenta lacking. **Inflorescence** 75–112 cm long, 70–100 cm wide, branched to 4 orders; axes cream to light green; prophyll c. 60 cm long, to c. 17 cm wide, glabrous; peduncular bract c. 58 cm long, to 15 cm wide, glabrous; subsequent peduncular bracts similar to primary peduncular bract, but diminishing in size to 3 cm wide, to 2 mm high; rameal bracts extremely small, either acute or wrinkled wavy; peduncle 8–13 cm long, 4–5 cm wide, 2–2.5 cm thick, green, with scattered small dark flat scales; rachis to 105 cm long, to 5 cm wide at the base, to 2.5 cm thick, cream to green, small flat brown scales scattered throughout, dense clusters of scales around bases of buds; basal branches c. half the diameter of rachis, alternate/spiral on the rachis, diminishing in thickness and evenly distributed along rachis; rachillae to 30 cm long, to 5 mm thick, terete, straight, green with scattered small dark scales. **Staminate flowers** to 11 mm long, to 4 mm wide in bud, opening to 25 mm wide at anthesis; sepals 5–6 mm long, 3–4 mm wide, cream-green, glabrous with scattered light-coloured scales; petals 9.8–10 mm long, to 5 mm wide, cream-green, glabrous; stamens to 10 mm long; filaments to 4 mm long; anthers 5.5–6.0 mm long, medifixed, versatile, white-cream; pistillode 8–15 mm long, basally rugose, apically attenuated. **Pistillate flowers** to 15 mm high to 10 mm wide at receptivity; sepals 6–7 mm long, 5–7 mm wide, green; petals 9–11 mm long, 6–8 mm wide, green; stigmas spreading to c. 5 mm at receptivity. **Fruit** 50–65 mm long, to 27–60 mm wide; stigmatic remains 8–10 mm long; epicarp c. 0.5 mm thick, red at maturity; mesocarp 2.5–3.0 mm thick, fibres 1–2 mm wide, compacted 6–8 mm thick; endocarp 2–3 mm thick; perianth not persistent on fruit. **Seed** to 32 mm long, to 26 mm wide; endosperm homogeneous; embryo 5 mm long at maturity. Figures 10.65, 10.66.

Foxtail palm

Distribution and ecology: Endemic to Melville Range, Cape York Peninsula, Queensland, in scattered small to moderate-sized groups in open forest among large granite boulders, as the dominant species. Occurring primarily on the lower slopes, 50–400 m asl (Fig. 10.67). Rainfall is seasonal but relatively low and there are localised high temperatures because of the radiative effect of the sun-heated granite boulders. The entire population is within the Cape Melville National Park. Flowering all months; fruiting all months. Conservation status – *Vulnerable* (Bostock and Holland 2007).

Figure 10.65 *Wodyetia bifurcata* in open woodland with granite boulders, Melville Range, north Queensland.

Typification: The holotype of *Wodyetia bifurcata* is *Irvine 2184* [CNS], collected in Nov. 1981, from Melville Range, Queensland. The collection consists of leaf portions with pinnae, inflorescence and infructescence portions and fruit (Fig. 10.67).

Etymology: 'outer endocarp with strongly forking, flattened, tough black fibres': With reference to the bifurcating fibres of the outer endocarp, from the Latin *bifurcus*, two-pronged or forked.

Notes: *Wodyetia bifurcata* is characterised by large plumose leaves with pinnae longitudinally divided with up to 17 segments, multi-stamen flowers (58–72) and large red fruit to 65 mm long with large black fibres, to 2 mm wide, in the mesocarp.

Normanbya

Normanbya F.Muell. ex Becc., *Ann. Jard. Bot. Buitenzorg* 2: 91 [170, 171] (1885); F.Mueller, *Fragm.* 11: 57 (1878) *nom. prop.*; *Saguaster* sect. *Normanbya* (F.Muell. ex Becc.) Kuntze, *Lex. Gen. Phan.*: 495 (1903). Type: *Cocos normanbyi* W.Hill = *Normanbya normanbyi* (W.Hill) L.H.Bailey.

Solitary, tall, erect, pleonanthic, monoecious palms. **Stems** slender to moderate; leafscars prominent; internodes long. **Leaves** paripinnate, plumose, cleanly deciduous; leafbases forming a tubular elongate crownshaft, eventually splitting opposite the petiole at abscission; petiole lacking or short, channelled on the adaxial surface, rounded on the abaxial surface, densely scaly, with whitish tomentum and scattered brown scales; rachis long, flat to angled or rounded on the adaxial surface, rounded on the abaxial surface, densely scaly. **Pinnae** in a single plane on the rachis, subopposite to opposite or clustered, reduplicate, divided longitudinally to base into 7–11 segments, apically truncate and praemorse, outer pinnae segments with thickened margins, some with midribs, some lacking midribs, uniseriate scales on the abaxial surface; ramenta lacking. **Inflorescence** infrafoliar, paniculate, branched to 3 orders, erect to pendulous in fruit, branches divaricate, protandrous; prophyll tubular, 2-keeled; peduncular bract tubular, apically exserted; peduncle short; rachis moderate; rachis bracts low, subtending lower branches which are angled and stout, rachillae terete, rigid. **Flowers** unisexual, sessile, spirally arranged in well-spaced triads of a single pistillate flower subtended by a pair of staminate flowers

Figure 10.66 *Wodyetia bifurcata*. **Top left:** Staminate flowers at anthesis. **Top right:** Pistillate flowers at anthesis. **Bottom left:** Mature fruit. **Bottom right:** Epicarp removed to reveal mesocarp fibres.

one either side, or only on the proximal portion and then with staminate flowers distally, in pairs or solitary. **Staminate flowers** symmetrical, bullet-shaped in bud; 3 imbricate rounded sepals; 3 thick rigid ovate valvate petals; stamens 24–40; filaments short; anthers long, apically bifid, dorsifixed, introrse; pistillode lageniform, longer than stamens, apically bulbous. **Pistillate flower** ovoid; 3 broad imbricate sepals; 3 broad imbricate petals; staminodes 3, dentiform; gynoecium ovoid, unilocular, uniovulate; style short, 3 large recurved stigmas, ovule pendulous. **Pollen** ellipsoidal; aperture monosulcate; exine tectate, scabrate. **Fruit** ovoid to obpyriform, beaked distally, 1-seeded; stigmatic remains apical; epicarp smooth, thin and semi-succulent; mesocarp thin, fleshy, moderately fibrous toward

Figure 10.67 *Wodyetia bifurcata*. **Left:** Distribution based on herbarium specimen data. **Right**. Type specimen. Australia. Queensland. Melville Range, 4 km north of Abbey Peak, 23 Nov. 1981, *A.K.Irvine 2184*; holotype CNS [1 of 10 sheets].

the endocarp, with thin tawny fibres adherent to the endocarp; endocarp thin; perianth not persistent on fruit. **Seed** ovoid, slightly pointed, laterally attached, terete in cross-section; hilum lateral; raphe long, unbranched; endosperm ruminate; embryo basal; germination adjacent-ligular; eophyll bifid.

Normanbya is a monotypic genus endemic to north-east Queensland. The relationships to other genera are unclear, but some phylogentic studies place it closest to *Drymophloeus*, or to *Carpentaria* and *Wodyetia* (Dransfield *et al.* 2008).

Etymology: '*Forse non va errato il Barone T. v. Mueller proponendo per essa il nuovo genere Normanbya*' (Beccari 1885, p. 91). [Perhaps this new taxon is best to be known by the name suggested by Baron Mueller as *Normanbya*]: Derived from the specific epithet used in W. Hill's protologue for the species (see Etymology for species below). The generic name *Normanbya* was first suggested by Mueller (1878) in reference to the possibility that with more collections the species *Ptychosperma normanbyi* may be sufficiently distinct to warrant placement in a new genus, but *Normanbya* was a proposed name and nomenclaturally illegitimate in that instance. Beccari (1885) is therefore the legitimate place of publication.

Normanbya normanbyi

Normanbya normanbyi (W.Hill) L.H.Bailey, *Gentes Herb.* 2: 188 (1930); *Cocos normanbyi* W.Hill, *Rep. Brisbane Bot. Gard. 1874*: 6 (1874); *Areca normanbyi* (W.Hill) F.Muell., *Fragm.* 8: 235 (1874); *Ptychosperma normanbyi* (W.Hill) F.Muell., *Fragm.* 11: 56 (1878); *Ptychosperma normanbyanum* Hook.f., *Gen. Pl.* 3: 893 (1880), *orth. var.*; *Drymophloeus normanbyi* (W.Hill) Hook.f. ex Becc., *Ann. Jard. Bot. Buitenzorg* 2: 168 (1885); *Normanbya muelleri* Becc., *Ann. Jard. Bot. Buitenzorg* 2: 171 (1885); *Saguaster normanbyi* (W.Hill) Kuntze, *Revis. Gen. Pl.* 2: 735 (1891); *Normanbya australis* (H.Wendl. & Drude) Baill., *Hist. Pl.* 13: 364 (1895); *Saguerus australis* H.Wendl. & Drude, *in* H.Baillon, *Histoire des Plantes.* Part 3, 13: 364 (1895). Type: Australia. Queensland. Daintree R., undated [1873?], *W.Hill s.n.*; holotype K [1 sheet, leaf only].

The last is included as a synonym of *Normanbya normanbyi* although Baillon indicated that it was a new combination for *Saguerus australis*, but perhaps meaning *Saguerus australasicus* which would be a synonym for *Arenga australasica*. Therefore *Normanbya australasicus* as used by various authors is incorrect. Govaerts and Dransfield (2005) argue that this is an incorrect spelling, not a new name, but if *australasica* is cited as the basionym then it could be treated as a new name. *Normanbya australasicus*

(H.Wendl. & Drude) Baill. is included as a synonym of *Arenga australasica* in Govaerts and Dransfield (2005), but this is not correct. Original spelling was *australis* and was placed under *Normanbya* not *Arenga*.

Stem to 30 m tall; trunk to 25 cm dbh; leafscars to 20 mm wide; internodes to 12 cm wide, green to grey. **Leaves** 7–10, to 2.5 m long, arching, 75–95 pinnae each side of rachis; leafsheath to 120 cm long, light green with white to green tomentum, and black scales densely aggregated at the apex and on the base of the petiole; petiole to 30 cm long, densely scaly; rachis to 220 cm long, densely scaly. **Pinnae** to 45 cm long, divided longitudinally to the base into 7–11 segments, 10–70 mm wide, broadening toward the apex, apex praemorse, distal pinnae twisted, dark green on the adaxial surface, blue-grey on the abaxial surface; midrib raised on adaxial surface; ramenta lacking. **Inflorescence** to 90 cm long, to 90 cm wide; prophyll to 90 cm long, to 20 cm wide, glaucous to grey-white; peduncular bract to 80 cm long, to 18 cm wide, glaucous to grey-white; peduncle to 7 cm long, to 7 cm wide, to 2.5 cm thick, green with scattered small dark scales; basal branches opposite, c. same thickness as rachis; rachis to 70 cm long, green with scattered small dark scales; rachillae to 32 cm long, to 4 mm thick, angled, curved, semi-erect to pendulous. **Flowers** in triads, spirally arranged. **Staminate flowers** to 15 mm long, to 7 mm wide in bud, opening to 30 mm wide at anthesis; sepals 5–7 mm long, 6–8 mm wide, dark green; petals 10–14 mm long, 6–9 mm wide, woody, thick, yellowish-green to straw coloured; stamens 8–11 mm long; filament c. 8 mm long; anthers to 9 mm long, glistening clear white; pistillode 18–22 mm long. **Pistillate flower** 12–15 mm long, 9–10 mm wide in bud; sepals 5–6 mm long, 7–8 mm wide, apple green; petals 9–11 mm long, 7–9 mm wide, light green; stigmas 2.5–3 mm long, strongly recurved, spreading 4–5 mm wide at receptivity, lucid cream. **Fruit** 35–50 mm long, 25–39 mm wide; stigmatic remains c. 5 mm long; epicarp to 0.25 mm thick, smooth, frequently glaucous, pink to scarlet-brown at maturity; mesocarp 5–10 mm thick, fibres to 0.5 mm thick; endocarp 2–3 mm thick; perianth not persistent on fruit. **Seed** to 35 mm long, to 25 mm wide; endosperm deeply ruminate; seed coat orange-red. Figures 10.68, 10.69.

Black palm

Distribution and ecology: Endemic to north-east Queensland from just south of Cooktown to near Mossman, in rainforest, swamp forest and mangrove margins, occurring in large to small populations and as scattered individuals as a canopy emergent on various soils, 0–700 m asl (Fig. 10.70). Flowering July–Dec; fruiting Aug.–Mar. Conservation status – *No present threats*.

Figure 10.68 *Normanbya normanbyi* in lowland rainforest, Daintree River, north Queensland.

The ecology of *Normanbya normanbyi* has been relatively well studied. The dispersal of fruits was found to be influenced by gravity and overland water flow; there was only minor dispersal by animal vectors, mainly rodents (Lott *et al.* 1995). Phenology was studied by Inkrot *et al.* (2007): flowering peaks in the dry season and fruit production coincides with the wet season, with a protandrous syndrome of a staminate phase of about 40 days, a period of quiescence of about nine days and a pistillate phase of about 14 days, for each active inflorescence. Fruit matures in about 21 weeks. At the population level, there was a greater degree of staminate synchrony than pistillate synchrony. Ants (*Leptomyrmex unicolor*) were found to be attracted to sugar-rich nectars and wound sap, but were not implicated in pollination (Blüthgen *et al.* 2004).

Typification: The holotype of *Normanbya normanbyi* (basionym *Cocos normanbyi*) is an unnumbered and

Figure 10.69 *Normanbya normanbyi*. **Top left:** Inflorescences. **Top right:** Staminate flower at anthesis. **Bottom left:** Pistillate flower at anthesis. **Bottom right upper:** Mature fruit. **Bottom right lower:** Fruit, whole and in longitudinal section, on abaxial surface of pinnae.

undated collection made by Walter Hill [K] and consists of a 'leaf of a young plant, Daintree River', as mentioned in the protologue ('In rich alluvial soil on the banks of the Daintree River'). The specimen also has a packet of staminate flowers labelled '*Pt normanbyana* Daintree River, Muell 1880', an illustration of fruit labelled '*Ptychosperma normanbyana,* 18th Nov 1880, Daintree river, Muell 1880, Fitzalan' and recent photos of fruit from an unknown source. This specimen includes the holotype to which further elements have subsequently been added (Fig. 10.70).

Etymology: 'This fine species of palm is named after the present Governor of Queensland' (Hill 1874, p. 6): Named for George Augustus Constantine Phipps (b.1819, d.1890), Marquess of Normanby, Governor of Queensland

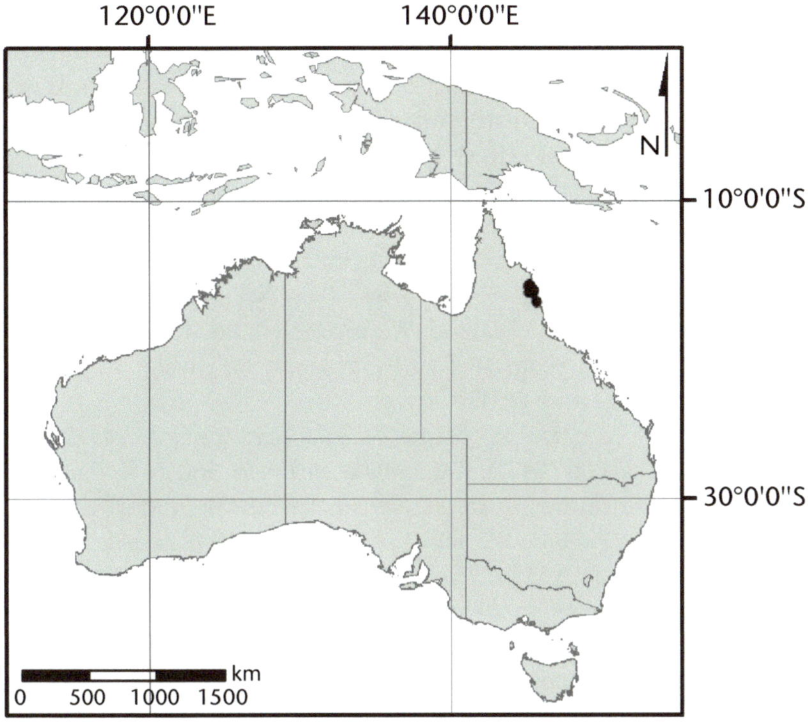

Figure 10.70 *Normanbya normanbyi*. **Left:** Distribution based on herbarium specimen data. **Right:** Type specimen. Australia. Queensland. Daintree River, undated [1873?], *W.Hill s.n.*; holotype K.

1871–74, Governor of New Zealand 1874–79 and Governor of Victoria 1879–84.

Notes: *Normanbya normanbyi* is distinguished by its solitary stem and tall emergent habit, plumose leaves with pinnae linearly divided into 7–9 segments and silver-grey on the abaxial surface, large flowers with numerous stamens (24–40), an extended pistillode to 12 mm long, and large pink to scarlet-brown ovoid to obpyriform fruit to 50 mm long.

■ RHOPALOSTYLIDINAE

Relationships

The Rhopalostylidinae resolves as monophyletic in most analyses (Pintaud 1999; Asmussen *et al.* 2006; Dransfield *et al.* 2008). The subtribe includes two genera, *Rhopalostylis* and *Hedyscepe*. Although resolved as monophyletic by molecular evidence, the genera are morphologically similar to others in the Archontophoenicinae.

Rhopalostylis

Rhopalostylis H.Wendl. & Drude, *Linnaea* 39: 180 [234] (1875). Type: *Areca baueri* Hook.f. = *Rhopalostylis baueri* (Hook.f.) H.Wendl. & Drude.

Kentia subgenus *Rhopalostylis* Becc., *Malesia* 1: 39 (1877). Type: *Kentia sapida* (Sol. ex G.Forst.) Mart. = *Rhopalostylis sapida* (Sol. ex G.Forst.) H.Wendl. & Drude.

Eora O.F.Cook, *J. Heredity* 18: 409 (1927). Type: *Eora sapida* (Sol. ex G.Forst.) O.F.Cook = *Rhopalostylis sapida* (Sol. ex G.Forst.) H.Wendl. & Drude.

Solitary, moderate to tall, erect, pleonanthic, monoecious palms. **Stems** slender to moderate, frequently with a swollen base; leafscars prominent; internodes short. **Leaves** paripinnate, cleanly deciduous; leafbases forming a tubular, short to long swollen crownshaft, eventually splitting opposite the petiole prior to abscission, scaly; petiole short, channelled on the adaxial surface, rounded on the abaxial surface, densely scaly; rachis long, erect to arching, horizontal or vertically oriented, angled to flat on the adaxial surface, rounded on the abaxial surface, densely scaly. **Pinnae** in a single plane on the rachis, subopposite to opposite, lanceolate, reduplicate, erect, rigid, apex tapered; midrib prominent, scaly on the abaxial surface in the proximal portion; parallel veinlets on abaxial surface; ramenta lacking. **Inflorescence** infrafoliar, paniculate, spreading, branched to 3 orders, erect to pendulous, branches divaricate, protogynous to subprotandrous or protandrous; prophyll tubular, bicarinate, long, apically pointed, fully enclosing the peduncular bract; peduncular bract inserted slightly above attachment of prophyll, tubular; bracts deciduous when flowers still in bud; rachis bracts prominent, tapered, subtending rachillae; peduncle stout, short, flattened; rachis longer than the peduncle, tapered; rachillae short, stout. **Flowers** unisexual, sessile, in well-spaced or congested triads of a single pistillate flower subtended by a pair of staminate flowers one either side, borne spirally on the rachillae in the proximal portion and with staminate flowers distally, in pairs or solitary; floral pits shallow, margins raised, sharp. **Staminate flowers** asymmetric in bud; 3 free, narrow, keeled sepals; 3 curved, pointed, thick valvate petals, same as or longer than the sepals; stamens 6; filaments inflexed in bud; anthers erect, linear, dorsifixed, basally bifid, latrorse; pistillode cylindrical, about as long as the stamens, tapered, apically lobed. **Pistillate flowers** symmetrical; 3 imbricate, pointed sepals; 3 imbricate petals, valvate at the apex; staminodes 0–3, filamentous; gynoecium ellipsoidal, unilocular, uniovulate; style obscure; 3 recurved stigmas; ovule large, laterally attached. **Pollen** ellipsoidal; aperture monosulcate or trichotomosulcate; exine tectate, foveolate to fossulate. **Fruit** ellipsoid to globose or subglobose, 1-seeded; stigmatic remains apical; epicarp thin, smooth; mesocarp thin, with fibres adhering to the endocarp; endocarp thin, fragile; perianth persistent on fruit. **Seed** ellipsoidal to globose, laterally attached; hilum wide; raphe fibres few to many, anastomosing; endosperm homogeneous; embryo basal; germination adjacent-ligular; eophyll bifid.

Rhopalostylis is a genus of two species, distributed on Norfolk Island and New Zealand (Kermadec Islands, North Island, South Island and Chatham Islands). Closely related to *Hedyscepe*, from which it differs in the protogynous to subprotandrous flowering syndrome (protandrous in *Hedyscepe*), as well as flowers with six stamens (multi-stamen flowers in *Hedyscepe*).

Etymology: '*Generi nomen tribuimus ex stylii rudimento clavaeformi in flor. ♂*,' (Wendland and Drude 1875, p. 180) [The generic name is derived from the rudimentary claviform pistillode in the male flowers]: From the Greek *rhopalo*, club-like, and *stylis*, style.

Rhopalostylis baueri

Rhopalostylis baueri (Hook.f.) H.Wendl. & Drude, *Linnaea* 39: 234, tab. 1, Fig. 2 (1875); *Areca baueri* Hook.f., *Fl. Novae-Zel.* 2: 262 (1853); *Kentia baueri* (Hook.f.) Seem., *Fl. Vit.* 8: 269 (1868); *Drymophloeus baueri* (Hook.f.) Scheff., *Natuur. Tijd. Ned. Indië* 32: 155 (1872) *nom. provis.*; *Eora baueri* (Hook.f.) O.F.Cook, *J. Heredity* 18: 409 (1927). Type: lectotype. Illustration of *Areca baueri* in Hook.f., *Bot. Mag.* 94: tab. 5735 (1868) here designated.

Rhopalostylis baueri var. *cheesemanii* (Becc. ex Cheeseman) Sykes, *Kermadec Is Fl.* (*Bull. New Zealand Dept. Sci. Industr. Res.*, No. 119) 184 (1977); *Rhopalostylis cheesemanii* Becc. ex Cheeseman, *Trans. Proc. N.Z. Inst.* 49: 47 (1917). Type: New Zealand. Kermadec Islands, Sunday (Raoul Island), Aug. 1887, *T.F.Cheeseman s.n.*; lecto; K, *fide* de Lange *et al.* (2005).

[*Areca sapida*, *in* mss, D. Solander, *Plantae Novae Hollandiae* 3 (1768–71); *Areca sapida auct. non.* Sol. ex G. Forst., *in* S.Endlicher, *Prodr. Fl. Norfolk.*: 26 (1833); *in* Martius, *Hist. Nat. Palm.* 3: 172, tab. 151, 152 (1838); *Kentia sapida* (Sol. ex G. Forst.) Mart., *Hist. Nat. Palm. Edn 1*, 3: 312 (1838)].

Stems to 18 m tall, to 15 cm dbh, enlarged at the base to 30 cm diam.; leafscars prominent, raised, 1–3 cm wide; internodes to 12 cm long, at first green, ageing to grey; subsurface stem saxophone-shaped from initial geotrophic growth of seedling. **Leaves** 9–12, 3–4 m long, arching, 42–60 pinnae each side of rachis; crownshaft 50–60 cm long, bulging, green with brown mealy scales, splitting shallowly opposite the petiole, margins smooth to coarsely fibrous; petiole to 40 cm long, green, dense dark mealy tomentum; rachis to 360 cm long, green, indumentum similar to petiole. **Pinnae** to 70 cm long, to 5 cm wide, coriaceous, dark green on both the adaxial and abaxial surfaces; midrib prominently raised on adaxial surface, moderately raised on abaxial surface, abaxial midrib with dense flat crustose scales in the proximal portion, 1 moderately prominent raised parallel vein on either side of midrib c. half-way to margin, parallel veinlets numerous; ramenta lacking. **Inflorescence** 30–90 cm long, to 60 cm wide, branched to 2 orders, axes cream-white at first, turning green in fruit; prophyll to 52 cm long, to 12 cm wide, beaked, green, glabrous; peduncular bract to 52 cm long, to 11 cm wide, beaked, green, glabrous; rudimentary rachis bracts on lower branches, elongate, pointed, thin; peduncle to 6 cm long, 5–7 cm wide, green, glabrous; rachis to 25 cm long, cream, glabrous; rachillae to 35 cm long, straight, angular, to 10 mm wide, glabrous; floral pits cupular. **Flowers** in triads, paired or single staminate flowers distally, moderately protogynous or synchronous, or protandrous. **Staminate flowers** 5–7 mm long, 4–5 mm wide in bud, opening 8–12 mm wide at anthesis; sepals 2–5 mm long, 0.5–2 mm wide, narrow, acuminate, filamentous, dark brown; petals 5–8 mm long, 3–5 mm wide, acuminate, white-cream; stamens 5–6 mm long; filaments 2–3 mm long; anthers 3–4 mm long, medifixed, versatile, orange; pistillode 6–7 mm long, cylindrical, straight, slightly expanded at the apex, club-like. **Pistillate flowers** 4–7 mm long, 4–6 mm wide in bud; sepals 3–4 mm long, 4–5 mm wide, thin, dark brown, margins minutely ciliolate; petals 5–6 mm long, 3.5–4 mm wide, white-cream, margins minutely ciliolate; stigmas to 1.5 mm long, strongly recurved, lucid cream, opening to 2.5 mm wide at receptivity; ovary oblong; stigmas recurved; staminodes 3, dentiform. **Fruit** globose to subglobose, 11–17 mm long, 10–13 mm wide; epicarp c. 0.2 mm thick, smooth, red at maturity; stigmatic remains low, to c. 2 mm wide; mesocarp 1.2–2 mm thick, fleshy, fibres to 1 mm thick, in rows; endocarp to 0.5 mm thick, cartilaginous, fragile; perianth persistent on fruit. **Seed** 10.5–14 mm long, 10–12 mm wide. Figures 10.71, 10.72.

Figure 10.71 *Rhopalostylis baueri* in *Araucaria* forest, Mt Pitt, Norfolk Island. Photo by Jeanne Price.

Norfolk Island palm

Distribution and ecology: On Norfolk Island and the Kermadecs (New Zealand), in moist forest on volcanic soils, most common at higher elevations, occurring in large populations and small groups, 0–300 m asl (Fig. 10.73). The Norfolk Island population is mostly restricted to conservation areas because of land clearance. The population on the Kermadecs was formerly known as *R. baueri* var. *cheesemanii* (Becc. ex Cheeseman) Sykes. Flowering; Sept.–Dec.; fruiting; Oct.–Mar. Conservation status – *No present threats*.

Typification: The first use of the name *Areca baueri* by Hooker (1853) did not include any specimen citations. Diagnostic features distinguishing it from *Areca sapida* were provided, thus supporting the proposal that it is the formal place of publication of the name. This assessment has been variously interpreted: Govaerts and Dransfield (2005) cited *A. baueri* Hook.f. ex Lem. in *Ill. Hort.* 15: 575 (1868) as formal publication of the name, while de Lange *et al.* (2005) cited *Kentia baueri* Seem. as the basionym, with *A. baueri* considered both a superfluous and illegitimate name. However, there is sufficient evidence to support Hooker's (1853) publication as formalisation of the name *A. baueri*.

The choice of a type for *A. baueri* is problematic as no specimens were cited in Hooker's (1853) publication. This does not exclude any specimens previously collected under the name of *A. sapida* as being a type of *A. baueri*; however, no such specimens have been located. Hooker (1868) subsequently provided an illustration in *Curtis's Botanical Magazine* drawn from plants cultivated in Kew, originating from seeds collected by Allan Cunningham from Norfolk Island, thus relating directly to the 1853 publication. In the absence of type specimens, this illustration is hereby designated as the lectotype (Fig. 10.73).

Etymology: 'Von Martius also separates them, but gives Forster's name to the Norfolk Island plant, whereas Forster figured the New Zealand one only, to which the name of sapida must remain attached, whilst that of *Baueri* may be given to the Norfolk Island species if it proves really distinct. There is as much difference between the narrow and broad pinnae of specimens of *A. sapida* growing in Kew Gardens, as between the latter and the pinnae of *A. baueri*' (Hooker 1853, p. 262). Named for Ferdinand Bauer (b.1760, d.1826), natural history artist, who accompanied Matthew Flinders on the *Investigator* 1801–03 in company with the botanist Robert Brown and the landscape painter William Westall, and spent August 1804 to March 1805 on Norfolk Island illustrating plants

Figure 10.72 *Rhopalostylis baueri*. **Top left:** Staminate flower at anthesis. **Top right:** Pistillate flower at anthesis. **Bottom left:** Crownshaft and infructescence with immature fruit. Photo by Jeanne Price. **Bottom right:** Mature fruit.

and animals. Bauer was born in Austria into a family of artists and illustrators, and was trained in Vienna. He developed a technique of colour-coding his subjects in the field, which allowed him to complete his illustrations in the studio. Many of Bauer's Australian plant illustrations remain unpublished, and are kept in the Natural History Museum, London.

Notes: *Rhopalostylis baueri* is distinguished by a solitary habit, horizontal leaves with linear pinnae, a divaricate inflorescence with variable flowering syndrome, staminate flowers with six stamens, and a claviform (club-like) pistillode swollen toward the apex, and globose to subglobose fruit to 17 mm long and red at maturity.

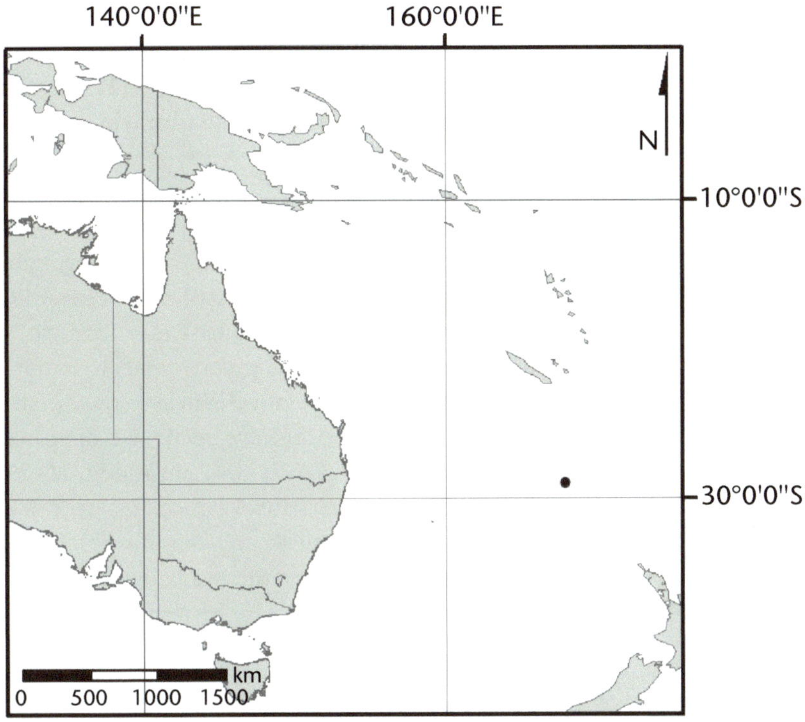

Figure 10.73 *Rhopalostylis baueri*. **Left:** Distribution based on herbarium specimen data. **Right:** Type specimen. Lectotype, illustration of *Areca baueri, in* J.D.Hooker, *Botanical Magazine* **94**: Tab. 5735 (1868).

Hedyscepe

Hedyscepe H.Wendl. & Drude, *Linnaea* 39: 178 [203] (June 1875); *Kentia* subgenus *Hedyscepe* (H.Wendl. & Drude) Becc., *Malesia* 1: 39 (1877). Type: *Kentia canterburyana* C.Moore & F.Muell. = *Hedyscepe canterburyana* (C.Moore & F.Muell.) H.Wendl. & Drude.

[*Hedyscepe* Drude & H.Wendl., *Nachr. Königl. Ges. Wiss. Georg-Augusts-Univ. 1875*: 55 (February, 1875), *nom illeg*. See *Laccospadix* for discussion about validity of this name].

Solitary, moderate, erect to leaning, pleonanthic, monoecious palms. **Stem** stout, expanded at the base, leafscars prominent, internodes short. **Leaves** paripinnate, cleanly deciduous; leafbases forming a tubular crownshaft, splitting opposite the petiole prior to abscission, densely tomentose; petiole short, channelled on the adaxial surface, rounded on the abaxial surface, moderately scaly; rachis long, arching, angled to flat on the adaxial surface, rounded on the abaxial surface. **Pinnae** in a single plane on the rachis, subopposite to opposite, lanceolate, reduplicate, apex acute, rigid, erect, glabrous on the adaxial surface, tomentose on the abaxial surface and margins; midrib and parallel veinlets raised, scaly; ramenta lacking. **Inflorescence** infrafoliar, paniculate, branched to 3 orders, spreading to pendulous, branches divaricate, thick, protandrous; prophyll 2-keeled, tubular, fully enclosing peduncular bract, inserted slightly above attachment of prophyll; peduncular bract similar to prophyll but not keeled, tubular; bracts deciduous immediately prior to floral anthesis; rameal bracts frequently subtend branches; peduncle short, horizontal; rachis longer than the peduncle; rachillae rigid, erect, slightly flexuous. **Flowers** unisexual, sessile, in triads of a single pistillate flower subtended by a pair of staminate flowers one either side, borne spirally on the rachillae, with staminate flowers distally, in pairs or solitary. **Staminate flowers** asymmetric in bud, 3 imbricate sepals, basally connate, long and narrow, keeled; 3 valvate petals, longer than sepals; stamens 9–12, solitary when antesepalous, antepetalous in pairs; filaments awl-shaped, inflexed in bud; anthers erect, versatile, medifixed, basally bifid, latrorse; pistillode cylindrical, longer than the stamens, tapered, apically lobed. **Pistillate flower** with 3 imbricate, rounded sepals; 3 imbricate, briefly valvate petals, longer than sepals; staminodes 3, dentiform; gynoecium unilocular, uniovulate; style short; 3 recurved stigmas; ovule basally attached. **Pollen** ellipsoidal; aperture monosulcate; exine tectate, finely reticulate.

Fruit ellipsoidal, large, 1-seeded; stigmatic remains apical; epicarp thick, smooth; mesocarp thick, with longitudinal fibres attached to the endocarp; endocarp thick, crustaceous, fragile; perianth persistent on mature fruit. **Seed** ellipsoidal, laterally attached, hilum elongate, raphe fibres anastomosing; endosperm homogeneous; embryo basal; germination adjacent-ligular; eophyll bifid.

Hedyscepe is a monotypic genus endemic to Lord Howe Island. Closely related to *Rhopalostyis*, from which it differs in the protandrous syndrome (protogynous to subprotandrous in *Rhopalostylis*), as well as multi-stamen flowers (six stamens in *Rhopalostylis*).

Etymology: '*Nomen genericum delegimus, quod incolae huic Palmae pulchrae nomen "umbrella" imposuerunt*' (Wendland and Drude 1875). The generic name is taken from that which has been assigned to it in its habitat, the umbrella palm; from the Greek *hedys*, pleasant, and *skepe*, a covering.

Hedyscepe canterburyana

Hedyscepe canterburyana (C.Moore & F.Muell.) H.Wendl. & Drude, *Linnaea* 39: 204, tab. 1, Fig. 4 (1875); *Kentia canterburyana* C.Moore & F.Muell., *Fragm.* 7: 101 (1870); *Veitchia canterburyana* F.Muell., *Fragm.* 7: 101 (1870) *nom. superfl.* Type: Australia. New South Wales. Lord Howe Island, Mt Lidgbird, undated, *C.Moore & R.D.Fitzgerald s.n.*; holotype MEL [2 sheets]; isotype BO.

Stem 3–15 m tall, 12–20 cm dbh, expanded at the base; leafscars 1–2 cm wide, stepped, irregular, light coloured; internodes 2–5 cm long, glaucous-green to grey. **Leaves** 8–12, 1.5–3 m long, arcuate with a moderate lateral twist, 40–60 pinnae per side; crownshaft 30–50 cm long, slightly to broadly bulging, glaucous-white to grey; petiole 12–20 cm long; rachis to 2.8 m long. **Pinnae** 20–30 cm long, to 3 cm wide, linear, acuminate, erect, rigid, dark green on the adaxial surface, lighter green on the abaxial surface; midrib prominently raised on the adaxial surface, less raised on the abaxial surface, 3–4 parallel veinlets each side of the midrib, dense coarse scales on the abaxial midrib; ramenta lacking. **Inflorescence** 30–80 cm long, to 80 cm wide, branched to 3 orders, axes initially cream, becoming dark green in fruit; prophyll to 30 cm long, to 12 cm wide, strongly winged, green, glabrous; peduncular bract to 30 cm long, to 10 cm wide, tubular, green, glabrous; peduncle 10–15 cm long, 5–8 cm wide, glabrous; rachis stout, 20–60 cm long, curving, basally 4–7 cm wide, tapered toward the apex, green, glabrous; rachillae 15–25 cm long, erect to semi-pendulous, regular and spirally arranged along rachis, angular; floral pits shallow, margins barely raised, smooth. **Flowers** in triads in proximal half of rachilla, paired or single staminate flowers in distal half. **Staminate flowers** to 12 mm long, to 5 mm wide in bud, opening to c. 10 mm wide at anthesis; sepals 3–4 mm long, 2–3 mm wide, acuminate, cream-ivory; petals 8–12 mm long, 5–6 mm wide, ovate, acute, cream-ivory; filaments c. 5 mm long, straight; anthers c. 5 mm long, cream-ivory; pistillode 7–9 mm long, cylindrical, twisted, apically rounded. **Pistillate flowers** to 6 mm long, 6 mm wide in bud; sepals to 4 mm long, to 5 mm wide imbricate, purplish-dark green; petals to 6 mm long, to 6 mm wide, imbricate, purplish-dark green; stigmas more or less sessile, c. 1 mm long, recurved, opening to c. 2 mm wide at anthesis. **Fruit** 38–42 mm long, 26–29 mm diam.; stigmatic remains not extended; epicarp c. 0.2 mm thick, smooth, dull red; mesocarp 2–5 mm thick, fibrous; endocarp c. 1 mm thick; perianth persistent on the fruit, c. 4 mm high. **Seed** ellipsoidal, 27–31 mm long, 19–21 mm diam. Figures 10.74, 10.75.

Figure 10.74 *Hedyscepe canterburyana* in moderate-altitude forest, Lord Howe Island. Photo by William J. Baker, Royal Botanic Gardens Kew.

Figure 10.75 *Hedyscepe canterburyana*. **Top left:** Staminate flowers at anthesis. Photo by Ian Hutton. **Top right:** Pistillate flowers at anthesis. Photo by Ian Hutton. **Bottom left:** Infructescence with mature fruit. Photo by Jeanne Price. **Bottom right:** Fruit in longitudinal section and seed, part of type specimen.

Big mountain palm

Distribution and ecology: Endemic to Lord Howe Island on Mt Gower and Mt Lidgbird, mainly on basalt soils as an emergent in evergreen closed forest, closed evergreen scrub and broad-leaved herb formations, and occurring in small populations or scattered groups or individuals, 335–830 m asl (Fig. 10.76). Flowering Mar.–July; fruiting Sept.–Jan. Conservation status – *Vulnerable* (Johnson 1996).

Typification: The holotype of *Hedyscepe canterburyana* (basionym *Kentia canterburyana*) is the unnumbered collection *Moore & Fitzgerald* [MEL] collected in June 1869 from high elevation on Mt Lidgbird, Lord Howe Island. The

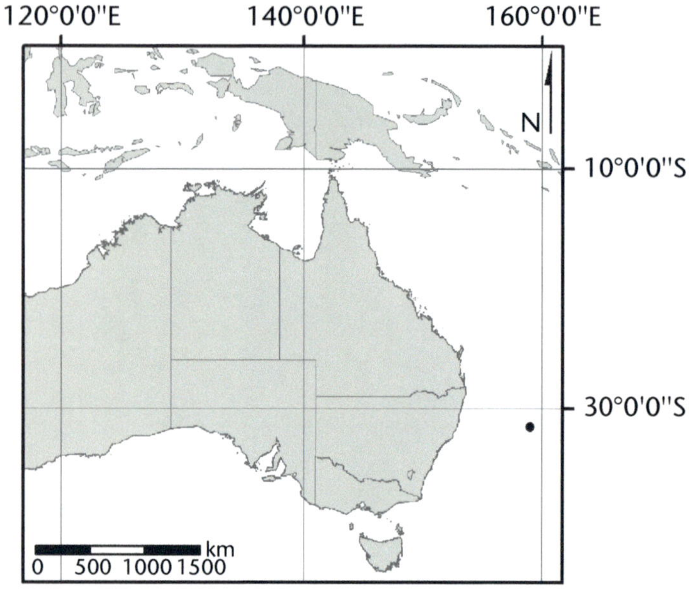

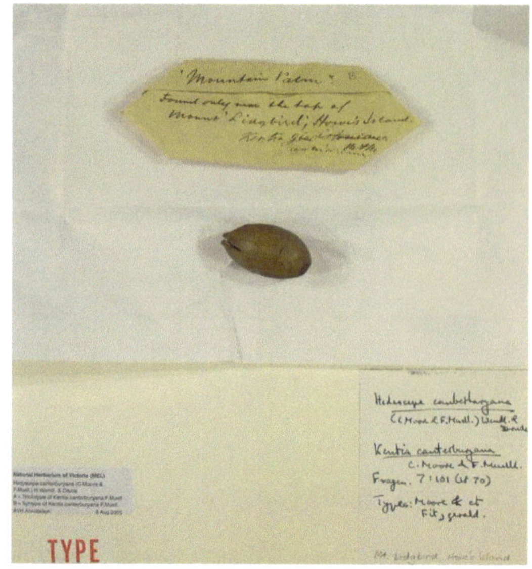

Figure 10.76 *Hedyscepe canterburyana*. **Left:** Distribution based on herbarium specimen data. **Right:** Type specimen. Australia. Lord Howe Island, Mt Lidgbird, undated, *C.Moore and R.D.Fitzgerald s.n.*; holotype MEL [1 of 2 sheets].

specimen consists of a packet of fruit and seeds, of which there is one whole fruit, one whole seed and assorted fragments (Fig. 10.76). The protologue described habit, leaf and inflorescence morphology, most probably based on Moore's field notes.

Etymology: '*Speciem autem tribuo nobili Comiti J.H.T. de Canterbury, nunc Coloniae Victoriae excellentissimo Gubernatori*' (Mueller 1870). [This species is named as a tribute to Viscount J.H.T. Canterbury, now Governor of the Colony of Victoria]: Named for Sir John Henry Thomas Manners-Sutton (b.1814, d.1877), 3rd Viscount Canterbury, Governor of Victoria 1866–73.

Notes: *Hedyscepe canterburyana* is distinguished by a solitary habit, arching leaves with erect pinnae, an inflorescence with a stout curving rachis, protandrous flowering syndrome, staminate flowers with 9–12 stamens, a cylindrical pistillode swollen toward the base, and ellipsoidal fruit to 42 mm long and dull red at maturity.

■ UNPLACED ARECEAE

Relationships

The relationships of some Areceae genera, including *Hydriastele*, have not been determined due to conflicting or inconclusive molecular evidence. *Hydriastele*, as a genus, is resolved as monophyletic (Asmussen *et al.* 2006; Loo *et al.* 2006; Norup *et al.* 2006; Dransfield *et al.* 2008) and its relationship appears to be as sister to other western Pacific Areceae (Lewis and Doyle 2002). A relationship to genera in the Arecinae (*Areca*, *Pinanga* and *Nenga*) may be suspected but at present it cannot be supported.

Hydriastele

Hydriastele H.Wendl. & Drude, *Linnaea* 39: 180 [208] (1875); *Kentia* subg. *Hydriastele* (H.Wendl. & Drude) Becc. *Malesia* 1: 39 (1877); *Hydriastele* sect. *Euhydriastele* Kuntze, *Lex. Gen. Phan.*: 289 (1903); *Hydriastele* H.Wendl. & Drude sect. *Hydriastele* Kuntze, *Lex. Gen. Phan.*: 289 (1903). Type: *Kentia wendlandiana* F.Muell. = *Hydriastele wendlandiana* (F.Muell.) H.Wendl. & Drude.

Gronophyllum Scheff., *Ann. Jard. Bot. Buitenzorg* 1: 135 [153] (1876); *Hydriastele* sect. *Gronophyllum* (Scheff.) Kuntze, *Lex. Gen. Phan.*: 289 (1903). Type: *Gronophyllum microcarpum* Scheff. = *Hydriastele microcarpa* (Scheff.) W.J.Baker & Loo.

Kentia Blume, *Bull. Sci. Phys. Nat. Néerl.* 1: 64 (1838) *nom. illeg.*, *non* Adans. (1763); *Kentia* Blume subg. *Kentia*, *Malesia* 1: 39: (1877). Type: *Kentia procera* Blume = *Hydriastele procera* (Blume) W.J.Baker & Loo.

Gulubia Becc., *Ann. Jard. Bot. Buitenzorg* 2: 128 [131, 134] (1885). Type: *Kentia moluccana* Becc. = *Hydriastele moluccana* (Becc.) W.J.Baker & Loo.

Paragulubia Burret, *Notizbl. Bot. Gard. Berlin–Dahlem* 13: 84 (1936). Type: *Paragulubia macrospadix* Burret = *Hydriastele macrospadix* (Burret) W.J.Baker & Loo.

NB: Generic synonyms which do not have a taxonomic association with Australian taxa are not included in this treatment. A full list of generic synonyms is provided in Dransfield *et al*. (2008).

Solitary or clustering, small to large, pleonanthic, monoecious palms. **Stems** slender to robust, leaning to erect; leafscars prominent to obscure, internodes short to long. **Leaves** infrequently simple or paripinnate, variously segmented, cleanly deciduous; leafsheaths forming a tubular elongate crownshaft, eventually splitting opposite the petiole prior to abscission, glabrous to densely scaly; petiole short to long, channelled on the adaxial surface, rounded on the abaxial surface, glabrous to densely scaly; rachis short to long, arching to straight, angled to flat on the adaxial surface, rounded on the abaxial surface, glabrous to densely scaly. **Pinnae** single or variously united, subopposite to opposite or grouped, linear to cuneate, reduplicate, erect to pendulous; apex truncate and praemorse or acuminate or acute, terminal pair broad; midrib prominent; abaxial glabrous or densely scaly on the midrib and veinlets; ramenta on abaxial midrib and veinlets. **Inflorescence** infrafoliar, paniculate, broomlike with branches ± parallel, spreading in fruit, branched 1–3 orders, or spicate, protogynous or protandrous; prophyll 2-keeled, fully enclosing the peduncular bract; peduncular bract inserted slightly above attachment of prophyll, or distant, tubular; flowers fully developed at bract fall; peduncle short; rachis short; rachillae attachment congested, straight or flexuous, arching to pendulous in fruit. **Flowers** unisexual, sessile, in congested opposite or decussate triads, or whorled, of a single pistillate flower subtended by a pair of staminate flowers one either side, with solitary or paired staminate flowers distally. **Staminate flowers** asymmetric in bud; 3 short triangular sepals, basally connate; 3 fleshy, acuminate, triangular, valvate petals, c. 4 times longer than the sepals, variously coloured; stamens 6–24, filaments short, epipetalous; anthers elongate, basifixed, latrorse; pistillode short and trifid, or lacking. **Pistillate flowers** globose or conical, smaller than the staminate; 3 rounded imbricate sepals; 3 rounded, imbricate, apically briefly valvate petals; staminodes 3–6, dentiform; gynoecium globose to ovoid or conical, unilocular, uniovulate; 3 small recurved stigmas, sessile or on a short style; ovule hemi-anatropous, laterally attached. **Pollen** ellipsoidal; aperture monosulcate or rarely trichotomosulcate; exine tectate, scabrate or reticulate. **Fruit** ellipsoidal, ovoid, cylindrical or globose, sometimes apically beaked; stigmatic remains apical; epicarp smooth or minutely pebbled; mesocarp thin, fleshy or dry with thin longitudinal fibres, sometimes longitudinally ribbed; endocarp thin or crustose, or obsolescent; perianth persistent or not, on mature fruit. **Seed** ellipsoid, ovoid or globose, laterally or basally attached, hilum elongate; raphe branches anastomosing; endosperm homogeneous or ruminate; embryo basal; germination adjacent-ligular; eophyll bifid, apices praemorse.

Hydriastele is a genus of c. 48 species distributed from Sulawesi, Indonesia, to Fiji, and with three species in Australia, two of which are endemic. The genus is highly diversified in New Guinea, where 32 species are recognised. Of the species occurring in Australia, *H. costata* is a common New Guinea species of which there are outlier populations in Cape York Peninsula; *Hydriastele wendlandiana* is most typical of the genus, being a moderate clustering species confined to moist areas in rainforest; and *Hydriastele ramsayi* represents a unique element, as it is the only species in the genus not occurring in rainforest, but in open woodland in a monsoonal climate with severe seasonal dryness and regular fires.

Etymology: *'Nomen derivatum a verbis γνοινς = nymphe, et σπήλη = columna, propter caudieum altitudinem prope fontes excelsam.'* (Wendland and Drude 1875). [The name is derived from the Greek words for water and pillar, on account of the stems growing in swamps]: Although not explained in the protologue, but possibly with reference to the species growing in inundated areas; from the Greek *hydria*, water vessel or fountain, and *stele*, tube or column.

Key to species of Hydriastele in Australia

1 Stems clustered; leaves with irregularly grouped pinnae; apex of pinnae praemorse .... **3. *H. wendlandiana***

1: Stem solitary; leaves regularly pinnate; apex of pinnae acuminate or acute **2**

2 Leaf rachis arching; pinnae held erect on the rachis; fruit ellipsoidal to cylindrical, to 15 mm long .... **2. *H. ramsayi***

2: Leaf rachis more or less straight; pinnae pendulous; fruit ovoid, to 10 mm long .... **1. *H. costata***

1 Hydriastele costata

Hydriastele costata F.M.Bailey, *Queensland Agric. J.* 2: 129 (1898). Type: Australia. Queensland. Somerset, Cape York Peninsula, Nov. 1897, *F.L.Jardine s.n.*; holotype BRI [1 sheet and parcel].

Kentia costata Becc., *Malesia* 1: 36 (1877); *Gulubia costata* (Becc.) Becc. var. *costata*, *Ann. Jard. Bot. Buitenzorg* 2: 134

(1885). Type: Indonesia. Papua. Aru Is., 1873, *O.Beccari 11152*; holotype FI.

Gulubia costata var. *minor* Becc., *Ann. Jard. Bot. Buitenzorg* 2: 135 (1885). Type: Indonesia. Papua. Arsus, *O.Beccari 11153*; holotype FI.

Gulubia costata var. *pisiformis* Becc., *Ann. Jard. Bot. Buitenzorg* 2: 136 (1885). Type: New Guinea, May 1878, *O.Beccari 11149*; holotype FI.

Pinanga pisiformis Teijsm. ex Becc., *Ann. Jard. Bot. Buitenzorg* 2: 136 (1885) *nom. inval.*

Kentia microcarpa (Lauterb. & K.Schum. ex K.Schum. & Lauterb.) Warb. ex K.Schum. & Lauterb., *Fl. Schutzgeb. Südsee*: 207 (1900 [1901]); *Kentia costata* var. *microcarpa* Lauterb. & K.Schum. ex K.Schum. & Lauterb., *Fl. Schutzgeb. Südsee*: 207 (1900 [1901]). Type: Papua New Guinea. Morobe Province. Finschhafen, 5 Aug. 1897, *Lauterbach 751*; lectotype L, isolectotype FI.

Gulubia affinis Becc., *Bot. Jahrb. Syst.* 58: 444 (1923). Type: Papua New Guinea. East Sepik Province. Hunstein Mts, undated, *C.L.Ledermann 8228*; holotype B, destroyed.

Gulubia costata var. *gracilior* Burret, *Notizbl. Bot. Gart. Berlin-Dahlem* 13: 81 (1936). Type: Papua New Guinea. Western Province. Oriomo R., Woroi, *L.J.Brass 5887*; holotype B destroyed; isotypes A, BO, BRI, NY.

Stem solitary, to 35 m tall, to 40 cm dbh, expanded at the base; leafscars to 35 mm wide; internodes to 25 cm long, smooth, grey. **Leaves** 10–15, to 4 m long, straight, but lower leaves tending to be below the horizontal, 80–120 pinnae each side of rachis; crownshaft to 150 cm long, cylindrical, light green suffused with lilac, with dense grey tomentum; petiole 15–40 cm long; rachis to 350 cm long. **Pinnae** regularly arranged, to 127 cm long, acuminate, laxly pendulous, green on both surfaces; midrib prominently raised on adaxial surface, less prominently raised on abaxial surface; 4–5 veinlets either side of midrib; ramenta on midrib on abaxial surface, to 5 mm long, twisted and chafy, basifixed or medifixed. **Inflorescence** to 125 cm long, to 40 cm wide, branched to 2 orders, axes green; prophyll to 120 cm long, to 20 cm wide, glabrous, cream to light green, basally flushed pink to red; peduncular bract to 100 cm long, to 15 cm wide, glabrous, cream to light green, basally flushed pink to red; peduncle to 15 cm long, to 10 cm wide at point of prophyll attachment, green; rachis to 20 cm long; rachillae to 62 cm long, to 7 mm thick, straight, 4-sided, green, straight, at first erect but pendulous with fruit; floral pits shallow, margins lipped, scarcely raised. **Flowers** in triads, paired or single staminate flowers distally, opposite, protogynous. **Staminate flowers** to 10 mm long in bud, not opening widely at anthesis; sepals 1–1.5 mm long, to 1–1.5 mm wide, green; petals 9–10 mm long, 3–5 mm wide, cream; stamens 5–7 mm long; filaments c. 0.5 mm long; anthers 2–3 mm long, bright yellow; pistillode lacking. **Pistillate flowers** 1.5–3 mm high, 2–3 mm wide in bud; sepals 0.2–0.5 mm long, to c. 0.5 mm wide, light green; petals 2–2.5 mm long, 2–2.5 mm wide, cream-green; stigmas to c. 0.2 mm long. **Fruit** ovoid, to 10 mm long, to 5 mm diam.; epicarp longitudinally ridged, smooth, blue-grey to deep maroon, with broad cream to pink longitudinal markings; mesocarp to 1 mm thick, fleshy; fibres dense and thick; endocarp crustose; perianth not persistent on fruit. **Seed** ovoid, to 4 mm long, to 4 mm diam., laterally attached; endosperm homogeneous. Figures 10.77, 10.78.

Gulubia palm

Distribution and ecology: Common in lowland New Guinea, but in Australia confined to Cape York Peninsula, Queensland, where it is uncommon, distributed from Lockerbie Scrub south to Iron Range, in rainforest and swamp forest in large and small colonies as a towering

Figure 10.77 *Hydriastele costata* in lowland swamp forest, Jacky Jacky Creek, Lockerbie Scrub, Cape York Peninsula, north Queensland.

Figure 10.78 *Hydriastele costata*. **Top left:** Inflorescence bracts, prophyll (left) and peduncular bract (right). **Top right:** Staminate and pistillate flowers at anthesis. **Bottom left:** Pistillate flowers post-anthesis. **Bottom right upper:** Mature fruit. **Bottom right lower:** Fruit, whole and in longitudinal section.

canopy emergent, 0–20 m asl (Fig. 10.79). Flowering July–Oct.; fruiting Sept.–Jan. Conservation status – *Vulnerable* (Bostock and Holland 2007).

Typification: The holotype of *Hydriastele costata* is an unnumbered collection by Frank Jardine [BRI], November 1897, from Somerset, Cape York, Queensland. The collection consists of a folded leaf and a packet of fruit (Fig. 10.79).

Etymology: 'Referring to the prominent ribs on the fruit' (Bailey 1898): From the Latin *costa*, rib.

Notes: *Hydriastele costata* has an intricate taxonomic history. The species was first described in 1877 by Beccari as *Kentia costata* and later transferred to *Gulubia*, as *G. costata*, the name by which it had until recently been known. In 1898, Bailey independently and apparently

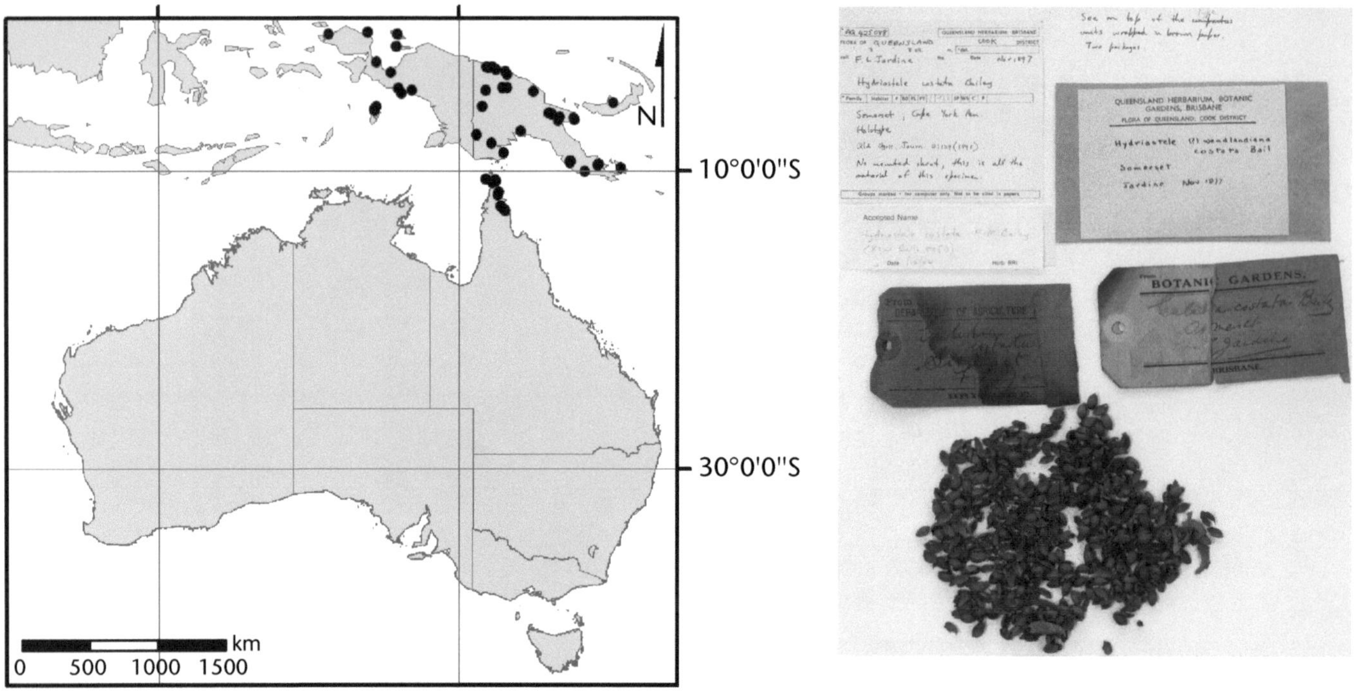

Figure 10.79 *Hydriastele costata*. **Left:** Distribution based on herbarium specimen data. **Right:** Type specimen. Australia. Queensland. Somerset, Cape York Peninsula, Nov. 1897, *F.L.Jardine s.n.*; holotype BRI.

without knowledge of Beccari's former use, chose the same specific epithet, *costata*, when describing *H. costata*, for a species of palm collected at Somerset on Cape York Peninsula. Bailey was unaware that his *H. costata* and Beccari's *G. costata* were the same species until it was pointed out to him by Beccari, who indicated that the two names were synonymous. This was later interpreted by Chapman to produce the authorship '(F.M. Bailey) Becc.', but this combination is nomenclaturally illegitimate as both species are represented by different types and therefore any combination is nomenclaturally illegitimate. Bailey, as author of the currently accepted name *H. costata*, has authorship priority.

Hydriastele costata is distinguished by a tall solitary habit, leaves held horizontal with pendulous pinnae, small ovoid fruits to 10 mm long, with maroon epicarp with longitudinal cream to pink markings, and homogeneous endosperm.

2 Hydriastele ramsayi

Hydriastele ramsayi (Becc.) W.J.Baker & Loo, *Kew Bull.* 59: 67(2004); *Gulubia ramsayi* Becc., *Webbia* 3: 159 (1910); *Kentia ramsayi* (Becc.) Becc., *Webbia* 4: 148 (1913); *Gronophyllum ramsayi* (Becc.) H.E.Moore, *Gentes Herb.* 9: 265 (1963). Type: Australia. Northern Territory. Port Essington, undated, *E.P.Ramsay s.n.*; holotype MEL [1 sheet]; isotype FI.

Stem solitary, to 30 m tall, to 35 cm dbh, moderately ventricose; leafscars slightly raised, to 20 mm wide; internodes to 20 cm long, green to grey. **Leaves** 9–12, to 2.5 m long, arching, with apex recurved, 50–60 pinnae each side of rachis; crownshaft to 70 cm long, yellow-green with white to grey tomentum; petiole to 80 cm long; rachis to 170 cm long. **Pinnae** to 90 cm long, 2–5 cm wide, erect, regularly arranged, opposite to subopposite, linear, acute, coriaceous, grey-green on both surfaces; midrib raised prominently on adaxial surface, raised moderately on abaxial surface, with scattered round black scales; 1–2 moderately prominent parallel veinlets each side of midrib, and 4–6 thin parallel veinlets; margins thickened; ramenta lacking. **Inflorescence** to 50 cm long, to 15 cm wide, branched to 2 orders, axes cream; prophyll to 52 cm long, to 6 cm wide, light green glaucous, internally yellow; peduncular bract to 50 cm long, to 3 cm wide; peduncle to 4 cm long, to 3 cm wide; rachis to 3 cm long; rachillae at first erect but pendulous with fruit; floral pits shallow, margins smooth, scarcely raised. **Flowers** in triads, paired or single staminate flowers distally, opposite, protandrous. **Staminate flowers** to 10 mm long, to 6 mm wide in bud, not opening widely at anthesis; sepals to

2 mm long, to 2 mm wide, cream; petals to 10 mm long, to 3 mm wide, cream-yellow; stamens to 7 mm long; filaments to 0.2 mm long; anthers to 5 mm long, bright yellow; pistillode lacking. **Pistillate flower** 1.5–5 mm high, to 4 mm wide in bud; sepals 1.5–3 mm long, 2–3 mm wide, cream; petals 5–8 mm long, 2–3 mm wide, green. **Fruit** ellipsoidal to cylindrical, to 15 mm long, to 8 mm diam.; epicarp smooth, to c. 0.5 mm thick, glossy red at maturity; mesocarp to c. 1 mm thick, dryish, densely fibrous; endocarp thin; perianth persistent on fruit. **Seed** elongate, to c. 14 mm long, to c. 6 mm wide; laterally attached; endosperm homogenous. Figures 10.80, 10.81.

Northern Kentia palm

Distribution and ecology: Endemic to Northern Territory, confined to the Top End, in tall coastal eucalypt forest in deep sandy soil, forming dense colonies, small groups or scattered individuals, also in sandstone areas, on escarpments and slopes, 5–90 m asl (Fig. 10.82). Large populations are on Cobourg Peninsula and Melville Island. Populations are subjected to regular fires. Flowering July–Oct.; fruiting Sept.–Jan. Conservation status – *No present threats*.

Figure 10.80 *Hydriastele ramsayi* in semi-open woodland, floodplain of East Branch, West Alligator River, Northern Territory.

Typification: The holotype of *Hydriastele ramsayi* (basionym *Kentia ramsayi*) is an unnumbered and undated collection by E.P. Ramsay [MEL], collected at Port Essington, Northern Territory. The specimen consists solely of a packet of fruit (Fig. 10.82). The protologue was a detailed description of the fruit, accompanied by an illustration of a single fruit in longitudinal view. Beccari (1910, p. 160) stated in the protologue that Mueller had suggested that the species was in the genus *Kentia*, but he had never published it.

Etymology: '*I frutti maturi di questa Palma vennero raccolti dal sig. Ramsay a Port Essington*'. [The mature fruits of this palm were collected by Mr Ramsay at Port Essington]: Named for collector of the type specimen Edward Pearson Ramsay (b.1842, d.1916), zoologist and ornithologist, Curator of the Australian Museum 1874–94, founder of the Linnean Society of New South Wales in 1874.

Notes: *Hydriastele ramsayi* is distinguished by a tall solitary habit with a ventricose stem, arching leaves with erect pinnae, ellipsoidal to cylindrical fruits to 15 mm long, glossy red at maturity, and homogeneous endosperm.

3 Hydriastele wendlandiana

Hydriastele wendlandiana (F.Muell.) H.Wendl. & Drude, *Linnaea* 39: 209 (1875); *Kentia wendlandiana* F.Muell., *Fragm.* 7: 102 (1870). Type: Australia. Northern Territory. Liverpool R., 25 May 1867, *B.Gulliver s.n*; holotype MEL [3 sheets]; isotypes BO, BRI.

Hydriastele wendlandiana var *microcarpa* H.Wendl. & Drude, *Linnaea* 39: 210, pl. 2, figs 6–7 (1875). Type: Australia. Queensland. O'Connell R., *J.Nernst s.n.*: holotype MEL.

Hydriastele douglasiana F.M.Bailey, *Queensland Agric. J.* 1: 232 (1897). Type: Australia. Queensland. Cape York Peninsula, Somerset, Polo Ck, June 1897, *F.M.Bailey 18*; holotype BRI.

Stems clustered, to 25 m tall, to 12 cm dbh; leafscars slightly raised, to 15 mm wide; internodes to 25 cm long, green ageing to cream-grey. **Leaves** 7–10, to 2 m long, erect to horizontal, 7–15 pinnae each side;

Figure 10.81 *Hydriastele ramsayi*. **Top left:** Crown with inflorescences. **Top right:** Staminate flowers pre-anthesis and pistillate flowers at anthesis. **Bottom left:** Crown with infructescence. **Bottom right:** Fruit, whole and in longitudinal section.

crownshaft to 50 cm long, green with moderate whitish tomentum, scattered dark scales; petiole 20–50 cm long, densely scaly; rachis to 1.5 m long, densely scaly. **Pinnae** irregularly arranged and grouped, some united in the middle part of the rachis, with extended bare portion distally, single pinnae linear, 20–50 cm long, 2.5–10 cm wide, apex truncate and praemorse, dark green on both surfaces; apical pinnae basally united; midrib thick, raised on adaxial surface; parallel veinlets almost as thick as midrib; ramenta medifixed to basifixed, to 2 mm long, on midribs and veinlets, orange to red-brown. **Inflorescence** 20–35 cm long, to 12 cm wide, branched to 1 order; axes light green, erect becoming pendulous with fruit; prophyll to 50 cm long,

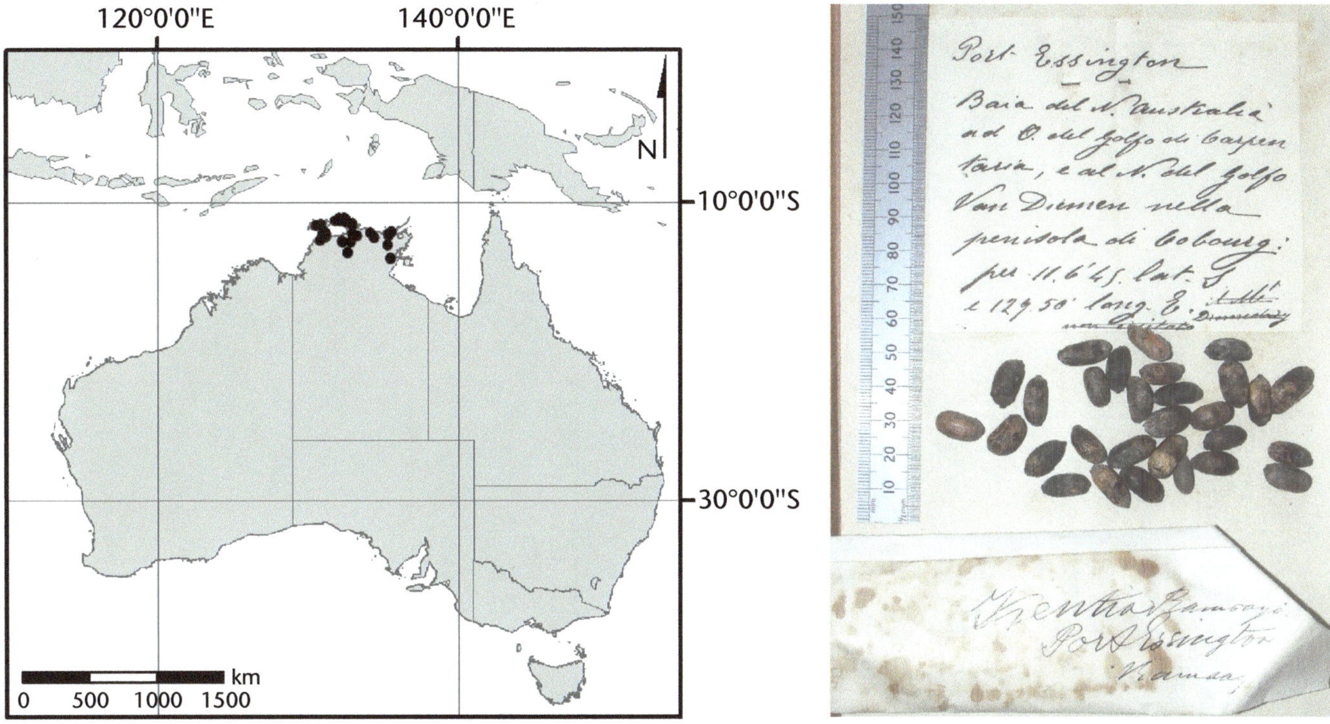

Figure 10.82 *Hydriastele ramsayi*. **Left:** Distribution based on herbarium specimen data. **Right:** Type specimen. Australia. Northern Territory. Port Essington, undated, *E.P.Ramsay s.n.*; holotype MEL.

to 10 cm wide, with a whitish bloom; peduncular bract to 45 cm long, to 7 cm wide, glabrous, light brown to green; peduncle to 7 cm long, glabrous, green; rachis to 4 cm long, glabrous, green; rachillae to 40 cm long, to 3 mm thick, straight, 4-sided, yellowish-green; floral pits shallow, lipped, margins scarcely raised, smooth. **Flowers** in triads, paired or solitary staminate flowers distally, decussate, protogynous. **Staminate flowers** to 9 mm long in bud, loosely open at bract fall and pollen release, opening to c. 1 cm wide at anthesis; sepals to 1 mm long, basally connate, narrow, apically pointed; petals to 9 mm long, to 4 mm wide, acuminate; stamens to 7 mm long; filaments to 1 mm long; anthers 3–4 mm long, basally narrowly sagittate, cream; pistillode lacking. **Pistillate flower** globose to ovate, to 3 mm high, to 2 mm wide in bud; sepals to 1 mm long, to 1 mm wide, light green; petals to 3 mm long, to 2 mm wide, light green to yellow-green; stigmas to c. 0.5 mm long. **Fruit** ovoid to globose, 8–10 mm long, 5–8 mm wide; epicarp smooth, 0.1– 0.2 mm thick, bright red at maturity; mesocarp 1–1.5 mm thick, fleshy, fibres thin; endocarp c. 0.1 mm thick, crustose; perianth not persistent on fruit. **Seed** globose, 5–8 mm diam., laterally attached; endosperm homogeneous or shallowly ruminate. Figures 10.83, 10.84.

Figure 10.83 *Hydriastele wendlandiana* in lowland rainforest, Licuala State Forest, Mission Beach, north-east Queensland.

Figure 10.84 *Hydriastele wendlandiana*. **Top left:** Pinnae with praemorse apex. **Top right:** Inflorescence with flowers in bud. **Bottom left:** Staminate and pistillate flowers at anthesis. **Bottom right:** Mature fruit.

Australian Kentia palm

Distribution and ecology: Occurring in Queensland and Northern Territory, in rainforest and seasonally inundated swamp forest, evergreen monsoon vine forest associated with permanent fresh water, in drier areas as a riparian element in various soil types, 0–200 m asl (Fig. 10.85). Flowering Jul.–Nov.; fruiting Oct.–Jan. Conservation status – *No present threats*.

Typification: The holotype of *Hydriastele wendlandiana* (basionym *Kentia wendlandiana*) is an unnumbered collection by B. Gulliver [MEL], collected in 1867 from Liverpool River, Northern Territory. The collection consists of a leaf, two inflorescences and a packet of fruit (Fig. 10.85).

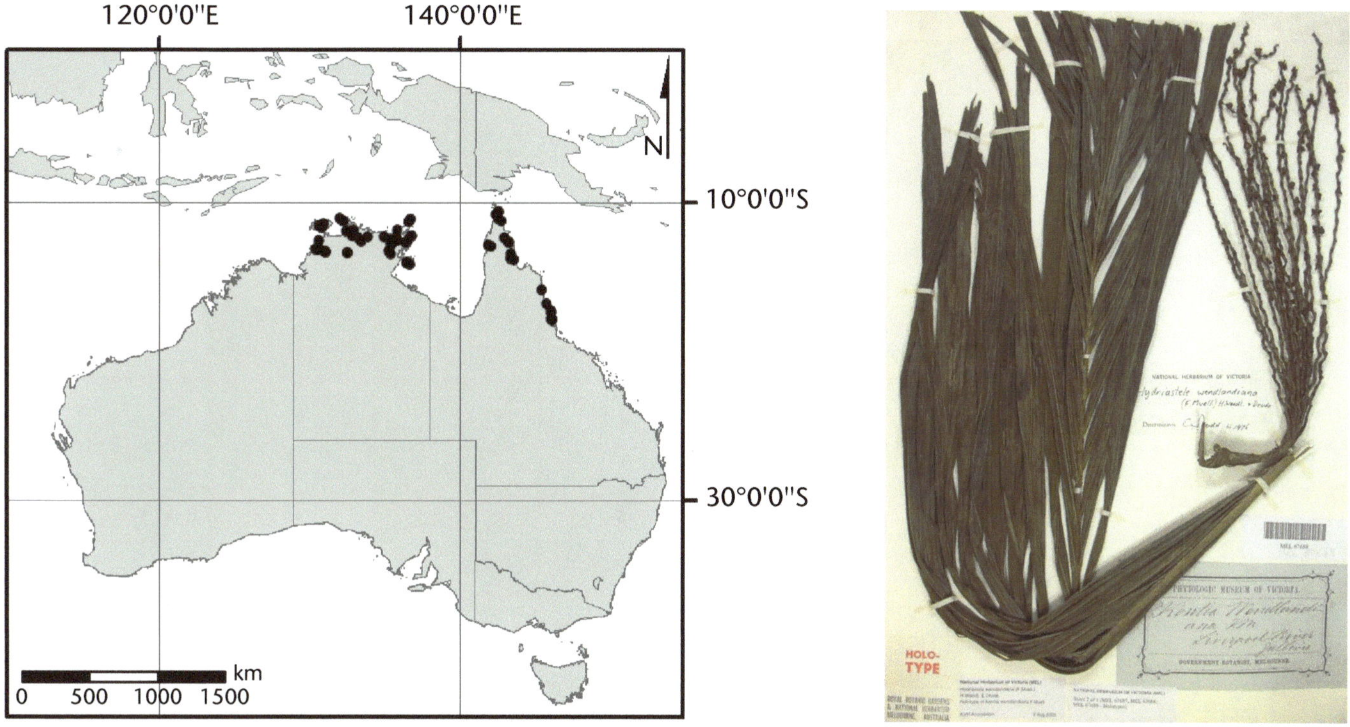

Figure 10.85 *Hydriastele wendlandiana*. **Left:** Distribution based on herbarium specimen data. **Right:** Type specimen. Australia. Northern Territory. Liverpool River, 25 May 1867, *B.Gulliver s.n*; holotype MEL [1 of 3 sheets].

Etymology: There was no explicit explanation in the protologue of the derivation of the epithet, but most likely to honour Hermann Wendland (b.1823, d.1903), botanist and horticulturist, Director of Royal Botanic Gardens Herrenhausen, Hannover, Germany, 1869–97. He made a specialty of the cultivation of palms and later of taxonomy, producing the first monograph of Australian palms, *Palmae Australasicae* (Wendland and Drude 1875).

Notes: *Hydriastele wendlandiana* var. *microcarpa* was described for a collection from 'O'Connell River' in North Queensland, made by J. Nernst [MEL], with fruit 8 mm long by 5.5 mm diameter, of which an illustration was produced in Wendland and Drude (1875). 'O'Connell River' is a former name for Proserpine Creek, which drains into Whitsunday Passage. There are no records of *Hydriastele* being collected from anywhere south of about Mission Beach, which is about 500 km to the north of Proserpine Creek, the southern limit of distribution. It must be assumed that the use of this place name is an error; it may relate to the place from which it was dispatched, rather than the actual place of collection.

Hydriastele douglasiana was described for a collection from Polo Creek, Somerset, Cape York Peninsula, labelled as collected by F.M. Bailey in 1897 but most likely collected by Frank Jardine's son, Douglas, for whom the species was named: 'Mr. Jardine's son brought from a few miles distant specimens of what may prove an hitherto undescribed palm' (Bailey 1897, p. 232). The only comparison that Bailey made with *H. wendlandiana* was included in his accompanying notes, in which he stated: 'This elegant palm is nearly allied to *H. wendlandiana*, but does not fully agree with the published description of that species' (Bailey 1897, p. 232). This taxon is referable to *H. wendlandiana*.

Hydriastele wendlandiana is distinguished by a clustering habit, leaves with irregular single or grouped pinnae, fruit ovoid to globose to 10 mm long, bright red at maturity, and endosperm homogeneous or shallowly ruminate.

11. Doubtful and excluded names

Actinophloeus bleeseri Burret, *Repert. Sp. Nov. Regni Veg.* 24: 266 (1928). Name cited as a possible alternative name for *Ptychosperma bleeseri* Burret, in a footnote to the protologue of that species. See *Carpentaria bleeseri* below.

Areca alicae F.Muell., *Gartenflora* 28: 199 [200] (1879). Type: Cultivation? Australia. Queensland. About 10 miles N of Trinity Bay, undated, *W.Hill s.n.*; holotype MEL; isotype FI. Refers to a probably cultivated plant of *Areca triandra* Roxb. ex Buch.-Ham. The description given by Mueller (1880) matches that species, as does the type specimen. The origin and history of the plant from which the type of *Areca alicae* was taken is not known. However, the area around Cairns was settled in the decades prior to the collection and historical records indicate that many exotic plants had been grown as part of acclimation programs or as garden ornamentals.

Borassus flabellifer L. *Sp. Pl.*: 1187 (1753). Type: '*Ampana*' and '*Carimpana*' Rheede, *Hort. malab.* 1: 11–14, pl. 9, 10 (1678). F.M.Bailey, *in Queensland Agric. J.* 2: 233 (1897). This refers to a single staminate plant cultivated near the now-abandoned homestead of F.L. Jardine at Somerset, Cape York Peninsula, Queensland. According to Bailey, the plant was acquired from 'a locality upon the Peninsula forty miles from the coast'. Jardine, in correspondence quoted by Bailey, described the locality as having 'quite a clump of palms, some having very thick stems'. Bailey quoted the palm as being 20–30 ft (6–9 m) high. The plant at Somerset was extant in November 2008, with a stem height of 18–20 m.

Carpentaria bleeseri Burret, *Repert. Sp. Nov. Regni Veg.* 24: 268 (1928). Name cited as a possible alternative name for *Carpentaria acuminata*, in discussion following the protologue of *Ptychosperma bleeseri* Burret.

Corypha juncea, *in* mss, D.Solander, *Plantae Novae Hollandiae* 3: 158 (1768–71). Name used in an unpublished manuscript, for a palm observed in eastern Australia.

Corypha nuda, *in* mss, D.Solander, *Plantae Novae Hollandiae* 3: 11, 158 (1768–71). Name used in an unpublished manuscript, and later related to *Livistona* spp. of eastern Australia (Vallance *et al.* 2001, p. 233).

Desmoncus minor R. *et* P., in W.Hill, *Catalog. Pl. Brisbane Bot. Gard.* 21 (1875). Type: not designated. Described as 'evergreen climber, Trinidad', and named for a plant cultivated in Brisbane Botanic Gardens, probably received with this 'tag' name from Trinidad Botanical Gardens. *Desmoncus* is a genus of palms from Central and South America.

Euterpe speciosa Hort. in *Blumeng.* 1: 1155 (1896). A name used as a synonym of *Howea belmoreana* (C.Moore & F.Muell.) Becc., *fide* Henderson and Galeano (1996).

Jubaea speciosa H.K., *in* W.Hill, *Catalog. pl. Brisbane bot. gard.* 23 (1875). Described as 'evergreen tree, Mauritius', and named for a plant cultivated in Brisbane Botanic Gardens. most likely a misspelling of *Jubaea spectabilis* Kunth, which is now a synonym of *Jubaea chilensis* (Molina) Baill., the cultivated Chilean wine palm.

Kentia australis Hort. ex anon., *Gard. Chron.* 6 (1873); H.Wendland *in* Kerchove, *Palmiers*: 68 (1878). The name *Kentia australis* was included in a list in the *Gard. Chron.* (*loc. cit.*), of the four palms from Lord Howe Island: 'introduced under the names of *Kentia* (*Veitchia*) *canterburyana* [*Hedyscepe canterburyana*], *K. australis*, *K. belmoreana* [*Howea belmoreana*], and *K. forsteriana* [*Howea forsteriana*], all obtained from Lord Howe's Island'. There was no description provided, so *Kentia australis* is a *nom. nud.* By a process of elimination, this could relate to the fourth species of palm from Lord Howe Island, *Lepidorrhachis mooreana* (F.Muell.) O.F.Cook.

Kentia rubricaulis Hort., *Gard. Chron. n.s.* 5: 603 (1876). The inadequate description and lack of type makes it impossible to place this taxon. This name is in APNI, but it is not conclusively an Australian plant, being listed with an explicitly Australian species of *Araucaria* and a *Pandanus* of unknown provenance. The brief description reads: 'with pinnate ovate leaves, having red stalks'. Govaerts and Dransfield (2005) place it as a synonym of *Chambeyronia macrocarpa* (Brong.) Vieill. ex Becc., with the authorship as *K. rubricaulis* Linden ex Salomon.

Licuala australasica Hort. ex H.Wendl., *in* Kerchove, *Palmiers*: 67, 249 (1878). This name was included in an account of Australian palms, but without a description or specimens it cannot be placed.

Licuala jeanenceyi André, *Rev. Hort.* 70: 263 (1898). This name was applied to a species 'from an unnamed locality within Australia or the Australian archipelago'. Govaerts and Dransfield (2005) placed it as a synonym of *Licuala ramsayi*, but this is unsupportable because of the lack of a specimen and the brief description.

Licuala muelleri var. *robusta* (Warb.) Becc., *in* P.Cuccuini and C.Nepi, *Quad. Bot. Amb. Appl.*, 17/1: 99 (2006). This name was included in a list of specimens conserved in FI. It is an invalid (though possibly proposed) name used only on a herbarium specimen sheet.

Livistona enervis Hort. ex anon., *Ill. Gart.-Zeitung* 16: 346 (1891). This name was applied to a palm from the nursery collection of William Bull. The leaves were described as delicately textured and with narrow segments. The description is inadequate for correct identification. It could be a misspelling of *Livistona inermis* R.Br., which was a name erroneously applied to *Livistona decora* (W.Bull) Dowe in Europe at that time.

Livistona filifera Hort. ex H.Wendl., *in* Kerchove, *Palmiers*: 250 (1878). This appears to be a horticultural tag name applied to *Livistona inermis auct. non.* R.Br., a name incorrectly used for *Livistona decora* (W.Bull) Dowe in horticulture in the late 1800s. Without a description or specimens it cannot be placed.

Livistona hoffmanni Hort. ex André, *Rev. Hort.* 77: 527 (1905). This name was applied to plants exhibited at a meeting of the Royal Horticultural Society of England. André quoted W. Watson, in the Gardeners' Chronicle (no publication details provided), suggesting that they were plants of *Livistona mariae*. Without specimens or a description, this cannot be substantiated, and the name remains a name of no standing.

Livistona humilis R.Br. var. *bloomfieldiensis*, *in* P.Cuccuini and C.Nepi, *Quad. Bot. Amb. Appl.*, 17/1: 104, 192 (2006). This name was included in a list of specimens conserved in FI. Cuccuini and Nepi (2006) suggested it was a synonym of *Livistona humilis* var. *sclerophylla* Becc. and *Livistona muelleri*. It is an invalid (though possibly proposed) name used only on a herbarium specimen sheet.

Livistona humilis var. *leichhardtii* (F.Muell.), *in* P.Cuccuini and C.Nepi, *Quad. Bot. Amb. Appl.*, 17/1; 104 (2006). This name was included in a list of specimens conserved in FI. It is an invalid (though possibly proposed) name used only on a herbarium specimen sheet.

Livistona mülleri Hort. ex H.Wendl., *in* Kerchove, *Palmiers*: 68, 250 (1878). The lack of description and specimens makes it impossible to place this taxon. This taxon appears not to be associated with *Livistona muelleri* F.M.Bailey, *Queensland Fl.* 5: 1683 (1902), but was a name included in a list of supposedly Australian palm species.

Livistona occidentalis Hort. ex Hook.f., *Report on the progress and condition of the Royal Gardens at Kew* 1882: 64 (1884). This name was used by Hooker in a list of palms cultivated in the Royal Gardens, Kew. He listed both *L. occidentalis* and *Thrinax tunicata* Hort. ex Verschaff. as synonyms of *Brahea dulcis*. With the raising of *Livistona mariae* subsp. *occidentalis* Rodd to specific level (Dowe and Jones 2004), the subspecies names would normally have become the specific epithet if the name had not been previously used. See *Livistona nasmophila* Dowe & D.L.Jones.

Phoenix canariensis Chabaud, *Provence Agric.* 19: 293 (1882). A non-Australian species treated in Conn and Walsh (1993) as naturalised in northern Victoria, New South Wales, South Australia and Western Australia. Indigenous to the Canary Islands and widely cultivated as an ornamental throughout the world.

Phoenix dactylifera L., *Sp. Pl.* 1188 (1753). A non-Australian species treated in Jessop and Toelken (1986) and Florabase (2009) as naturalised in semi-arid to arid regions of Australia, along watercourses and near springs (Noack 1994). Of unknown exact origin, but possibly from the north African to middle East region. Widely cultivated and of major economic importance as a source of food.

Pritchardia maideniana Becc., *Webbia* 4: 213, fig.12 (1913). Type: Cultivation. Royal Botanic Gardens, Sydney, NSW, 1912, *Boorman*; lecto: FI (*fide* Hodel 2007). This species was described from a single plant growing in the Royal Botanic Gardens, Sydney. It apparently originated from seeds collected in the Hawaiian Islands. Hodel (2007) recognised the conspecifity of *P. maideniana* and *P. affinis* Becc. which occurs on Kona, Kau and Puna islands, Hawaii. The name *P. maideniana* has priority over *P. affinis*.

Ptychosperma drudei H.Wendl., *in* G.Bentham and J.D.Hooker, *Gen. Pl.* 3: 892 (1883); *Saguaster drudei* (H.Wendl.) Kuntze, *Revis. Gen. Pl.* 2: 735 (1891). This name was placed as a synonym under *Archontophoenix alexandrae* (F.Muell.) H.Wendl. & Drude without explanation by Martelli (1935), and reiterated by Essig (1978). The brief description, apart from *'fructum 2-spermum'* (Bentham and Hooker 1883, p. 892), and lack of extant specimens, prevent this taxon being placed. Govaerts and Dransfield (2005) tentatively placed it as a synonym of *Archontophenix alexandrae*.

Sagus blackalli W.Hill, *Catalog. Pl. Brisbane Bot. Gard.*: 21 (1875). The description given to this palm by Hill was 'evergreen tree, Cape York'. The lack of specimens and detailed description makes it impossible to place this taxon. See Dowe (2004) for further discussion of this name.

Seaforthia nobilis Lhotsky, *Jour. Austral. Alps*: 59 (1835). A name used for one of two palm species observed by Lhotsky in the vicinity of Sydney. It most likely refers to *Archontophoenix cunninghamiana* (H.Wendl.) Wendl. & Drude, but cannot be nomenclaturally linked to that species.

Seaforthia robusta Hort. ex H.Wendl., *in* Kerchove, *Palmiers*: 257 (1858). This taxon has been placed as a synonym of *Rhopalostylis baueri* but this is unsupportable in the absence of a description or specimens.

Seaforthia veitchii Hort. ex anon., *Gard. Chron.*: 37 (1870). Apart from it being described as 'a fine Australian species', the lack of description and specimens makes it impossible to place this taxon.

Syagrus romanzoffianum (Cham.) Glassman, *Fieldiana, Bot.* 31: 382 (1968). A non-Australian species treated in Stanley and Ross (1989) as naturalised in southern Queensland and the Atherton Tableland, and widely cultivated as an ornamental palm throughout the warmer and moister areas of Australia. Occurs naturally in South America from central Brazil to eastern Uruguay.

12. Field identification of Australian palm species

This key to Australian palm species is based on characters of mature specimens, that may be observed in the field. However, because of the limited vegetative variation among some species, the presence of flowers and/or fruits is necessary for successful use of the key. The key includes species that occur on Australia's off-shore territories (six spp.) as well as those from continental Australia (54 spp.). The locations of off-shore species are indicated in the key.

1 Leaves costapalmate or palmate (fan palms) 2
1: Leaves bipinnate or pinnate (fishtail, feather and/or climbing palms) 21
2 Leaf with wedge-shaped segments, narrowest at the centre of the leaf, and widening toward the leaf margins .... ***Licuala ramsayi***
2: Leaf with linear to lanceolate segments, tapering toward the leaf margin 3
3 Leafbases cleft at attachment to stem; inflorescences terminal on stem .... ***Corypha utan***
3: Leafbases not cleft at attachment; inflorescences interfoliar 4
4 Divisions between leaf segments of mature leaves >60% of the length of the segments; segment apices deeply bifid, pendulous and commonly ribbon-like 5
4: Divisions between leaf segments of mature leaves <60% of the length of the segments; segment apices most commonly semi-pendulous, or rigid, infrequently pendulous 10
5 Proximal portion of petiole and leafbase purple .... ***Livistona drudei***
5: Proximal portion of petiole and leafbase green, infrequently with brown discoloration with age 6
6 Stem with leafbases remaining attached and with c. half the petiole remaining intact or deteriorating with age or fire .... ***Livistona benthamii***
6: Stem with leafbases remaining attached only in lower portions and with none or only small amount of petiole remaining 7
7 Inflorescences with diameter of the peduncle >6 cm .... ***Livistona decora***
7: Inflorescences with diameter of the peduncle <6 cm 8
8 Petiole lacking marginal spines or with indistinct spines or prickles, largest to 5 mm long .... ***Livistona inermis***
8: Petiole with marginal spines >5 mm long 9
9 Robust tall palms to 30 m high, dbh 24–35 cm .... ***Livistona concinna***
9: Moderate palm to 15 m high, dbh 8–20 cm .... ***Livistona lorophylla***
10 Abaxial surface of leaf with coppery brown floccose tomentum; apex of the segment with a cleft <4 cm long .... ***Livistona fulva***
10: Abaxial surface of leaf lacking coppery tomentum; apex of the segment with a cleft >4 cm long 11
11 Fruit obovoid 12
11: Fruit ellipsoid or globose 13

12 Inflorescences sexually dimorphic, on separate plants, female plants having a single partial inflorescence at the apex of the inflorescence, male plants with 4–7 partial inflorescences placed regularly along the primary axis .... ***Livistona humilis***

12: Inflorescences not sexually dimorphic, both female and male plants with 5–6 partial inflorescences placed regularly along the primary axis .... ***Livistona eastonii***

13 Fruit ellipsoidal; leaf segments with rigid apices .... ***Livistona muelleri***

13: Fruit globose; leaf segments pendulous **14**

14 Fruit >25 mm diam. **15**

14: Fruit <25 mm diam. **16**

15 Inflorescence bracts densely tomentose-lanuginose throughout .... ***Livistona lanuginosa***

15: Inflorescence bracts mostly glabrous with scattered patches of tomentum .... ***Livistona alfredii***

16 Adaxial surface of leaf with a covering of wax easily removed with fingernail; segment apices semi-pendulous **17**

16: Adaxial surface of leaf lacking wax; segment apices pendulous **20**

17 Inflorescence bracts tomentose throughout **18**

17: Inflorescence bracts glabrous or with scattered patches of tomentum, usually in the proximal portion **19**

18 Lamina deeply folded into a narrow 'v' .... ***Livistona rigida***

18: Lamina not deeply folded, being a broad 'v' .... ***Livistona mariae***

19 Lamina pruinose, grey-green to bluish-green; fruit with glossy epicarp .... ***Livistona victoriae***

19: Lamina glossy, grey-green; fruit with semi-glossy epicarp .... ***Livistona nasmophila***

20 Fruit reddish-brown to dull black, occasionally glaucous .... ***Livistona australis***

20: Fruit glossy jet black, not glaucous .... ***Livistona nitida***

21 Leaves bipinnate (fishtail palms) .... ***Caryota albertii***

21: Leaves pinnate (feather and/or climbing palms) **22**

22 Climbing palms (rattans, lawyer vines or wait-a-whiles) **23**

22: Non-climbing palms **30**

23 Leaf rachis extending into a cirrus, i.e. a whip-like structure with grapnel-like hooks on the abaxial surface and used as a climbing organ **24**

23: Leaf rachis not extending into a cirrus **26**

24 Stem solitary .... ***Calamus vitiensis***

24: Stems clustering **25**

25 Leafsheaths densely spiny .... ***Calamus warburgii***

25: Leafsheaths lacking spines or with widely scattered single spines .... ***Calamus aruensis***

26 Pinnae cuneate, fishtail-like in shape, apices praemorse, i.e. appearing chewed off .... ***Calamus caryotoides***

26: Pinnae linear to lanceolate; apices acuminate **27**

27 Pinnae without bristly spines on the adaxial surface .... ***Calamus australis***

27: Pinnae with bristly spines on the adaxial surface **28**

28 Spines on leafsheaths arranged in oblique combs; spines yellowish to green-brown; leaves stiffly arched .... ***Calamus moti***

28: Spines on leafsheaths randomly distributed; spines reddish-brown; leaves not stiffly arched **29**

29 Stem diam. 5–16 mm; leaves 1 m or less with 7–10 pinnae each side of rachis .... ***Calamus muelleri***

29: Stem diam. >20 mm; leaves to 3 m with 30–55 pinnae each side of rachis .... ***Calamus radicalis***

30 Leafbases not forming a tubular crownshaft **31**

30: Leafbases forming a tubular crownshaft **44**

31 Inflorescence a solitary spike, or with 3–8 spikes that have a common base and a common prophyll **32**

31: Inflorescence not as above **39**

32 Seed with ruminate endosperm; inflorescence a solitary spike .... ***Laccospadix australasicus***

32: Seed with homogeneous endosperm; inflorescence either a solitary spike or multi-spicate **33**

33 Tall solitary palms >10 m tall; inflorescence a solitary spike or multi-spicate; peduncular bract attached to inflorescence within the portion covered by the prophyll; peduncular bract persistent **34**

33: Diminutive or moderate, solitary or clustering palms <6 m tall; inflorescence always a solitary spike; peduncular bract attached to the inflorescence distally of the portion covered by the prophyll; peduncular bract deciduous **35**

34 Inflorescence a solitary spike; leaves ascending and arching, pinnae erect (Lord Howe Island) ..... ***Howea belmoreana***

34: Inflorescence 3–8-spiked; leaf horizontal to descending; pinnae horizontal to semi-pendulous (Lord Howe Island) ..... ***Howea forsteriana***

35 Stem solitary ..... ***Linospadix monostachyos***

35: Stems clustering **36**

36 Fruit globose-turbinate to subglobose, 5–9 mm long ..... ***Linospadix microcaryus***

36: Fruit ellipsoid, or elongate-cylindrical, 8–18 mm long **37**

37 Petiole lacking or rarely to 2 cm long; leaf usually undivided, occasionally with few divisions with united segments ..... ***Linospadix apetiolatus***

37: Petiole 4–51 cm long; leaf always variously divided with both single and united segments in various arrangements **38**

38 Staminate flowers bullet-shaped in bud, widely opening at anthesis; petals green; stamens 7–20 ..... ***Linospadix minor***

38: Staminate flowers globose in bud, barely opening at anthesis; petals orange-yellow; stamens 6–9 ..... ***Linospadix palmerianus***

39 Inflorescences with male flowers in catkin-like branches and female flowers in a terminal head; mangrove habitat ..... ***Nypa fruticans***

39: Inflorescences paniculate; terrestrial habitat **40**

40 Peduncular bracts 3–5; pinnae with silver-coppery felt-like tomentum on abaxial surface; dioecious palms ..... ***Oraniopsis appendiculata***

40: Peduncular bract solitary; pinnae with silver-grey tomentum or lacking tomentum on abaxial surface; monoecious palms **41**

41 Fruit >10 cm diam.; endosperm with large central cavity partially containing liquid; pinnae with acuminate apices ..... ***Cocos nucifera***

41: Fruit <3 cm diam.; endosperm solid; pinnae with praemorse apices **42**

42 Stem solitary (Christmas Island) ..... ***Arenga listeri***

42: Stems clustered **43**

43 Large palm to 20 m tall; pinnae with dense grey indumentum on abaxial surface; fruit 8–28 mm wide ..... ***Arenga australasica***

43: Moderate palm to 7 m tall; pinnae with indistinct grey indumentum on abaxial surface; fruit 15–18 mm wide ..... ***Arenga microcarpa***

44 Leaf with plumose (fox-tail like) appearance; pinnae longitudinally split into numerous segments **45**

44: Leaf with pinnae in a single plane; pinnae not longitudinally split **46**

45 Abaxial surface of pinnae blue-grey; pinnae longitudinally split into 7–11 segments; mesocarp with 0.5 mm wide fibres; endosperm ruminate ..... ***Normanbya normanbyi***

45: Abaxial surface of pinnae green; pinnae longitudinally split into 2–18 segments; mesocarp with 1–2 mm wide fibres; endosperm homogeneous ..... ***Wodyetia bifurcata***

46 Seed with five lobes in cross-section **47**

46: Seed terete (round) in cross-section **48**

47 Stem solitary; endosperm deeply ruminate ..... ***Ptychosperma elegans***

47: Stems clustering; endosperm homogeneous ..... ***Ptychosperma macarthurii***

48 Stems clustering ..... ***Hydriastele wendlandiana***

48: Stem solitary **49**

49 Staminate flowers with 6 stamens 50
49: Staminate flowers with >6 stamens 53
50 Inflorescences broom-like with rachillae ± parallel 51
50: Inflorescences are divaricating panicles 52
51 Leaves arching; pinnae erect on the rachis .... ***Hydriastele ramsayi***
51: Leaves with a ± straight rachis; pinnae laxly pendulous .... ***Hydriastele costata***
52 Moderate palms <4 m tall; crownshaft split opposite the petiole; seed with an operculum covering the embryo (Lord Howe Island) .... ***Lepidorrhachis mooreana***
52: Tall palms >10 m high; crownshaft tubular; seed without an operculum covering the embryo (Norfolk Island) .... ***Rhopalostylis baueri***
53 Fruit >35 mm long; crownshaft densely tomentose, whitish-grey-bluish (Lord Howe Island) .... ***Hedyscepe canterburyana***
53: Fruit <20 mm long; crownshaft not as above 54
54 Leaves recurved, not twisted along the rachis; pinnae with praemorse apices, i.e. appearing chewed off; pinnae grouped at the base of the leaf .... ***Carpentaria acuminata***
54: Leaves moderately arched, twisted along the rachis; pinnae with acuminate apices; pinnae not grouped at the base of the leaf 55
55 Pinnae with ramenta, i.e. twisted hair-like scales up to 6 mm long, on the midrib on the abaxial surface 56
55: Pinnae lacking ramenta on the midrib on the abaxial surface 57
56 Abaxial surface of pinnae with dense silver-grey scales; fruit 20–26 mm long .... ***Archontophoenix purpurea***
56: Abaxial surface of pinnae light green; fruit 10–15 mm long .... ***Archontophoenix cunninghamiana***
57 Fruit with <0.5 mm wide fibres in the mesocarp .... ***Archontophoenix myolensis***
57: Fruit with >0.5 mm wide fibres in the mesocarp 58
58 Fruit 15–25 mm long; inflorescences branched to 3 orders .... ***Archontophoenix tuckeri***
58: Fruit 8–15 mm long; inflorescences branched to 4 orders 59
59 Inflorescences 80–140 cm long, peduncle to 12 cm wide; leaf with 90 pinnae each side of rachis; pinnae to 90 cm long .... ***Archontophoenix maxima***
59: Inflorescences 50–100 cm long, peduncle <10 cm wide; leaf with 60–80 pinnae each side of the rachis; pinnae to 75 cm long .... ***Archontophoenix alexandrae***

Checklist of genera and species

***Calamus* L.**
Calamus aruensis Becc.
Calamus australis Mart.
Calamus caryotoides A.Cunn. ex Mart.
Calamus moti F.M.Bailey
Calamus muelleri H.Wendl.
Calamus radicalis H.Wendl. & Drude
Calamus vitiensis Warb. ex Becc.
Calamus warburgii K.Schum.

***Nypa* Steck**
Nypa fruticans Wurmb

***Livistona* R.Br.**
Livistona alfredii F.Muell.
Livistona australis (R.Br.) Mart.
Livistona benthamii F.M.Bailey
Livistona concinna Dowe & Barfod
Livistona decora (W.Bull) Dowe
Livistona drudei F.Muell. ex Drude
Livistona eastonii C.A.Gardner
Livistona fulva Rodd
Livistona humilis R.Br.
Livistona inermis R.Br.
Livistona lanuginosa Rodd
Livistona lorophylla Becc.
Livistona mariae F.Muell.
Livistona muelleri F.M.Bailey
Livistona nasmophila Dowe & D.L.Jones
Livistona nitida Rodd
Livistona rigida Becc.
Livistona victoriae Rodd

***Licuala* Thunb.**
Licuala ramsayi (F.Muell.) Domin

***Caryota* L.**
Caryota albertii F.Muell. ex H.Wendl.

***Arenga* Labill. ex DC**
Arenga australasica (H.Wendl. & Drude) S.T.Blake ex H.E.Moore
Arenga listeri Becc. ex Oliver
Arenga microcarpa Becc.

***Corypha* L.**
Corypha utan Lam.

***Oraniopsis* (Becc.) J.Dransf., A.K.Irvine & N.W.Uhl**
Oraniopsis appendiculata (F.M.Bailey) J.Dransf., A.K. Irvine & N.W.Uhl

***Cocos* L.**
Cocos nucifera L.

***Archontophoenix* H.Wendl. & Drude**
Archontophoenix alexandrae (F.Muell.) H.Wendl. & Drude
Archontophoenix cunninghamiana (H.Wendl.) H.Wendl. & Drude
Archontophoenix maxima Dowe
Archontophoenix myolensis Dowe
Archontophoenix purpurea Hodel & Dowe
Archontophoenix tuckeri Dowe

***Lepidorrhachis* (H.Wendl. & Drude) O.F.Cook**
Lepidorrhachis mooreana (F.Muell.) O.F.Cook

***Linospadix* H.Wendl.**
Linospadix apetiolatus Dowe & A.K.Irvine
Linospadix microcaryus (Domin) Burret
Linospadix minor (W.Hill) Burret
Linospadix monostachyos (Mart.) H.Wendl.
Linospadix palmerianus (F.M.Bailey) Burret

***Laccospadix* H.Wendl. & Drude**
Laccospadix australasicus H.Wendl. & Drude

***Howea* Becc.**
Howea belmoreana (C.Moore & F.Muell.) Becc.
Howea forsteriana (C.Moore & F.Muell.) Becc.

***Ptychosperma* Labill.**
Ptychosperma elegans (R.Br.) Blume
Ptychosperma macarthurii (H.Wendl, ex H.J.Veitch) H.Wendl. ex Hook.f.

***Carpentaria* Becc.**
Carpentaria acuminata (H.Wendl. & Drude) Becc.

***Wodyetia* A.K.Irvine**
Wodyetia bifurcata A.K.Irvine

***Normanbya* F.Muell. ex Becc.**
Normanbya normanbyi (W.Hill) L.H.Bailey

***Rhopalostylis* H.Wendl. & Drude**
Rhopalostylis baueri (Hook.f.) H.Wendl. & Drude

***Hedyscepe* H.Wendl. & Drude**
Hedyscepe canterburyana (C.Moore & F.Muell.) H.Wendl. & Drude

***Hydriastele* H.Wendl. & Drude**
Hydriastele costata F.M.Bailey
Hydriastele ramsayi (Becc.) W.J.Baker & Loo
Hydriastele wendlandiana (F.Muell.) H.Wendl. & Drude

Glossary

abaxial (abaxially) side of an organ that faces away from the axis bearing it

acaulescent lacking an above-ground stem

acropetal maturation from a lower to a higher level on an axis in flowering

acuminate gradually tapered to a point

acute distinctly and sharply pointed

adaptation process of evolutionary modification that results in improved survival and reproduction efficiency

adaptive radiation diversification of a species or single ancestral type into several forms, each adaptively specialised to a specific environmental niche

adaxial (adaxially) side of an organ that faces the axis bearing it

adjacent-ligular with the new shoot developing next to the side of the seed and enclosed in a ligule in seed germination

adnate with one organ joined to another

allopatric of populations occurring in separate, non-overlapping geographic areas and unable to cross-breed

anastomosing forming a network

anatropous bent parallel to its stalk with the micropyle adjacent to the hilum in an ovule

anther pollen-bearing chamber of a stamen

anthesis time of flower opening

aperture an opening in a pollen grain through which the pollen tube emerges

apex (apical) end portion, borne at the top

appressed lying flat or pressed closely

arcuate curved like a bow

autecology (autecological) the branch of ecology that deals with the biological relationship between an individual organism or individual species and its environment

autochthonous a species having evolved *in situ*

axil (axillary) upper angle of the junction between the petiole and stem

basal at the base of an organ or structure

basifixed attached at the base

basionym the original or first validly described name for a species

basipetal maturation from a higher to a lower level on an axis, characteristic of some species in the tribe Caryoteae

bicarinate with two keels

bicentric (bicentricity) having two centres

bifid divided into two ± equal parts

bifurcate forked

biogeography study of the geographic distribution of organisms

bipinnate doubly pinnate, a characteristic of the leaves of species of *Caryota*

bract a modified enclosing leaf

bracteole a small bract on the flower stalk

caducous falling away early

campanulate of three-dimensional shape, bell-shaped

carinate keeled

carpel a single unit of the gynoecium

Cenozoic geological era from the end of the Cretaceous to the present day, from 65 million years ago

chartaceous papery

ciliate fringed with hairs

cincinnus arranged with each successive flower arising in the axil of a bracteole on the stalk of the previous flower

cirrus (cirrate) a whip-like extension of the leaf rachis armed with reflexed spines used as a climbing organ, characteristic of rattans in the subfamily Calamoideae

clade a monophyletic group or branch of a phylogenetic tree

clustered (clustering) basally composed of two to many stems

colliculose (colliculate) covered with small rounded projections

concolorous with each surface of the same colour

conic (conical) having the form of a cone

connate joined

connective tissue joining the locules of one anther to another

coriaceous leathery

costapalmate describing a palmate leaf with a strong rib (costa) extending through the lamina, characteristic of species of *Livistona*

Cretaceous geological period between the Jurassic and the Tertiary, from 144 to 66.4 million years ago

crownshaft cylinder formed by the leafbases, characteristic of most species in the subfamily Arecoideae

crustaceous dry and brittle

cucullate hooded or hood-shaped

cuneate in two-dimensional shape, wedge-shaped and widest distally

cupule (cupular) resembling a cup

decumbent a reclining stem with the apex ascending
deflexed bent downwards
demography the study of populations
dentate (dentiform) toothed or tooth-like in appearance
dichotomous equally forking
dimorphic occurring in two different forms
dioecious (dioecy) when plants bear either staminate or pistillate flowers
discolorous with each surface of a different colour
distal farthest from the point of attachment
distichous arranged in two rows
disulcate with two apertures in pollen
divaricate spreading widely
dyad in pairs
echinate prickly or with low broad spines
ecological amplitude the limits of environmental conditions within which an organism can live and function
edaphic of or relating to soil, especially as it affects living organisms
ellipsoid(al) in three-dimensional shape, elliptical in reach rotational plane
elliptic(al) in two-dimensional shape, oblong with regularly rounded ends
embryo the rudimentary plant within a seed
endemic restricted in natural distribution to a specific geographic region
endocarp the innermost layer of the fruit wall
endosperm nutrient source for the developing embryo in a seed
Eocene subdivision of the Tertiary period, from 58 to 36 million years ago
eophyll the first functional leaf of a seedling
epicarp the outermost layer of the fruit wall
epipetalous united with or borne on petals
excrescence an outgrowth of hardened wart-like tissue
exine the outer layer of pollen
exserted projected beyond
extant presently in existence, not lost or extinct
falcate sickle-shaped
filament the stalk bearing the anther
fimbriate fringed with fine hairs or thread-like segments
flabellate fan-shaped
flagellum a whip-like climbing organ armed with reflexed spines, adnate to or emergent from the leafsheath, and characteristic of rattans in the subfamily Calamoideae
flexuous wavy, bent alternately in opposite directions
fossulate a furrowed or minutely trenched surface
foveolate a minutely pitted surface
free organs that are not fused
frugivore (frugivorous) an animal that feeds primarily on fruit
fusiform in three-dimensional shape, spindle-shaped, tapering at each end
gibbous pouched, more convex in one place than another
glabrous a smooth surface lacking indumentum
glaucous a surface reflecting a white, pale-blue or greyish sheen, usually by wax or finely divided particles
globose in three-dimensional shape, spherical
Gondwana southern supercontinent which comprised the now-separated continents of South America, Africa, Antarctica and Australia, as well as the subcontinent of India and the smaller landmasses of Madagascar, New Zealand, parts of Arabia and probably parts of south-east Asia
granular a surface appearing grainy
grapnel series of single or clustered recurved hooks associated with the abaxial surface of the leaf rachis, cirrus, flagellum or inflorescences of rattans in the subfamily Calamoideae, used as a climbing or support aid
gynoecium the ovule-bearing organ consisting of one to many carpels, each with an ovary, a style and stigmas
habitat the kind of environment in which a plant lives
hapaxanthic the process of an individual plant (or stem of a cluster) flowering once then dying, characteristic of species in the tribe Caryoteae
hastula a flap of tissue borne at the insertion of the lamina on the petiole of a palmate leaf
head compact arrangement of sessile flowers and fruits, a capitulum
hemianatropous the ovule turned so that the micropyle is at right angles to its stalk
hermaphrodite (hermaphroditic) bisexual, male and female together in one flower
hilum the attachment scar on a seed
holotype a single specimen or illustration designated by the author of a plant name at the time of original publication, which fixes the application of the name
homogeneous uniform and similar throughout
hyaline translucent, thin
imbricate overlapping
imparipinnate a pinnate leaf with a terminal pinna
indumentum a covering of scales or hairs
induplicate V-shaped in cross-section
inflexed bent inwards
inflorescence collective term for the aggregation of flowers
infrafoliar borne below the leaves
infructescence the grouping in which fruits are borne on a plant
interfoliar borne among the leaves
internode portion between two successive leafscars on a stem
introrse opening toward the centre of the flower, of anthers
isolectotype duplicate of a lectotype
isotype a specimen that is a duplicate of the holotype
keel(ed) a ridge, resembling a boat keel
Kimberley Region that part of Western Australia lying north of 19°S
knee a swelling at the junction of the leafsheath and the petiole, characteristic of rattans in the subfamily Calamoideae
lacerate jagged, appearing torn or irregularly cleft

lageniform in three-dimensional shape, swollen at the base and narrow at the apex
lamina the blade of a leaf
lanceolate narrow and tapering at both ends
lanuginose a surface with a dense cover of long, soft woolly hairs
lateral on or at the side
laterite(ic) a hard, iron-rich, rock-like soil layer formed by heavy leaching of soil under markedly seasonal tropical conditions
latrorse opening lateral to the filament in anthers
leafbase portion of leaf proximal to the petiole, narrowly or broadly clasping the stem as typical of many Coryphoideae genera, or forming a closed tubular crownshaft as in many Arecoideae genera
leafscar the remnants or impression of the former attachment of a leaf on a stem
lectotype a specimen that serves in place of a lost or unplaced holotype
lepidote a surface covered in scales
ligule a usually fibrous appendage associated with the leafbase
linear long and narrow with ± parallel sides
lobe projection or division of a leaf or other organ
locule the cavity that contains the ovary in the gynoecium
macrofossil a fossil large enough to be examined by direct inspection
Malesia biogeographical province extending from Malaysia east to New Guinea
marine transgression when an influx of the sea covers areas of previously exposed land
medifixed attached at the middle
megathermal high-temperature as in humid subtropical or tropical climates, with the coldest monthly mean temperature above 18°C
membranous thin and translucent
meristem undifferentiated plant tissue from which new cells are formed
mesic being neither wet nor arid
mesocarp middle layer of the fruit wall
mesothermal moderate temperatures in the range 20–30°C
microfossil a fossil which requires magnification for it to be examined, usually by microscope
micropyle a minute opening in the ovule of a seed plant through which the pollen tube usually enters
midrib the main vein of a pinna or segment
Miocene a division of the Tertiary, from 23 to 5 million years ago
molecular phylogenetics a process of determining the rates and patterns of change occurring in DNA and proteins in order to reconstruct the evolutionary history of genes and organisms
monocarpic flowering and fruiting once, then dying
monoecious plant bearing both staminate and pistillate flowers
monogeneric with one genus
monophyly (monophyletic) a branch of a phylogenetic tree that includes all descendants of a common ancestor
monosulcate pollen with a single aperture (sulcus)
monotypic with one species
monsoon(al) a wind regime that brings heavy summer rainfall to northern Australia
mucilage (inous) gelatinous-like substance
neotype a specimen selected in place of a holotype in the absence of original material
obovoid in three-dimensional shape, egg-shaped and broader distally
obpyriform in three-dimensional shape, pear-shaped and attached at the narrower end
Oligocene a division of the Tertiary, from 36.6 to 23.7 million years ago
ovate in two-dimensional shape, egg-shaped, if attached by the broad end
ovoid in three-dimensional shape, egg-shaped
ovule structure in which the female cycle of reproduction occurs in flowers
Palaeocene division of the Tertiary, from 66.4 to 57.8 million years ago
palmate with segments radiating from a central point resembling a fan in a leaf
panicle (paniculate) inflorescence with iterative branching patterns (racemes) from a central axis
pappilate (pappilose) a surface with minute, rounded elongated projections
paripinnate a pinnate leaf without a terminal pinna
partial inflorescence the first-order branch and the branches it carries, characteristic of some inflorescence types in the subfamilies Calamoideae and Coryphoideae
pedicel the stalk of each individual flower or fruit
peduncle the proximal unbranched portion of an inflorescence
peduncular bract(s) a single bract or multiple bracts on the main axis of the inflorescence, attached between the prophyll and rachis bracts
pendulous hanging pendant
perianth the sepals and petals together
persistent parts not shed after flowering
petal a unit of the corolla
petiole (petiolate) the stalk between attachment and the lowest pinnae or leaf segments on a leaf
phylogeny (phylogenetic) evolutionary relationships between groups of living things, inferred using a variety of techniques to establish the relative importance of various shared features
phytogeography (phytogeographical) study of the geographic distribution of plants
phytoliths microscopic silica-based structures formed within and between the living cells of most monocotyledons and some dicotyledons, and often preserved as microfossils
Pilbara a region of Western Australia bounded by the Tropic of Capricorn and 19°N, extending from the coast almost to the Northern Territory border

pinna(e) segment(s) of a pinnate leaf
pinnate pinnae arranged in a feather-like arrangement on a leaf, i.e. with similar number of pinnae on each side of the rachis
pinnule the pinnae of a bipinnate leaf, characteristic of species of *Caryota*
pistillate bearing a pistil
pistillode a sterile gynoecium
plate tectonics theory that explains the global distribution of geological phenomena such as seismicity, volcanism, continental drift and mountain-building in terms of the formation, destruction, movement and interaction of the earth's lithospheric plates
Pleistocene division of the Quaternary, from 1.6 million years ago to the last 10 000 years
pleonanthic the process of an individual plant flowering repeatedly
Pliocene division of the Tertiary, from 5.3 to 1.6 million years ago
plumose segments feathered or plume-like in a leaf
pollen a mature microspore and carrier of genetic information in a staminate flower
polygamodioecious having bisexual and male flowers on some plants and bisexual and female flowers on others
polygamomonoecious having bisexual and unisexual flowers on the same plant
polygamous having bisexual and unisexual flowers on the same or different plants
praemorse with the apex appearing torn-off or bitten, characteristic of the pinnae apices in subtribe Ptychospermatinae
pre-Linnean informal plant names prior to 1753 and the commencement of the Linnean classification system
pro parte [Latin] in part
propagule any structure that can give rise to a new plant, either vegetative portions, seeds or spores
prophyll the first bract borne on the inflorescence
protandrous when pollen shed precedes stigma receptivity on a single inflorescence
protogynous when stigma receptivity precedes pollen shed on a single inflorescence
proximal nearest to the point of attachment
pseudomonomerous a gynoecium with one fertile carpel and one locule but with parts of two other carpels present
punctiform a surface with small dots
pyriform in three-dimensional shape, resembling a pear, attached at the broader end
Quaternary the last 1.6 million years of the geological scale
rachilla(e) the flower-bearing branch of an inflorescence
rachis the axis beyond the petiole or beyond the peduncle of the leaf or of an inflorescence
radiation the process of dispersal of organisms across a landscape
ramenta variously shaped ragged-edged scales, large and irregular, sometimes present on the midrib and veins on the abaxial surface of pinnae
raphe a ridge or depression that is the source of vascular branches on a seed
rattan a climbing palm of the subfamily Calamoideae
reduplicate inverse V-shaped in cross-section
reflexed folded or turned back
relictual species species that are persistent examples of floras now mainly vanished
remote-ligular the young plant joined to the seed by a petiolar cotyledonary tube and bearing a ligule in seed germination
remote-tubular the young plant joined to the seed by a petiolar cotyledonary tube but lacking a ligule in seed germination
reniform in two-dimensional shape, kidney-shaped
reticulate large rounded or angular holes through the tectum of pollen
retrorse bent backward or downward
rhombic in two-dimensional shape, quadrilateral with obtuse lateral angles and acute basal and apical angles
rugose with a wrinkled surface
ruminate appearing streaked due to intrusion of the seed coat into the endosperm
sagittate arrow-shaped
sarcotesta(al) a fleshy layer developed from the outer seed coat, characteristic of most fruit in the subfamily Calamoideae
scabrate a roughened surface
sclerophyllous vegetation characterised by thick hard foliage due to well-developed sclerenchymatous tissue
segment a single division of a palmate or costapalmate leaf
sepal a unit of the calyx
sessile flat against, lacking a stalk
sister the closest relative in a phylogenetic tree
solitary with a single stem, not basally clustered
speciation the process of the evolutionary formation of new biological species, usually by the division of a single species into two or more genetically distinct species
spicate an unbranched inflorescence
stamen the stalk bearing the anther in a flower
staminate a flower that bears stamens
staminode vestigial stamens, often reduced to tooth-like flaps of tissue
sterile lacking functional sex organs
stigma (stigmatic) the portion of the pistil that receives the pollen
stigmatic remains the usually dried remnants of the stigma and perianth on the surface of the fruit
striate a lined surface
style the part of the pistil connecting the ovary and the stigma
subdistichous arranged in two rows but with items not perfectly aligned

subulate in two-dimensional shape, narrowing abruptly, awl-shaped
sulcus (sulci) an elongate aperture on one of the axes of a pollen grain
suprafoliar borne above the leaves
sympatric related species that occur in the same geographical range
sympodial changing direction by frequent replacement of the growing apex with a lateral growing point below it
systematics the process of organising biological information in a taxonomic or phylogenetic manner
taphonomy (taphonimic potential) study of the conditions and process by which organisms become fossilised
taxonomy the science of organising and classifying living things into groups reflecting their natural phylogenetic relationships, and naming them
tectate (tectum) the uppermost layer of the exine of pollen
terete cylindrical, circular in transverse section
Tertiary geological period from 66.4 to 1.6 million years ago
tomentum (tomentose) a surface with a cover of short hairs, scales, wool or down
Top End part of the Northern Territory north of about Mataranka
triad in threes, flowers grouped with a central pistillate and two lateral staminate flowers, characteristic of most species in the subfamily Arecoideae
trichotomosulcate having a three-armed sulcus in pollen
trilocular with three chambers, each bearing an ovary or seed in the gynoecium
triovulate bearing three ovules in the gynoecium, one in the locule of each carpel
trullate in two-dimensional shape, broadest below the middle, rhombic, with the two equal lower sides shorter than the two equal upper sides, shaped like a bricklayer's trowel
truncate as though cut off at the end
turbinate in three-dimensional shape, obconical, top-shaped
type specimen a specimen upon which the name was established and to which it is bound
ultramafic type of rock that is rich in heavy metals and that usually supports a distinctive flora
unisexual with one functional sexual role, i.e. with pistils or stamens only
vagile (vagility) able to move about or disperse freely in a given environment
valvate margins of segments meeting neatly without overlapping
ventricose swollen or inflated
verrucate (verrucose) a wart-like element of the exine of pollen
versatile when the anther appears hinged and is movable on the filament
vestigial diminished in size, structure or function
vicariance the process whereby an ancestral species splits as a result of the imposition of a barrier within the original population
vicariant species closely related and ecologically equivalent species that tend to be mutually exclusive, occupying disjunct geographical areas
Wallace's Line the boundary that marks unrelated biological realms east and west of a line drawn approximately through central Malesia. Thomas Huxley coined the term, based on the work of Alfred R. Wallace
zonasulcate with a ring-like meridional or equatorial aperture in pollen

Abbreviations

µm	micron (one-millionth of a metre)
AFLP	amplified fragment length polymorphism
alt.	altitude
Apr.	April
asl	above sea-level
auct. non.	(Latin) *auctorum non*, not of the author; used to show that the preceding name has been misapplied by some authors and is not used in the sense of the original author
Aug.	August
b.	borne
bp	before present
c.	(Latin) *circa*, about, approximately
Ck	creek
cm	centimetres
cpDNA	chloroplast DNA (deoxyribonucleic acid)
d.	died
dbh	diameter at breast height
Dec.	December
diam.	diameter
e.g.	(Latin) *exempli gratia*, for example
et. al.	(Latin) *et alii*, and others
f.	(Latin) *filius*, son of
Feb.	February
Fig.(s)	figure(s)
ft	feet
hort.	(Latin) *hortorum*, of gardens
hwy	highway
i.e.	(Latin) *id est*, that is
ICBN	International Code of Botanical Nomenclature
in litt.	(Latin) *in litteris*, in correspondence
ined.	(Latin) *ineditus*, unpublished
Is.	Island
Jan.	January
Jul.	July
Jun.	June
k	one thousand
km	kilometres
m	metres
Mar.	March
mm	millimetres (one-thousandth of a metre)
mss	manuscript
Mt	Mount, Mountain
My	million years
Mya	million years ago
N	north
NE	north-east
n.s.	new series
nom. alt.	(Latin) *nomen alternativum*, alternate name
nom. con.	(Latin) *nomen conservandum*, conserved name
nom. illeg.	(Latin) *nomen illegitimum*, illegitimate name under the rules of the International Code of Botanical Nomenclature
nom. inval.	(Latin) *nomen invalidum*, name not validly published
nom. nud.	(Latin) *nomen nudum*, name published without a description
nom. prov.	(Latin) *nomen provisorium*, provisional name
nom. rej.	(Latin) *nomen rejiciendum*, rejected name
Nov.	November
Oct.	October
orth. var.	orthographic variation
p.	page
pers. comm.	personal communication
pro syn.	(Latin) *pro synonyma*, an invalid name originally proposed as a synonym of an accepted name
R.	River
rd	road
s. lat.	(Latin) *sensu lato*, in a broad sense
s.n.	(Latin) *sine numero*, without number
SE	south-east
Sept.	September
ser.	series
sp.	species (singular)
spp.	species (plural)
sub.	(Latin) *subter*, under
subsp.	subspecies
tab.	tabula (plate)
var.	variety

Herbarium acronyms

A Arnold Arboretum, Harvard University, USA
AAU Aarhus University, Department of Systematic Botany, Denmark
B Berlin, Botanischer Garten und Botanisches Museum, Germany
BH Bailey Hortorium, Cornell University, USA
BM [British Museum] Natural History Museum, London, UK
BO Bogor, Herbarium Bogoriense, Indonesia
BRI Brisbane, Queensland Herbarium, Australia
CANB Australian National Botanic Gardens, Canberra, Australia
CBG Canberra, Australian National Herbarium, Australia
CNS Cairns, Australian Tropical Herbarium, James Cook University, Australia
DNA Darwin, Northern Territory Herbarium, Australia
FI Herbarium Universitatis Florentinae, Museo di Storia Naturale dell'Università, Italy
FTG Fairchild Tropical Botanic Gardens, Miami, Florida, USA
GOET Universität Göttingen, Abteilung Systematische Botanik, Germany
JCT James Cook University of North Queensland, Townsville, Australia
K Royal Botanic Gardens Kew, UK
L Leiden, Rijksherbarium, Netherlands
LAE Lae, Papua New Guinea Forest Research Institute, Papua New Guinea
LZ Universität Leipzig, Institut für Botanik, Germany
M Botanische Staatssamlung München, Germany
MEL Melbourne, National Herbarium of Victoria, Australia
MO Missouri Botanical Garden, Saint Louis, USA
NOU Nouméa, Institut de Recherche pour le Développement, New Caledonia
NSW National Herbarium of New South Wales, Sydney, Australia
NY New York Botanical Garden, USA
PERTH Perth, Western Australian Herbarium, Australia
PR National Museum in Prague, Department of Botany, Czech Republic
PVNH Port Vila, ORSTOM, Vanuatu
U Nationaal Herbarium Nederland, Leiden, Netherlands
UC University of California, Berkeley, USA
US United States National Herbarium, Washington, USA

References

Albert RM, Bamford MK and Cabanes D (2009) Palaeoecological significance of palms at Olduvai Gorge, Tanzania, based on phytolith remains. *Quaternary International* **193**, 41–48.

Alley NF and Clarke JDA (1992) Stratigraphy and palynology of Mesozoic sediments from the Great Australian Bight area, southern Australia. *BMR Journal of Australian Geology and Geophysics* **13**, 113–129.

Anon (1877) *Kentia Mac Arthuri*: Nouvelle-Caledonie. *Belgique Horticole* **17**, 241.

Asmussen CB and Chase MW (2001) Coding and noncoding plastid DNA in palm systematics. *American Journal of Botany* **88**, 1103–1117.

Asmussen CB, Baker WJ and Dransfield J (2000) Phylogeny of the palm family (Arecaceae) based on *rps*16 intron and *trn*L-*trn*F plastid DNA sequences. In *Monocots: Systematics and Evolution*. (Eds KL Wilson and DA Morrison) pp. 525–537. CSIRO Publishing, Melbourne.

Asmussen CB, Dransfield J, Deickmann V, Barfod AS, Pintaud J-C and Baker WJ (2006) A new subfamily classification of the palm family (Arecaceae): evidence from plastid DNA phylogeny. *Botanical Journal of the Linnean Society* **151**, 15–38.

Atkins KJ (2008) *Declared Rare and Priority Flora List for Western Australia, 26 February 2008*. Department of Environment and Conservation. Perth, Western Australia.

Bachman S, Baker WJ, Brummitt N, Dransfield J and Moat J (2004) Elevational gradients, area and tropical island diversity: an example from the palms of New Guinea. *Ecography* **27**, 299–310.

Bailey FM (1889) Botany of the Bellenden-Ker Expedition, by Fredk. Manson Bailey, Colonial Botanist. In *Report by A Meston on the Government Scientific Expedition to the Bellenden-Ker Range (Wooroonooran), North Queensland*. (Ed. A Meston), *Queensland, Votes and Proceedings of the Legislative Assembly* Vol. 4, pp. 14–29 (repaginated as 1218–1233 in bound copy). James C Beal, Government Printer, Brisbane.

Bailey FM (1890) Palmae. In *A Synopsis of the Queensland Flora, Third Supplement*, pp. 77–79. James C Beal, Government Printer, Brisbane.

Bailey FM (1891) Final supplement to the report of the botany of the Bellenden-Ker Expedition. In *Annual Report of the Department of Agriculture for the year 1890–1891. Queensland, Votes and Proceedings of the Legislative Assembly* Vol. 4, pp. 21–22 (repaginated as 639–640 in bound copy). James C Beal, Government Printer, Brisbane.

Bailey FM (1896) Contributions to the Queensland Flora. Order Palmae. *Queensland Botany Bulletin* **13**, 13–14.

Bailey FM (1897) Contributions to the Queensland Flora. Order Palmae. *Queensland Agricultural Journal* **1**, 228–233.

Bailey FM (1898) Contributions to the Queensland Flora. Order Palmae. *Queensland Agricultural Journal* **2**, 129–131.

Bailey FM (1902) Palmae. *Queensland Flora* **5**, 1670–1688. Government Printer, Brisbane.

Bailey FM (1909) Contributions to the Flora of Queensland. Order Palmae. *Queensland Agricultural Journal* **23**, 35.

Bailey FM (1912) *Comprehensive Catalogue of Queensland Plants*. Government Printer, Brisbane.

Baillon H (1895) *Histoire des Plantes*. Part 3. Librairie Hachette and Cie, Paris.

Baker WJ and Dransfield J (2000) Towards a biogeographic explanation of the Calamoid palms. In *Monocots: Systematics and Evolution*. (Eds KL Wilson and DA Morrison) pp. 545–553. CSIRO Publishing, Melbourne.

Baker WJ and Dransfield J (2006) *Field Guide to the Palms of New Guinea*. Kew Publishing, Royal Botanic Gardens, Kew.

Baker WJ and Hutton I (2006) *Lepidorrhachis*. *Palms* **50**, 33–38.

Baker WJ, Coode MJE, Dransfield J, Dransfield S, Harley MM, Hoffmann P and Johns RJ (1998) Patterns of distribution of Malesian vascular plants. In *Biogeography and Geological Evolution of SE Asia*. (Eds R Hall and JD Holloway) pp. 243–258. Backhuys Publishers, Leiden.

Baker WJ, Dransfield J and Hedderson TA (2000a) Phylogeny, character evolution, and a new classification of the Calamoid palms. *Systematic Botany* **25**, 297–322.

Baker WJ, Hedderson TA and Dransfield J (2000b) Molecular phylogenetics of *Calamus* (Palmae) and related rattan genera based on 5S nrDNA spacer sequence data. *Molecular Phylogenetics and Evolution* **14**, 218–231.

Baker WJ, Bayton RP, Dransfield J and Maturbongs RA (2003) A revision of the *Calamus aruensis* (Arecaceae) complex in New Guinea and the Pacific. *Kew Bulletin* **58**, 351–370.

Ballance PF, Gregory MR and Gibson GW (1981) Coconuts in Miocene turbidites in New Zealand: possible evidence for tsunami origin of some turbidity currents. *Geology* **9**, 592–595.

Banks J (1962) *The Endeavour Journal of Joseph Banks 1768–1771: Edited by J.C. Beaglehole*. Public Library of New South Wales/Angus and Robertson, Sydney.

Barfod AS and Dowe JL (2005) The enigmatic Australian fan-palm *Licuala ramsayi*. *Palms* **49**, 15–23.

Barfod AS, Banka R and Dowe JL (2001) Field guide to palms in Papua New Guinea. *AAU Reports* **40**, 1–79.

Barham AJ (1999) The local environmental impact of prehistoric populations on Saibai Island, northern Torres Strait, Australia: enigmatic evidence of Holocene swamp lithostratigraphic records. *Quaternary International* **59**, 71–105.

Barker NP, Weston PH, Rutschmann F and Sauquet H (2007) Molecular dating of the 'Gondwanan' plant family Proteaceae is only partially congruent with the timing of the break-up of Gondwana. *Journal of Biogeography* **34**, 2012–2027.

Barlow BA (1981) The Australian flora: its origin and evolution. *Flora of Australia* **1**, 25–75.

Bauhin J (1650–51) *Palma indica nucifera*. *Historia Plantarum Universalis* **1**, 375–376. Ebroduni.

Bayton RP (2007) A revision of *Borassus* (Arecaceae). *Kew Bulletin* **62**, 561–586.

Beccari O (1877) Le specie di palme raccolte alla Nuova Guinea do O Beccari e dal medesimo adesso desritte, con note suule species dei paesi circonvicini. *Malesia* **1**, 9–102.

Beccari O (1885) Reliquiae Schefferianae. *Annales du Jardin Botanique Buitenzorg* **2**, 77–171.

Beccari O (1886) Nuovi studi sulle palme asiatiche. *Malesia* **3**, 58–149.

Beccari O (1889) Palmae. In *Die Flora von Kaiser Wilhelms Land*. (Eds K Schumann and M Hollrung) pp.15–17. Asher and Co., Berlin.

Beccari O (1908) Asiatic palms – Lepidocaryeae. Part I. The species of *Calamus*. *Annals of the Royal Botanic Garden, Calcutta* **11**, 1–518.

Beccari O (1910) Palme Australasiche Nuove o Poco Note. *Webbia* **3**, 131–165.

Beccari O (1921) Recensione delle Palme del vecchio mondo appartenenti alla tribu delle Corypheae, con descrizione della specie e varieta nuove che vi appartengono. *Webbia* **5**, 5–70.

Beccari O (1931) Asiatic palms – Corypheae. *Annals of the Royal Botanic Garden, Calcutta* **13**, 1–356.

Beccari O and Pichi-Sermolli REG (1955) Subfamiliae Arecoidearum palmae gerontogeae tribuum et generum conspectus. *Webbia* **11**, 1–187.

Bellingham PJ (2008) Cyclone effects on Australian rain forests: an overview. *Austral Ecology* **33**, 580–584.

Bentham G (1878) Palmae. *Flora Australiensis* **7**, 132–147.

Bentham G and Hooker JD (1883) *Genera Plantarum: ad exemplaria imprimis in herbariis Kewensibus servata definita*. L Reeve and Co., London.

Berry EW (1914) The Upper Cretaceous and Eocene floras of South Carolina, Georgia. *US Geological Survey, Professional Paper* **84**, 1–200.

Berry EW (1926) *Cocos* and *Phymatocaryon* in the Pliocene of New Zealand. *American Journal of Science, Series 5* **12**(69), 181–184.

Bjorholm S, Svenning J-C, Skov F and Balslev H (2005) Environmental and spatial controls of palm (Arecaceae) species richness across the Americas. *Global Ecology and Biogeography* **14**, 423–429.

Bjorholm S, Svenning J-C, Baker WJ, Skov F and Balslev H (2006) Historical legacies in the geographical diversity patterns of New World palm (Arecaceae) subfamilies. *Botanical Journal of the Linnean Society* **151**, 113–125.

Blackburn DT and Sluiter IRK (1994) The Oligo-Miocene coal floras of southeastern Australia. In *History of the Australian Vegetation: Cretaceous to Recent*. (Ed. RS Hill) pp. 328–367. Cambridge University Press, Cambridge, UK.

Blake ST (1954) Botanical contributions of the Northern Australian Regional Survey. *Australian Journal of Botany* **2**, 99–140.

Blüthgen N, Gottsberger G and Fiedler K (2004) Sugar and amino acid composition of ant-attended nectar and honeydew sources from an Australian rainforest. *Austral Ecology* **29**, 418–429.

Borchsenius F and Skov F (1997) Ecological amplitudes of Ecuadorian palms. *Principes* **41**, 179–183.

Bostock PD and Holland AE (Eds) (2007) *Census of the Queensland Flora 2007*. Queensland Herbarium, Environmental Protection Agency, Brisbane.

Bowman DMJS (2000) *Australian Rainforests: Islands of Green in a Land of Fire*. Cambridge University Press, Cambridge, UK.

Brennan K (1997) The impact of wet season burning on herbaceous plants in savanna woodlands in the Jabiru area. In *Proceedings of Bushfire '97*. (Eds B McKaige and RJ Williams) p. 24. CSIRO Tropical Ecosystems Research Centre, Winnellie, Northern Territory.

Bridson GDR and Smith ER (Eds) (1991) *B-P-H, Botanico-Periodicum-Huntianum Supplement*. Hunt Institute for Botanical Documentation, Carnegie Mellon University, Pittsburgh.

Brierley OW (1848) *Journal on HMS Rattlesnake*. Unpublished manuscript ZA505. Mitchell Library, Sydney.

Brown R (1810) Palmae. *Prodromus florae Novae Hollandiae et Insulae van Diemen* **1**, 266–268. Taylor, London.

Brown R (1814) General remarks, geographical and systematical, on the botany of *Terra Australis*. In *A Voyage to Terra Australis; Undertaken for the Purpose of Completing the*

Discovery of that Vast Country, and Prosecuted in the Years 1801, 1802, and 1803, in His Majesty's Ship the Investigator. (Ed. M Flinders) Vol. 2, pp. 533–594. G and W Nicol, London.

Brummitt RK and Powell CE (1992) *Authors of Plant Names.* Royal Botanic Gardens, Kew.

Buckley R and Harries HC (1984) Self-sown wild-type coconuts from Australia. *Biotropica* **16**, 148–151.

Bull W (1887) *Catalogues of New Beautiful and Rare Plants.* London.

Burbidge NT (1960) The phytogeography of the Australian region. *Australian Journal of Botany* **8**, 75–212.

Burret M (1928) Beiträge zur Kenntnis der Palmen von Malesia, Papua und der Südsee. *Repertorium Specierum Novarum Regni Vegetabili* **24**, 253–296.

Burret M (1935) Neue palmen aus Neuguinea II. *Notizblatt des Botanischen Gartens und Museums zu Berlin-Dahlem* **12**, 309–348.

Burret M (1936) Palmae chinenses. *Notizblatt des Botanischen Gartens und Museums zu Berlin-Dahlem* **13**, 582–606.

Burret M (1939) Palmae gesammelt in Neu Guinea von LJ Brass. *Journal of the Arnold Arboretum* **20**, 187–212.

Burret M (1943) Neue Palmen aus der Gruppe der Lepidocaryoideae II. *Notizblatt des Botanischen Gartens und Museums zu Berlin-Dahlem* **15**, 796–805.

Burrett C, Duhig N, Berry R and Varne R (1991) Asian and south-western Pacific continental terranes derived from Gondwana, and their biogeographical significance. *Australian Systematic Botany* **4**, 13–24.

Calvert GA, Lokkers C and Cumming R (2005) *Rare and Threatened Plants of the Townsville–Thuringowa Region.* Coastal Dry Tropics Landcare Inc., Townsville.

Cameron DG (1987) Temperate rainforests of East Gippsland. In *The Rainforest Legacy: Australian National Rainforests Study. Vol. 1: The Nature, Distribution and Status of Rainforests Types.* pp. 33–46. Australian Government Publishing Service, Canberra.

Carlile N and Priddel D (2009) Mortality of adult *Livistona australis* on Cabbage Tree Island, Australia. *Palms* **53**, 46–50.

Chapman AD (1990) Domin and Danes in Java and Australia 1909–1910. In *History of Systematic Botany in Australasia.* (Ed. PS Short) pp. 159–163. Australian Systematic Botany Society, Melbourne.

Chapman AD (1991) *Australian Plant Name Index.* Australian Flora and Fauna Series 11. Australian Government Printing Service, Canberra.

Chase MW, Soltis DE, Soltis PS, Rudall PJ, Fay MF, Hahn WJ, Sullivan S, Joseph J, Molvray M, Kores PJ, Givnish TJ, Sytsma KJ and Pires JC (2000) Higher-level systematics of the monocotyledons: an assessment of current knowledge and a new classification. In *Monocots: Systematics and Evolution.* (Eds KL Wilson and DA Morrison) pp. 3–16. CSIRO Publishing, Melbourne.

Chivas AR, García A, van der Kaars S, Couapel MJJ, Holt S, Reeves JM, Wheeler DJ, Switzer AD, Murray-Wallace CV, Banerjee D, Price DM, Wang SX, Pearson G, Edgar NT, Beaufort L, De Deckker P, Lawson E and Cecil CB (2001) Sea-level and environmental changes since the last interglacial in the Gulf of Carpentaria, Australia: an overview. *Quaternary International* **83–85**, 19–46.

Christophel DC and Greenwood DR (1988) A comparison of Australian tropical rainforests and Tertiary fossil leaf-beds. *Proceedings of the Ecological Society of Australia* **15**, 139–148.

Churchill DM (1973) The ecological significance of tropical mangroves in the Early Tertiary floras of southern Australia. *Geological Society of Australia Special Publication* **4**, 79–86.

Clarke PJ and Kerrigan RA (2000) Do forest gaps influence the population structure and species composition of mangrove stands in northern Australia? *Biotropica* **32**, 642–652.

Clifford HT (1981) Key to families of flowering plants. *Flora of Australia* **1**, 113–168.

Collinson ME (1993) Taphonomy and fruiting biology of recent and fossil *Nypa*. *Special Papers in Palaeontology* **49**, 165–180.

Comeau PL, Comeau YS and Johnson W (2003) *The Palm Book of Trinidad and Tobago including the Lesser Antilles.* International Palm Society, Lawrence, Kansas.

Conn BJ and Walsh NG (1993) Arecaceae. *Flora of Victoria.* (Eds NG Walsh and TJ Entwistle) pp. 165–167. Inkata Press, Melbourne.

Conran JG and Rozefelds AC (2003) A permineralised Oligocene palm trunk from near Springsure, southeastern Queensland. *Alcheringa* **27**, 125–134.

Cookson IC and Eisenack A (1965) Microplankton from the Dartmoor Formation, SW Victoria. *Proceedings of the Royal Society of Victoria* **79**, 133–137.

Cookson IC and Eisenack A (1967) Some Early Tertiary microplankton and pollen grains from a deposit near Strahan, western Tasmania. *Proceedings of the Royal Society of Victoria* **80**, 131–140.

Couper RA (1952) The spore and pollen flora of the *Cocos*-bearing beds, Mangonui, North Auckland. *Transactions of the Royal Society of New Zealand* **79**(3–4), 340–348.

Couper RA (1953) Upper Mesozoic and Cainozoic spores and pollen grains from New Zealand. *New Zealand Geological Survey Palaeontological Bulletin* **22**, 1–77.

Couper RA (1960) New Zealand Mesozoic and Cainozoic plant microfossils. *New Zealand Geological Survey Palaeontological Bulletin* **32**, 1–87.

Covacevich J (1981) Distribution of the *Nypa* palm in Australia. *Principes* **25**, 184–188.

Cowie ID, Short PS and Osterkamp Madsen M (2000) *Floodplain Flora: A Flora of the Coastal Floodplains of the Northern Territory, Australia.* Flora of Australia Supplementary Series No. 10. ABRS, Canberra/PWCNT, Darwin.

Crisp M, Cook L and Steane D (2004) Radiation of the Australian flora: what can comparisons of molecular phylogenies across multiple taxa tell us about the evolution of diversity in present-day communities? *Philosophical Transactions of the Royal Society, London B* **359**, 1551–1571.

Crisp MD, West JG and Linder HP (1999) Biogeography of the terrestrial flora. *Flora of Australia*, 2nd edn, **1**, 321–367.

Crowley GM (1995) *Fire on Cape York Peninsula*. Cape York Peninsula Land Use Strategy, Office of the Co-ordinator General of Queensland, Brisbane/Dept of the Environment, Sport and Territories, Canberra.

Crowley GM and Gagan MK (1995) Holocene evolution of coastal wetlands in wet-tropical northeastern Australia. *The Holocene* **5**, 385–399.

Cuccuini P and Nepi C (2006) The palms of Odoardo Beccari. *Quaderni di Botanica Ambientale e Applicata* **17**, 1–251.

Cunningham A (1827) A few general remarks on the vegetation of certain coasts of *Terra Australis*, and more especially of its north-western shores. In *Narrative of a Survey of the Intertropical and Western Coasts of Australia Performed Between the Years 1818 and 1822*. (Ed. PP King) pp. 497–533. John Murray, London.

Curran TJ, Gersbach LN, Edwards W and Krockenberger AK (2008) Wood density predicts plant damage and vegetative recovery rates caused by cyclone disturbances in tropical rainforest tree species of north Queensland, Australia. *Austral Ecology* **33**, 442–450.

Curry S, Maslin B and Maslin J (2002) *Allan Cunningham: Australian Collecting Localities*. Australian Biological Resources Study, Canberra.

Daghlian CP (1981) A review of the fossil record of monocotyledons. *Botanical Review* **47**, 517–555.

De Deckker P (2001) Late Quaternary cyclic aridity in tropical Australia. *Palaeogeography, Palaeoclimatology, Palaeoecology* **170**, 1–9.

De Lange PJ, Gardner RO, Sykes WR, Crowcroft GM, Cameron EK, Stalker F, Christian ML and Braggins JE (2005) Vascular flora of Norfolk Island: some additions and taxonomic notes. *New Zealand Journal of Botany* **43**, 563–596.

Dettmann ME and Clifford HT (2000) Monocotyledon fruits and seeds, and an associated palynoflora from Eocene-Oligocene sediments of coastal central Queensland, Australia. *Review of Palaeobotany and Palynology* **110**, 141–173.

Dingle H (2004) The Australo–Papuan bird migration system: another consequence of Wallace's Line. *Emu* **104**, 95–108.

Dixon D, Cowie I and Kerrigan R (2003) *Ptychosperma macarthurii* or *P. bleeseri*? The taxonomic status of *P. bleeseri* reconsidered. *The Beagle, Records of the Museums and Art Galleries of the Northern Territory* **19**, 81–86.

Domin K (1915) Beiträge zur flora und pflanzengeographie Australiens. Palmae. *Bibliotheca Botanica* **85**, 498–501.

Doutch HF (1972) The paleogeography of northern Australia and New Guinea and its relevance to the Torres Strait area. In *Bridge and Barrier: The Natural and Cultural History of Torres Strait*. (Ed. D Walker) pp. 1–10. Australian National University, Canberra.

Dowe J (1992) Robert Tucker: a contribution. *Mooreana* **2**(2), 14.

Dowe JL (1995) A preliminary review of the biogeography of Australian palms. *Mooreana* **5**, 7–22.

Dowe JL (1998) Regional priorities in Australia and the Pacific islands. In *Status Survey and Conservation Action Plan: Palms, Their Conservation and Sustained Utilization*. (Ed. D Johnson) pp. 79–85. IUCN/SSC Palm Specialist Group, Gland, Switzerland.

Dowe JL (2001) *Studies in the Genus Livistona (Coryphoideae: Arecaceae)*. PhD thesis. James Cook University, Townsville.

Dowe JL (2004) Taxonomic notes on palms (Arecaceae) in catalogues of the Brisbane Botanic Garden, Australia, of 1875 and 1885. *Austrobaileya* **6**, 967–971.

Dowe JL (2005) Observations of palms made by the botanist and explorer Ludwig Leichhardt, during the Australian Overland Expedition of 1844–1845. *Palms* **49**, 167–182.

Dowe JL (2007a) *Ptychosperma macarthurii*: discovery, horticulture and taxonomy. *Palms* **51**, 85–96.

Dowe JL (2007b) F.M.Bailey's palm botany and his ascent of Mt Bellenden-Ker, Queensland, Australia. *Palms and Cycads* **97**, 12–18.

Dowe JL (2007c) Notes on endangered and vulnerable Queensland palms. *Palms and Cycads* **95**, 8–16.

Dowe JL (2009a) A taxonomic account of *Livistona* R.Br. (Arecaceae). *Gardens Bulletin Singapore* **60**, 185–344.

Dowe JL (2009b) A preliminary note on the impact of Cyclone Larry on populations of *Archontophoenix alexandrae* (Arecaceae), north Queensland, Australia. *Palms and Cycads* **102**, 4–8.

Dowe JL and Barfod AS (2001) New species of *Livistona* R.Br. (Arecaceae) from north Queensland and Papua New Guinea. *Austrobaileya* **6**, 165–174.

Dowe JL and Broughton AD (2007) F.M.Bailey's ascent of Mt Bellenden-Ker in 1889, and notes on the publication priority of new vascular plant species from the expedition. *Austrobaileya* **7**, 555–566.

Dowe JL and Hodel DR (1994) A revision of *Archontophoenix* H Wendl. and Drude (*Arecaceae*) *Austrobaileya* **4**, 227–244.

Dowe JL and Irvine AK (1997) A revision of *Linospadix* in Australia, with the description of a new species. *Principes* **41**, 192–197, 211–217.

Dowe JL and Jones DL (2004) Nomenclatural changes for two Australian species of *Livistona* R.Br. (Arecaceae). *Austrobaileya* **6**, 979–981.

Dowe JL and Jones DL (in prep.) *Arecaceae. Flora of Australia* **39**.

Dowe JL and Smith LT (2002) A brief history of the coconut palm in Australia. *Palms* **46**, 134–138.

Dransfield J (1974) Notes on *Caryota no* Becc. and other Malesian *Caryota* spp. *Principes* **18**, 87–93.

Dransfield J (1978) Growth forms of rainforest palms. In *Tropical Trees as Living Systems*. (Eds PB Tomlinson and MH Zimmermann) pp. 247–268. Cambridge University Press, Cambridge, UK.

Dransfield J (1979) *A Manual of the Rattans of the Malay Peninsula*. Forest Department, West Malaysia.

Dransfield J (1981) Palms and Wallace's Line. In *Wallace's Line and Plate Tectonics*. (Ed. TC Whitmore) pp. 43–56. Oxford University Press, London.

Dransfield J (1984) The genus *Areca* (Palmae; Arecoideae) in Borneo. *Kew Bulletin* **39**, 1–22.

Dransfield J (1987) Bicentric distribution in Malesia as exemplified by palms. In *Biogeographical Evolution of the Malay Archipelago*. (Ed. TC Whitmore) pp. 60–72. Clarendon Press, Oxford.

Dransfield J and Beentje HJ (1995) *The Palms of Madagascar*. Royal Botanic Gardens, Kew/International Palm Society, Lawrence, Kansas.

Dransfield J, Ferguson IK and Uhl NW (1990) The coryphoid palms: patterns of variation and evolution. *Annals of the Missouri Botanical Garden* **77**, 802–815.

Dransfield J, Irvine AK and Uhl NW (1985) *Oraniopsis appendiculata*, a previously misunderstood Queensland palm. *Principes* **29**, 56–63.

Dransfield J, Uhl NW, Asmussen CB, Baker WJ, Harley MM and Lewis CE (2008) *Genera Palmarum: The Evolution and Classification of Palms*. Kew Publishing, Kew.

Drude O (1893) Uber die australischen *Livistona*–Arten. *Botanische Jahrbücher fur Systematik Pflanzengeschichte und Pflanzengeographie* **39**, 5–12.

Duke NC (1991) *Nypa* in the mangroves of central America: introduced or relicts? *Principes* **35**, 127–132.

Duke NC (2006) *Australia's Mangroves. The Authoritative Guide to Australia's Mangrove Plants*. University of Queensland Press, Brisbane.

Du Puy DJ and Telford IRH (1993) Arecaceae. *Flora of Australia* **50**, 430–434.

Egge RG (1979) Martius, the father of palms. *Principes* **23**, 158–170.

Endt D and Hayward B (1997) Modern relatives of New Zealand's fossil coconuts from high altitude South America. *New Zealand Geological Society Newsletter* **113**, 67–70.

Environment Australia (2000) *Revision of the Interim Biogeographic Regionalisation of Australia (IBRA) and the Development of Version 5.1 – Summary Report*. Dept of Environment and Heritage, Canberra.

Essig FB (1978) A revision of the genus *Ptychosperma* Labill. (Arecaceae). *Allertonia* **1**, 415–478.

Essig FB and Hernandez N (2002) A systematic histological study of palm fruits. V. Subtribe Archontophoenicinae (Arecaceae). *Brittonia* **54**, 65–71.

Essig FB, Manka T and Bussard L (1999) A systematic histological study of palm fruits. III. Subtribe Iguanurinae (Arecaceae). *Brittonia* **51**, 307–325.

Ettingshausen CF (1887) Beiträge zur Kenntniss der Fossilen Flora Neuseelands. *Denkschriften der Kaiserlichen Akademie der Wissenschaften. Mathematisch–naturwissenschaftliche Classe* **53**, 143–192.

Ettingshausen CF (1895) *Beiträge zur Kenntniss der Kreideflora Australiens*. F. Tempsky, Vienna.

Ewart AJ, White J, Rees B and Wood B (1912) Contributions to the Flora of Australia, No.18. *Proceedings of the Royal Society of Victoria* **24**, 255–259.

Findlay E (1998) *Arcadian Quest: William Westall's Australian Sketches*. National Library of Australia, Canberra.

Flinders M (1814) *A Voyage to Terra Australis: Undertaken for the Purpose of Completing the Discovery of that Vast Country, and Prosecuted in the Years 1801, 1802, and 1803, in His Majesty's Ship the Investigator*. G and W Nicol, London.

Florabase (2009) *Phoenix dactylifera*. The Western Australia Flora. Dept of Environment and Conservation/Western Australian Herbarium. florabase.dec.wa.gov.au, accessed 11 February 2009.

Floyd AG (1990) *Australian Rainforests in New South Wales. Vol. 1*. Surrey Beatty and Sons, Sydney.

Frakes LA (1999) Evolution of Australian environments. *Flora of Australia*, 2nd edn, **1**, 163–203.

Frohlich J, Hyde K and Guest DI (1997) Fungi associated with leaf spots of palms in north Queensland. *Mycological Research* **101**, 721–732.

Furtado CX (1956) Palmae Malesicae – XIX, The genus *Calamus* in the Malayan Peninsula. *Gardens Bulletin Singapore* **15**, 32–265.

Gardner CA (1923) Principes. Palmae. *Forest Department Bulletin* **32**, 36–37.

Gee CT (2001) The mangrove palm *Nypa* in the geologic past of the New World. *Wetlands Ecology and Management* **9**, 181–194.

George AS (1981) The background to the Flora of Australia. *Flora of Australia* **1**, 3–24.

George AS (1999) *William Dampier in New Holland*. Bloomings Books, Melbourne.

Gorman RM (1996) Demographic trends in *Licuala ramsayi* (F.Muell.) Domin (Arecaceae), Tam O'Shanter State Forest, Mission Beach, north Queensland. *Mooreana* **6**, 63–78.

Govaerts R and Dransfield J (2005) *World Checklist of Palms*. Royal Botanic Gardens, Kew.

Green PS (1994) Arecaceae. *Flora of Australia* **49**, 407–412.

Greenwood DR (1996) Eocene monsoon forests in central Australia? *Australian Systematic Botany* **9**, 95–112.

Greenwood DR and Conran JG (2000) The Australian Cretaceous and Tertiary monocot fossil record. In *Monocots: Systematics and Evolution*. (Eds KL Wilson and DA Morrison) pp. 52–59. CSIRO Publishing, Melbourne.

Gregg DR (1975) Type and figured specimens of fossil plants in the Canterbury Museum. *Records of the Canterbury Museum* **9**, 259–276.

Gregory MR and Campbell KA (2003) A 'Phoebichnus look-alike': a fossilised root system from Quaternary coastal dune sediments, New Zealand. *Palaeogeography, Palaeoclimatology, Palaeoecology* **192**, 247–258.

Greuter W, Barrie FR, Burdet HM, Chaloner WG, Demoulin V, Hawksworth DL, Jørgensen PM, Nicolson DH, Silva PC, Trehane P and McNeill J (1994) International Code of Botanical Nomenclature (Tokyo Code). *Regnum Vegetabile* **131**. Koeltz Scientific Books, Königstein.

Griffith MP, Noblick LR, Dowe JL, Husby CE and Calonje MA (2008) Cyclone tolerance in New World Arecaceae: biogeographic variation and abiotic natural selection. *Annals of Botany* **102**, 591–598.

Grove SJ, Turton SM and Siegenthaler DT (2000) Mosaics of canopy openness induced by tropical cyclones in lowland rain forest with contrasting management histories in northeastern Australia. *Journal of Tropical Ecology* **16**, 883–894.

Gunn BF (2004) The phylogeny of the Cocoeae (Arecaceae) with emphasis on *Cocos nucifera*. *Annals of the Missouri Botanical Garden* **91**, 505–522.

Hahn WJ (2002a) A phylogenetic analysis of the Arecoid line of palms based on plastid DNA sequence data. *Molecular Phylogenetics and Evolution* **23**, 189–204.

Hahn WJ (2002b) A molecular phylogenetic study of the Palmae (Arecaceae) based on *atp*B, *rbc*L, and 18S nrDNA sequences. *Systematic Biology* **51**, 92–112.

Hahn WJ and Sytsma KJ (1999) Molecular systematics and biogeography of the south east Asian genus *Caryota* (Palmae). *Systematic Botany* **24**, 558–580.

Harley MM (2006) A summary of fossil records for Arecaceae. *Botanical Journal of the Linnean Society* **151**, 39–67.

Harley MM and Morley RJ (1995) Ultrastructural studies of some fossil and extant palm pollen, and the reconstruction of the biogeographical history of subtribes Iguanurinae and Calaminae. *Review of Palaeobotany and Palynology* **85**, 153–182.

Harley MM, Kurmann MH and Ferguson IK (1991) Systematic implications of comparative morphology in selected Tertiary and extant pollen from the Palmae and the Sapotaceae. In *Pollen and Spores*. (Eds S Blackmore and SH Barnes) pp. 225–238. Clarendon Press, Oxford.

Hartwich S, Conran JG, Bannister JM and Lee DE (in prep.) Calamus-like (Arecaceae) fossil palm leaves and fruits from Late Eocene Otago, New Zealand. *Australian Systematic Botany*.

Hedley C (1893) On the relation of the fauna and flora of Australia to those of New Zealand. *Natural Science* **3**, 187–191.

Hekel H (1972) Pollen and spore assemblages from Queensland Tertiary sediments. *Publication of the Geological Survey of Queensland* (Palaeontology Paper 30) **355**, 1–33.

Helman C (1987) Rainforest in southern New South Wales. In *The Rainforest Legacy: Australian National Rainforests Study. Vol. 1: The Nature, Distribution and Status of Rainforest Types*. pp. 47–70. Australian Government Publishing Service, Canberra.

Henderson A and Galeano G (1996) *Euterpe, Prestoea*, and *Neonicholsonia* (Palmae: Euterpeinae). *Flora Neotropica* **72**, 1–90.

Henderson A, Galeano G and Bernal R (1995) *Field Guide to the Palms of the Americas*. Princeton University Press, Princeton.

Hilbert DW, Graham A and Hopkins MS (2007) Glacial and interglacial refugia within a long-term rainforest refugium: the wet tropics bioregion of NE Queensland, Australia. *Palaeogeography, Palaeoclimatology, Palaeoecology* **251**, 104–118.

Hill W (1874) Appendix. Description of some plants discovered during the north-east coast expedition. *Report on the Brisbane Botanic Gardens 1874*, p. 6.

Hnatiuk RJ (1977) Population structure of *Livistona eastonii* Gardn., Mitchell Plateau, Western Australia. *Australian Journal of Ecology* **2**, 461–466.

Hodel DR (1992) *Chamaedorea Palms: The Species and their Cultivation*. International Palm Society, Lawrence, Kansas.

Hodel DR (2007) A review of the genus *Pritchardia*. *Palms* **51** (Supplement), S1–S52.

Hodel DR and Pintaud J-C (1998). *The Palms of New Caledonia*. Kampon Tansasha, Nong Nooch Tropical Garden, Pattaya, Thailand.

Hodel DR and Vatcharakorn P (1998) The palms of Thailand. In *The Palms and Cycads of Thailand*. (Ed. DR Hodel) pp. 1–177. Nong Nooch Tropical Garden, Pattaya, Thailand.

Holmgren PK and Holmgren NH (1998 [continuously updated]) *Index Herbariorum: A Global Directory of Public Herbaria and Associated Staff.* New York Botanical Garden's Virtual Herbarium. http://sweetgum.nybg.org/ih/.

Hoogland RD (1972) Plant distribution patterns across the Torres Strait. In *Bridge and Barrier: The Natural and Cultural History of Torres Strait*. (Ed. D Walker) pp. 131–152. Australian National University, Canberra.

Hooker JD (1853) *The Botany of the Antarctic Voyage of HM Discovery Ships Erebus and Terror. II. Flora Novae-Zelandiae* Vol. 1. Reeve and Co., London.

Hooker JD (1857) *Seaforthia elegans. Curtis's Botanical Magazine* Tab. 4961.

Hooker JD (1859) *The Botany of the Antarctic Voyage of HM Discovery Ships Erebus and Terror in the years 1839–1843. Part III. Flora Tasmaniae. Vol. 1. Dicotyledones*. Reeve and Co., London.

Hooker JD (1868) *Areca baueri. Curtis's Botanical Magazine* **94,** Tab. 5735.

Hooker JD (1883) Palmae. In *Genera Plantarum*. (Eds G Bentham and JD Hooker) **3**, 870–948. Lovell Reeve and Co., London.

Hooker JD (1884) *Report on the Progress and Condition of the Royal Gardens at Kew during the Year 1882*. Eyre and Spottiswoode, London.

Hope GS (1994) Quaternary vegetation. In *History of the Australian Vegetation: Cretaceous to Recent*. (Ed. RS Hill) pp. 368–389. Cambridge University Press, Cambridge, UK.

Hopkins MS (1990) Disturbance: the forest transformer. In *Australian Tropical Rainforests: Science – Values – Meaning*. (Eds LJ Webb and J Kikkawa) pp. 40–52. CSIRO Publishing, Melbourne.

Huang YS, Chen SS, Lin TP and Chen YS (2002) Growth strain in coconut palm trees. *Tree Physiology* **22**, 261–266.

Humphreys WF, Brooks RD and Vines B (1990) Rediscovery of the palm *Livistona alfredii* on the North West Cape

Peninsula. *Records of the Western Australian Museum* **14**, 647–650.

Inkrot D, Sattler D, Geyer C and Morawetz W (2007) Flowering and fruiting phenology of *Normanbya normanbyi* (W.Hill) L.H.Bailey (Arecaceae), a palm endemic to the lowland tropical rainforest of north-eastern Australia. *Austral Ecology* **32**, 21–28.

Irvine A (1983) *Wodyetia*, a new arecoid genus from Australia. *Principes* **27**, 158–167.

IUCN (2008) *2008 IUCN Red List of Threatened Species*. www.iucnredlist.org. Downloaded on 28 March 2009.

Jaffre T and Veillon J-M (1988) *Morphologie, distribution et écologie des palmiers de Nouvelle Calédonie*. Rapports Scientifiques et Techniques, Sciences de la Vie, Botanique No. 2. ORSTOM, Noumea.

Jessop JP and Toelken HR (1986) *Flora of South Australia. Part IV. Alismataceae – Orchidaceae*. South Australian Government Printing Division, Adelaide.

Johnson RW (1981) The Queensland palm flora. *Queensland Naturalist* **22**, 10–20.

Johnson D (Ed.) (1996) *Palms: Their Conservation and Sustained Utilization. Status Survey and Conservation Action Plan*. IUCN, Gland, Switzerland.

Johnson D (2004) *The Geology of Australia*. Cambridge University Press, Cambridge, UK.

Johnston TH (1916) Presidential address: Frederick Manson Bailey, CMG, FLS, etc. *Proceedings of the Royal Society of Queensland* **28**, 3–10.

Jones DL (1996) *Palms in Australia*. 3rd edn. Reed Books, Melbourne.

Jones JG and McDougall (1973) Geological history of Norfolk and Phillip Islands, southwest Pacific Ocean. *Australian Journal of Earth Sciences* **20**, 239–254.

Kahn F and de Castro A (1985) The palm community in a forest of Central Amazonia, Brazil. *Biotropica* **17**, 210–216.

Kemp EM and Harris WK (1977) The palynology of Early Tertiary sediments, Ninetyeast Ridge, Indian Ocean. *Special Paper in Palaeontology* **19**, 1–69.

Kershaw AP, Martin HA and McEwen Mason JRC (1994) The Neogene: a period of transition. In *History of the Australian Vegetation: Cretaceous to Recent*. (Ed. RS Hill) pp. 299–327. Cambridge University Press, Cambridge, UK.

Kershaw P, Quilty PG, David B, van Huet S and McMinn A (2000) Palaeobiogeography of the Quaternary of Australasia. *Memoir of the Association of Australasian Palaeontologists* **23**, 471–516.

Kershaw PA, Clark JS, Gill AM and D'Costa DM (2002) A history of fire in Australia. In *Flammable Australia: The Fire Regimes and Biodiversity of a Continent*. (Eds RA Bradstock, AM Gill and DM D'Costa) pp. 3–25. Cambridge University Press, Cambridge, UK.

Khan AM (1976) Palynology of Tertiary sediments from Papua New Guinea. I. New form genera and species from upper Tertiary sediments. *Australian Journal of Botany* **24**, 753–781.

King PP (1827) *Narrative of a Survey of the Intertropical and Western Coasts of Australia: Performed Between the Years 1818 and 1822*. J Murray, London.

Kroenke LW (1996) Plate tectonic development of the western and southwestern Pacific Mesozoic to the present. In *The Origin and Evolution of Pacific Island Biotas, New Guinea to Eastern Polynesia: Patterns and Processes*. (Eds A Keast and E Miller) pp. 19–34. Academic Publishing, Amsterdam.

Kuo-Huang LL, Huang YS, Chen SS and Huang YR (2004) Growth stresses and related anatomical characteristics in coconut palm trees. *IAWA Journal* **25**, 297–310.

Labillardière C [J–JH de] (1800) Mémoire sur un nouveau genre de palmier nommé *Arenga*, par le C. Labillardière. *Bulletin des sciences, par la société philomathique, Paris* **45**, 161–162.

Labillardière J–JH de (1804–06) *Novae Hollandiae Plantarum Specimen*. Huzard, Paris.

Labillardière M [J–JH de] (1809) Mémoire sur un nouveau genre de palmier. *Mémoires de la classe des sciences mathématiques et physiques de l'institut national de France 1808 (2)* **9**, 251–256.

Ladd PG (1978) Vegetation history at Lake Curlip in lowland eastern Victoria, from 5200 b.p. to present. *Australian Journal of Botany* **26**, 393–414.

Lamarck JBAPM (1786) Encyclopedie méthodique. *Botanique* **2**, 130–131.

Lange RT (1982) Australian Tertiary vegetation: evidence and interpretation. In *A History of Australasian Vegetation*. (Ed. JMB Smith) pp. 44–89. McGraw-Hill, Sydney.

Latz PK (1975) Notes on the relict palm *Livistona mariae* F.Muell. in central Australia. *Transactions of the Royal Society of South Australia* **99**, 189–195.

Laurance WF and Curran TJ (2008) Impacts of wind disturbance on fragmented tropical forest: a review and synthesis. *Austral Ecology* **33**, 399–408.

Leach GJ (1992) *Livistona mariae* F.Muell. (Family Arecaceae). In *Threatened Australian Plants: Overview and Case Studies*. (Eds JH Leigh and JD Briggs) p. 24. Commonwealth of Australia, Canberra.

Lee DE, Lee WG and Mortimer N (2001) Where and why have all the flowers gone? Depletion and turnover in the New Zealand Cenozoic angiosperm flora in relation to palaeogeography and climate. *Australian Journal of Botany* **49**, 341–356.

Lee DE, Lindqvist JK, Bannister JM, Cieraad E, Raine JI, Kennedy EM and Conran JG (2009) Late Eocene climate of southern New Zealand: insights from the *in situ* Pikopiko Fossil Forest. In *Climatic and Biotic Events of the Paleogene, Wellington, NZ, 12–15 Jan. 2009, Program and Abstracts*. (Eds CP Strong, EM Crouch and C Hollis) p. 112. GNS Science Miscellaneous Series 16. GNZ, Wellington.

Leichhardt L (1847) *Journal of an Overland Expedition in Australia, from Moreton Bay to Port Essington, a Distance of upwards of 3000 Miles, during the years 1844–1845*. T and W Boone, London.

Lewis CE and Doyle JJ (2002) A phylogenetic analysis of tribe Areceae (Arecaceae) using low-copy nuclear genes. *Plant Systematics and Evolution* **236**, 1–17.

Liddle DT, Taylor SM and Larcombe DR (1996) Population changes from 1990 to 1995 and management of the endangered rainforest palm *Ptychosperma bleeseri* Burret (Arecaceae). In *Back from the Brink: Refining the Threatened Species Recovery Process.* (Eds S Stephens and S Maxwell) pp. 110–113. Surrey Beatty and Sons, Sydney.

Liddle DT, Brook BW, Matthews J, Taylor SM and Caley P (2006) Threat and response: a decade of decline in a regionally endangered rainforest palm affected by fire and introduced animals. *Biological Conservation* **132**, 362–375.

Linnaeus C (1753) *Species Plantarum*. Laurentii Salvii, Stockholm.

Lister JJ (1888) On the natural history of Christmas Island, in the Indian Ocean. *Proceedings of the Zoological Society London* **56**, 512–531.

Lonsdale WN and Braithwaite RW (1991) Assessing the effects of fire on vegetation in tropical savannas. *Austral Ecology* **16**, 363–374.

Loo AHB, Dransfield J, Chase MW and Baker WJ (2006) Low copy nuclear DNA, phylogeny and the evolution of dichogamy in the betel nut palms and their relatives (Arecinae, Arecaceae). *Molecular Phylogenetics and Evolution* **39**, 598–618.

Loope L, Duever M, Herndon A, Snyder J and Jansen D (1994) Hurricane impact on uplands and freshwater swamp forest. *Bioscience* **44**, 238–246.

Lothian TRN (1958) The Livistonas of Australia with particular reference to the central Australian cabbage palm. *Principes* **2**, 92–94.

Lothian TRN (1959) Further notes concerning the central Australian cabbage palm – *Livistona mariae. Principes* **3**, 53–63.

Lott RH, Harrington GN, Irvine AK and McIntyre S (1995) Density-dependent seed predation and plant dispersion of the tropical palm *Normanbya normanbyi. Biotropica* **27**, 87–95.

Lourensz RS (1981) *Tropical Cyclones in the Australian Region July 1909 to June 1980.* Bureau of Meteorology, Melbourne.

Mabberley D (1999) *Ferdinand Bauer: The Nature of Discovery.* Merrell Holberton/Natural History Museum, London.

Macarthur W Sir (1855–86) *Horticultural Correspondence, 1855–1886, with J. Veitch & Sons.* Macarthur Papers, Vol. 49, ML A2945. Mitchell Library Sydney.

MacGillivray J (1852) *Narrative of the Voyage of HMS Rattlesnake.* T and W Boone, London.

Macphail MK, Jordan GJ and Hill RS (1993) Key periods in the evolution of the flora and vegetation in western Tasmania. I. The Early–Middle Pleistocene. *Australian Journal of Botany* **41**, 673–707.

Macphail MK, Alley NF, Truswell EM and Sluiter IRK (1994) Early Tertiary vegetation: evidence from spores and pollens. In *History of the Australian Vegetation: Cretaceous to Recent.* (Ed. RS Hill) pp. 189–261. Cambridge University Press, Cambridge, UK.

Martelli U (1935) Sinonimia delle palme gerontogee della tribù delle Areceae. *Nuovo Giornale Botanico Italiano* **42**, 17–88.

Martin ARH (1982) Proteaceae and the early differentiation of the central Australian flora. In *Evolution of the Flora and Fauna of Arid Australia.* (Eds WP Barker and PJM Greenslade) pp. 77–83. Peacock Publications/Australian Systematic Botany Society/ANZAAS, South Australian Division.

Martin HA (1990) Tertiary climate and phytogeography in southeastern Australia. *Review of Palaeobotany and Palynology* **65**, 47–55.

Martin HA (1993) The palaeovegetation of the Murray basin, Late Eocene to Mid Miocene. *Australian Systematic Botany* **6**, 491–531.

Martin HA (1994) Australian Tertiary phytogeography. In *History of the Australian Vegetation: Cretaceous to the Recent.* (Ed. RS Hill) pp. 104–159. Cambridge University Press, Cambridge, UK.

Martius CFP von (1838) *Historia Naturalis Palmarum.* 1st edn. Vol. 3. Leipzig.

McDougall, I, Embleton BJJ and Stone DB (1981) Origin and evolution of Lord Howe Island, southwest Pacific Ocean. *Australian Journal of Earth Science* **28**, 155–176.

McGowran B, Archer M, Bock P, Darragh TA, Godthelp H, Hageman S, Hand SJ, Hill R, Li Q, Maxwell PA, McNamara KJ, Macphail M, Mildenhall D, Partridge AD, Richardson J, Shafik S, Truswell EM and Warne M (2000) Australasian palaeobiogeography: the Palaeogene and Neogene record. *Memoir of the Association of Australasian Palaeontologists* **23**, 405–470.

McIntyre DJ (1968) Further new pollen species from New Zealand Tertiary and uppermost Cretaceous deposits. *New Zealand Journal of Botany* **6**, 177–204.

McMinn WG (1970) *Allan Cunningham: Botanist and Explorer.* Melbourne University Press, Melbourne.

McNeill J, Barrie FR, Burdet HM, Demoulin V, Hawksworth DL, Marhold K, Nicolson DH, Prado J, Silva PC, Skog JE, Wiersema JH and Turland NJ (2006) *International Code of Botanical Nomenclature (Vienna Code). Regnum Vegetabile* **146**. ARG Gantner Verlag KG.

Metcalfe I (2002) Bridging Tectonic history of the SE Asian–Australian region. In *Bridging Wallace's Line: The Environmental and Cultural History and Dynamics of the SE Asian–Australian Region.* (Eds P Kershaw, B David, N Tapper, D Penny and J Brown) pp. 29–48. Catena Verlag, Reiskirchen.

Metcalfe DJ, Bradford MG and Ford AJ (2008) Cyclone damage to tropical rain forests: species- and community-level impacts. *Austral Ecology* **33**, 432–441.

Mildenhall DC (1980) New Zealand Late Cretaceous and Cenozoic plant biogeography: a contribution. *Palaeogeography, Palaeoclimatology, Palaeoecology* **31**, 197–233.

Mildenhall DC and Pocknall DT (1989) Miocene–Pleistocene spores and pollen from Central Otago, South Island, New Zealand. *New Zealand Geological Survey Palaeontological Bulletin* **59**, 1–128.

Milne LA (1988) Palynology of a Late Eocene lignitic sequence from the western margin of the Eucla Basin, Western Australia. *Memoir of the Association of Australasian Paleontologists* **5**, 285–310.

Mogea JP (1999) Relationships and phylogeny of the species of the genus *Arenga* (Palmae) based on morphology using the polarity method and the NTSYS program. *Memoirs of the New York Botanical Garden* **83**, 169–177.

Moore C (1871) Remarks on the botany of Lord Howe's Island. *Transactions of the Royal Society of New South Wales* **5**, 29–34.

Moore C (1884) *A Census of the Plants of New South Wales*. Government Printer, Sydney.

Moore HE Jr (1963) The types and lectotypes of some palm genera. *Gentes Herbarum* **9**, 245–274.

Moore HE Jr (1973) The major groups of palms and their distribution. *Gentes Herbarum* **11**, 27–141.

Moore HE Jr (1981) Odoardo Beccari (1843–1920). *Principes* **25**, 29–35.

Moore C and Betche E (1893) *Handbook of the Flora of New South Wales*. Charles Potter, Sydney.

Moore HE and Dransfield J (1979) The typification of Linnaean palms. *Taxon* **28**, 59–70.

Moraes MR (1996) Diversity and distribution of palms in Bolivia. *Principes* **40**, 75–85.

Moran GF, Muona O and Bell JC (1989) *Acacia mangium*: a tropical forest tree of the coastal lowlands with low genetic diversity. *Evolution* **43**, 231–235.

Morley RJ (2002) Tertiary vegetational history of southeast Asia, with emphasis on the biogeographical relationships with Australia. In *Bridging Wallace's Line: The Environmental and Cultural History and Dynamics of the SE Asian–Australian Region*. (Eds P Kershaw, B David, N Tapper, D Penny and J Brown) pp. 49–60. Catena Verlag, Reiskirchen.

Morton SR, Short J and Barker RD (1995) Refugia for biological diversity in arid and semi-arid Australia. Biodiversity Series, Paper No. 4. Dept of the Environment, Sport and Territories, Canberra.

Mueller F (1865) Palmae. *Fragmenta Phytographiae Australiae* **5**, 47–49.

Mueller F (1866) A systematic arrangement of the plants noticed around the Gulf of Carpentaria, from the Roper to the Gilbert River. In *Landsborough's Exploration of Australia from Carpentaria to Melbourne*. (Ed. JS Laurie) pp. 109–123. London.

Mueller F (1870) Palmae. *Fragmenta Phytographiae Australiae* **7**, 99–104.

Mueller F (1874a) Palmae. *Fragmenta Phytographiae Australiae* **8**, 221–222.

Mueller F (1874b) *Livistona leichhardtii*. *Fragmenta Phytographiae Australiae* **8**, 283.

Mueller F (1878) *Palmae. Fragmenta Phytographiae Australiae* **11**, 54–58.

Mueller F (1880) *Areca alicae*. *Gartenflora* **28**, 199–201.

Mueller F (1881) *Pandanaceae. Fragmenta Phytographiae Australiae* **11**, 127–128.

Mueller F (1882) A new palm from Queensland. *Chemist and Druggist. Australasian Supplement* **4** (Feb.), 7.

Mueller F (1889) *Second Systematic Census of Australian Plants, with Chronologic, Literary and Geographic Annotations*. McCarron Bird and Co., Melbourne.

Mueller F (1892) Note on the West Australian fan-palm. *Victorian Naturalist* **9**, 112.

Muller J (1968) Palynology of the Pedawan and Plateau Sandstone Formations (Cretaceous–Eocene) in Sarawak, Malaysia. *Micropaleontology* **14**, 1–37.

Muller J (1972) Palynological evidence for change in geomorphology, climate, and vegetation in the Mio–Pliocene of Malesia. In *The Quaternary Era in Malesia*. (Eds P Ashton and M Ashton) pp. 6–16. Dept of Geography, Hull/Institute for South-east Asian Biology, University of Aberdeen.

Nix HA and Kalma JD (1972) Climate as a dominant control in the biogeography of northern Australia and New Guinea. In *Bridge and Barrier: The Natural and Cultural History of Torres Strait*. (Ed. D Walker) pp. 61–91. Australian National University, Canberra.

Noack D (1994) Date palms and the exotic flora of Dalhousie Springs. *South Australian Geographical Journal* **93**, 81–89.

Norst MJ (1989) *Ferdinand Bauer: The Australian Natural History Drawings*. Lothian, Melbourne.

Norup MV, Dransfield J, Chase MW, Barfod AS, Fernando ES and Baker WJ (2006) Homoplasious character combinations and generic delimitation: a case study from the Indo-Pacific Arecoid palms (Arecaceae: Areceae). *American Journal of Botany* **93**, 1065–1080.

NRETAS (Natural Resources, Environment, the Arts and Sport) (2009) Threatened species list. NRETAS, Darwin. http://www.nt.gov.au/nreta/wildlife/animals/threatened/specieslist.html.

Oliver D (1891) *Arenga listeri*. *Hooker's Icones Plantarum* 3rd Series, **10**, Plate 1985.

Orscheg CK and Parsons RF (1996a) Ecology of *Livistona australis* (R.Br.) Martius (Arecaceae) at its southern limit, south-eastern Australia. Part I. Distribution and genetic variation. *Mooreana* **6**, 8–17.

Orscheg CK and Parsons RF (1996b) Ecology of *Livistona australis* (R.Br.) Martius (Arecaceae) at its southern limit, south-eastern Australia. Part II. Phenology, germination, herbivory, regeneration and fire effects. *Mooreana* **6**, 36–50.

Ostertag R, Silver WL and Lugo AE (2005) Factors affecting mortality and resistance to damage following hurricanes in rehabilitated subtropical moist forest. *Biotropica* **37**, 16–24.

Palmer E (1884) Notes on some Australian tribes. *Journal of the Royal Anthropological Institute of Great Britain* **13**, 276–346.

Palmer E (1903) *Early Days in North Queensland*. Angus and Robertson, Sydney.

Pan AD, Jacobs BF, Dransfield J and Baker WJ (2006) The fossil history of palms in Africa and new records from the Late Oligocene (~28–27 Myr) of northwestern

Ethiopia. *Botanical Journal of the Linnean Society* **151**, 69–81.

Parsons RF (2007) The southernmost limits for palms. *New Zealand Journal of Botany* **45**, 477–478.

Pettit NE and Dowe JL (2003) Distribution and population structure of the vulnerable riparian palm *Livistona lanuginosa* AN Rodd (Arecaceae) in the Burdekin River catchment, north Queensland. *Pacific Conservation Biology* **9**, 207–214.

Pickard J (1983) Vegetation of Lord Howe Island. *Cunninghamia* **1**, 133–266.

Pintaud J-C (1999) A cladistic analysis of the Archontophoenicinae (Palmae: Areceae) based on morphological and anatomical characters. *Memoirs of the New York Botanical Garden* **83**, 279–284.

Pintaud J-C and Baker WJ (2008) A revision of the palm genera (Arecaceae) of New Caledonia. *Kew Bulletin* **63**, 61–73.

Pintaud J-C, Jaffre T and Veillon J-M (1999) Conservation status of New Caledonia palms. *Pacific Conservation Biology* **5**, 9–15.

Pocknall DT (1985) Palynology of Waikato Coal Measures (Late Eocene–Late Oligocene) from the Raglan area, North Island, New Zealand. *New Zealand Journal of Geology and Geophysics* **28**, 329–349.

Pohlman CL, Goosem M and Turton SM (2008) Effects of severe Tropical Cyclone Larry on rainforest vegetation and understorey microclimate near a road, powerline and stream. *Austral Ecology* **33**, 503–515.

Pole MS (1993) Early Miocene flora of the Manuherikia Group, New Zealand. IV. Palm fossils. *Journal of the Royal Society of New Zealand* **23**, 283–288.

Pole MS (1998) Early Eocene estuary at Strahan, Tasmania. *Australian Journal of Earth Sciences* **45**, 979–985.

Pole M (1999) Latest Albian–earliest Cenomanian monocotyledonous leaves from Australia. *Botanical Journal of the Linnean Society* **129**, 177–186.

Pole MS (2003) New Zealand climate in the Neogene and implications for global atmospheric circulation. *Palaeogeography, Palaeoclimatology, Palaeoecology* **193**, 69–84.

Pole M (2007) Monocot macrofossils from the Miocene of southern New Zealand. *Palaeontologia Electronica* **10**, 1–21.

Pole MS and Macphail MK (1996) Eocene *Nypa* from Regatta Point, Tasmania. *Review of Palaeobotany and Palynology* **92**, 55–67.

Potztal E (1958) Professor Dr Max Burret. *Principes* **2**, 87–91.

Prebble M, Sim R, Finn J and Fink D (2005) A Holocene pollen and diatom record from Vanderlin Island, Gulf of Carpentaria, lowland tropical Australia. *Quaternary Research* **64**, 357–371.

Prider JN and Christophel DC (2000) Distributional ecology of *Gymnostoma australianum* (Casuarinaceae), a putative palaeoendemic of Australian wet tropic forests. *Australian Journal of Botany* **48**, 427–434.

Prior LD, Murphy BP and Russell-Smith J (2009) Environmental and demographic correlates of tree recruitment and mortality in north Australian savannas. *Forest Ecology and Management* **257**, 66–74.

Puotinen ML, Done TJ and Skelly WC (1997) *An Atlas of Tropical Cyclones in the Great Barrier Reef Region, 1969–1997.* CRC Reef Research Centre, Technical Report No. 19. CRC Reef Research Centre, Townsville.

Putz FE (1990) Growth habits and trellis requirements of climbing palms (*Calamus* spp.) in north-eastern Queensland. *Australian Journal of Botany* **38**, 603–608.

Raine JI, Mildenhall DC and Kennedy EM (2008) New Zealand fossil spores and pollen: an illustrated catalogue. 3rd edn. GNS Science Miscellaneous Series No. 4. http://www.gns.cri.nz/what/earthhist/fossils/spore_pollen/catalog/index.htm.

Rawling TJ and Lister GS (1999) Oscillating modes of orogeny in the Southwest Pacific and the tectonic evolution of New Caledonia. *Geological Society, London, Special Publications* **154**, 109–127.

Reeves JM, Chivas AR, García A, Holt S, Couapel MJJ, Jones BG, Cendón DI and Fink D (2008) The sedimentary record of palaeoenvironments and sea-level change in the Gulf of Carpentaria, Australia, through the last glacial cycle. *Quaternary International* **183**, 3–22.

Renner C, Foreman DB and Murray D (2000) Timing transantarctic disjunctions in the Atherospermataceae (Laurales): evidence from coding and noncoding chloroplast sequences. *Systematic Biology* **49**, 579–591.

Rickett HW and Stafleu FA (1959) Nomina generica conservanda et rejicienda spermatophytorum. *Taxon* **8**, 213–243.

Rigby JF (1995) A fossil *Cocos nucifera* L fruit from the latest Pliocene of Queensland, Australia. In *Global Environment and Diversification of Plants through Geological Time: Proceedings of the International Conference, Allahabad, India: Birbal Sahni Centenary Volume.* (Ed. DD Pant) pp. 379–381. Allahabad.

Rodd AN (1983) Arecales. Palmae. In *Flowering Plants in Australia.* (Eds BD Morley and HR Toelken) pp. 392–395. Rigby, Sydney.

Rodd AN (1998) Revision of *Livistona* (*Arecaceae*) in Australia. *Telopea* **8**, 49–153.

Rowe C (2007) Vegetation change following mid-Holocene marine transgression of the Torres Strait shelf: a record from the island of Mua, northern Australia. *The Holocene* **17**, 927–937.

Rumphius GE (1741) *Herbarium Amboinense* 1. J Burmann, Meinard Uytwerf, Amsterdam.

Sanderson R (2007) Many beautiful things: colonial botanists' accounts of the north Queensland rainforests. *Historical Records of Australian Science* **18**, 1–18.

Sarukhán J (1978) Studies on the demography of tropical trees. In *Tropical Trees as Living Systems.* (Eds PB Tomlinson and MH Zimmerman) pp. 163–184. Cambridge University Press, Cambridge, UK.

Sattler PS and Williams RD (1999) *The Conservation Status of Queensland's Bioregional Ecosystems.* Environmental Protection Agency, Brisbane.

Savolainen V, Anstett M-C, Lexer C, Hutton I, Clarkson JJ, Norup MV, Powell MP, Springate D, Salamin N and

Baker WJ (2006) Sympatric speciation in palms on an oceanic island. *Nature* **441**, 210–213.

Scheffer RHCC (1876) Sur quelques palmiers du groupe des Arécinées. *Annales du Jardin Botanique de Buitenzorg* **1**, 103–164.

Shafik S (1978) Palaeocene and Eocene nannofossils from Kings Park. *Bureau of Mineral Resources, Australian Bulletin* **192**, 165–172.

Shapcott A (1998a) The patterns of genetic diversity in *Carpentaria acuminata* (Arecaceae), and rainforest history in northern Australia. *Molecular Ecology* **7**, 833–847.

Shapcott A (1998b) The genetics of *Ptychosperma bleeseri*, a rare palm from the Northern Territory, Australia. *Biological Conservation* **85**, 203–209.

Shapcott A (2000) Conservation and genetics in the fragmented monsoon rainforest in the Northern Territory, Australia: a case study of three frugivore-dispersed species. *Australian Journal of Botany* **48**, 397–407.

Solander D (1768–71) *Plantae Novae Hollandiae, 1768–1771.* Palmae, Vol. 3. Unpublished handwritten manuscript. Natural History Museum, London.

Soltis DE, Soltis PS, Endress PK and Chase MW (2005) *Phylogeny and Evolution of Angiosperms.* Sinauer, Sunderland, Massachusetts.

Stafleu FA and Cowan RS [with Mennega E] (1976–98) *Taxonomic Literature: A Selective Guide to Botanical Publications and Collections with Dates, Commentaries and Types. 2nd edn.* IDC Publishers/International Association of Plant Taxonomy, Leiden.

Stanley TD and Ross EM (1989) Arecaceae. *Flora of South-eastern Queensland* **3**, 268–270. Queensland Department of Primary Industries, Brisbane.

Stork NE (2007) Australian tropical forest canopy crane: new tools for new frontiers. *Austral Ecology* **32**, 4–9.

Stover LE and Partridge AD (1973) Tertiary and Late Cretaceous spores and pollen from the Gippsland Basin, southeastern Australia. *Proceedings of the Royal Society of Victoria* **85**, 237–286.

Svenning J-C (2001) Environmental heterogeneity, recruitment limitation and the mesoscale distribution of palms in a tropical montane rain forest (Maquipucuna, Ecuador). *Journal of Tropical Ecology* **17**, 97–113.

Thunberg CP (1782) *Licuala*, Et nytt Palm-flagte. *Kongliga Vetenskaps Academiens Nya Handlingar* **3**, 284–287.

Tomlinson PB (1979) Systematics and ecology of the Palmae. *Annual Review of Ecology and Systematics* **10**, 85–107.

Tomlinson PB (1990) *The Structural Biology of Palms.* Clarendon Press, Oxford.

Tracey JG and Webb LJ (1975) *Key to the Vegetation of the Humid Tropical Region of North Queensland, with 15 maps at 1:100,000 scale.* CSIRO Long Pocket Laboratories, Indooroopilly, Queensland.

Tralau H (1964) The genus *Nypa* van Wurmb. *Kungliga Svenska Vetenskapsakademiens Handlingar* **10**, 1–29.

Trénel P, Gustafsson MHG, Baker WJ, Asmussen-Lange CB, Dransfield J and Borchsenius F (2007) Mid-Tertiary dispersal, not Gondwana vicariance explains distribution patterns in the wax palm family (Ceroxyloideae: Arecaceae). *Molecular Phylogenetics and Evolution* **45**, 272–288.

Truswell EM (1990) Australian rainforests: the 100 million year record. In *Australian Tropical Rainforests: Science–Values–Meaning.* (Eds LJ Webb and J Kikkawa) pp. 7–22. CSIRO Publishing, Melbourne.

Truswell EM (1993) Vegetation changes in the Australian Tertiary in response to climatic and phytogeographic forcing factors. *Australian Systematic Botany* **6**, 533–557.

Truswell EM and Harris WK (1982) The Cainozoic palaeobotanical record in arid Australia: fossil evidence for the origin of an arid adapted flora. In *Evolution of the Flora and Fauna of Arid Australia.* (Eds WR Barker and PJM Greenslade) pp. 67–76. Peacock Publications, Adelaide.

Truswell EM and Marchant NG (1986) Early Tertiary pollen of probable Droseracean affinity from central Australia. *Special Papers in Palaeontology* **35**, 163–178.

Truswell EM and Owen JA (1988) Eocene pollen from Bungonia, New South Wales. *Memoirs of the Association of Australian Palaeontologists* **5**, 259–284.

Truswell EM, Sluiter IR and Harris WK (1985) Palynology of the Oligocene–Miocene sequence in the Oakvale-1 corehole, western Murray Basin, South Australia. *BMR Journal of Australian Geology and Geophysics* **9**, 267–295.

Truswell EM, Kershaw AP and Sluiter IR (1987) The Australian–South-east Asian connection: evidence from the palaeobotanical record. In *Biogeographic Evolution of the Malay Archipelago.* (Ed. TC Whitmore) pp. 32–49. Oxford University Press, Oxford.

Tucker R (1988) *Palms of Subequatorial Queensland.* Palm and Cycad Societies of Australia, Milton.

Tucker R (1994) Studies in Queensland Pandanaceae. III. *Pandanus* L.f. subgenus *Acrostigma* (Kurz) Stone. *Mooreana* **4**, 49–65,

Turton SM (2008) Landscape-scale impacts of Cyclone Larry on the forests of northeast Australia, including comparisons with previous cyclones impacting the region between 1858 and 2006. *Austral Ecology* **33**, 409–416.

Uhl NW and Dransfield J (1987) *Genera Palmarum: A Classification of Palms based on the Work of HE Moore Jr.* LH Bailey Hortorium/International Palm Society. Allen Press, Lawrence, Kansas.

Vallance TG, Moore DT and Groves EW (2001) *Nature's Investigator: The Diary of Robert Brown in Australia, 1801–1805.* Australian Biological Resource Study, Canberra.

Vandermeer J (1994) Effects of Hurricane Joan on the palms of the Caribbean coast rainforest of Nicaragua. *Principes* **38**, 182–189.

Van Steenis-Kruseman MJ (1950) Malaysian plant collectors and collections. *Flora Malesiana* **1**, 1–639.

Veevers JJ (1991) Phanerozoic Australia in the changing configuration of proto-Pangea through Gondwanaland and Pangea to the present dispersed continents. *Australian Systematic Botany* **4**, 1–11.

Veitch HJ (1879) *Catalogue of Plants including Novelties for 1879.* James Veitch and Sons/Pollet and Co., London.

Wallis LA (2001) Environmental history of northwest Australia based on phytolith analysis at Carpenter's Gap 1. *Quaternary International* **83–85**, 103–117.

Wallis LA (2003) An overview of phytolith production patterns in selected northwestern Australian flora. *Review of Palaeobotany and Palynology* **125**, 210–248.

Waterhouse JT and Quinn CJ (1978) Growth patterns in the stem of the palm *Archontophoenix cunninghamiana. Botanical Journal of the Linnean Society* **77**, 73–93.

Watts P, Pomfrett JA and Mabberley D (1997) *An Exquisite Eye: The Australian Flora and Fauna Drawings 1801–1820 of Ferdinand Bauer.* Historic Houses Trust of New South Wales, Sydney.

Webb LJ (1958) Cyclones as an ecological factor in tropical lowland rainforest, north Queensland. *Australian Journal of Botany* **6**, 220–228.

Webb LJ (1959) A physiognomic classification of Australian rain forests. *Journal of Ecology* **47**, 551–570.

Webb LJ and Tracey JG (1972) An ecological comparison of vegetation communities on each side of Torres Strait. In *Bridge and Barrier: The Natural and Cultural History of Torres Strait.* (Ed. D Walker) pp. 109–129. Australian National University, Canberra.

Webb LJ and Tracey JG (1981) Australian rainforests: patterns and change. In *Ecological Biogeography of Australia.* (Ed. A Keast) pp. 605–694. Dr W Junk, The Hague.

Webb LJ, Tracey JG and Jessup LW (1986) Recent evidence for autochthony of Australian tropical and subtropical elements. *Telopea* **2**, 575–589.

Wendland H (1858) *Ptychosperma cunninghamiana. Botanische Zeitung* **16**, 346–347.

Wendland H and Drude O (1875) *Palmae Australasicae. Linnaea* **39**, 153–238.

Werner PA (2005) Impact of feral water buffalo and fire on growth and survival of mature savanna trees: an experimental field study in Kakadu National Park, northern Australia. *Austral Ecology* **30**, 625–647.

White CT (1950) FM Bailey: his life and work. *Proceedings of the Royal Society of Queensland* **61**, 105–114.

Williams RJ, Cook GD, Gill AM and Moore PHR (1999) Fire regime, fire intensity and tree survival in a tropical savanna in northern Australia. *Australian Journal of Ecology* **24**, 50–59.

Williams RJ, Griffiths AD and Allan GE (2002) Fire regimes and biodiversity in the savannas of northern Australia. In *Flammable Australia: The Fire Regimes and Biodiversity of a Continent.* (Eds RA Bradstock, JE Williams and AM Gill) pp. 281–304. Cambridge University Press, Cambridge, UK.

Williams RJ, Gill AM and Moore PHR (2003) Fire behaviour. In *Fire in Tropical Savannas: The Kapalga Experiment.* (Eds AN Andersen, GD Cook and RJ Williams) pp. 33–46. Springer-Verlag, New York.

Willis M (1949) *By their Fruits: A Life of Ferdinand von Mueller, Botanist and Explorer.* Angus and Robertson, Sydney.

Winter JW and Atherton RG (1987) An introduction to Australian rainforests. In *The Rainforest Legacy: Australian National Rainforests Study. Vol. 1: The Nature, Distribution and Status of Rainforest Types.* pp. 1–7. Australian Government Publishing Service, Canberra.

Wischusen JDH, Fifield LK and Cresswell RG (2004) Hydrogeology of Palm Valley, central Australia: a Pleistocene flora refuge? *Journal of Hydrology* **293**, 20–46.

Wyrwoll K-H (1993) An outline of Late Cenozoic palaeoclimatic events in the Cape Range region. *Records of the Western Australian Museum Supplement* **45**, 39–49.

Zona S (1990) A monograph of *Sabal* (Arecaceae: Coryphoideae). *Aliso* **12**, 583–666.

Zona S (1999) New perspectives on generic limits and relationships in the Ptychospermatinae (Palmae: Arecoideae). *Memoirs of the New York Botanical Garden* **83**, 255–263.

Index

www.ingramcontent.com/pod-product-compliance
Lightning Source LLC
LaVergne TN
LVHW070947210926
848482LV00002B/128
9780643096158